Optical Sensors for Structural Health Monitoring

Optical Sensors for Structural Health Monitoring

Editors

Paulo Antunes
Humberto Varum

MDPI • Basel • Beijing • Wuhan • Barcelona • Belgrade • Manchester • Tokyo • Cluj • Tianjin

Editors
Paulo Antunes
Aveiro University and Institute for Nanostructures, Nanomodelling and Nanofabrication
Instituto de Telecomunicações
Portugal

Humberto Varum
CONSTRUCT—LESE Faculty of Engineering of the University of Porto
Portugal

Editorial Office
MDPI
St. Alban-Anlage 66
4052 Basel, Switzerland

This is a reprint of articles from the Special Issue published online in the open access journal *Sensors* (ISSN 1424-8220) (available at: https://www.mdpi.com/journal/sensors/special_issues/optic_sens_SHM).

For citation purposes, cite each article independently as indicated on the article page online and as indicated below:

LastName, A.A.; LastName, B.B.; LastName, C.C. Article Title. *Journal Name* **Year**, *Volume Number*, Page Range.

ISBN 978-3-0365-0172-7 (Hbk)
ISBN 978-3-0365-0173-4 (PDF)

Contents

About the Editors

Paulo Antunes (Ph.D.) received a Physics Engineering degree in 2005, an M.Sc. in Applied Physics in 2007 and a Ph.D. in Physics Engineering in 2011 from the Universidade de Aveiro, Portugal. Currently, he is an Assistant Professor at the Physics Department of the Aveiro University, Portugal. As a researcher, he is also with the I3N-Aveiro (Institute of Nanostructures, Nanomodelling and Nanofabrication) and the Instituto de Telecomunicações. His research interests include the study and simulation of optical fiber sensors based on silica and polymeric fibers for static and dynamic measurements, data acquisition, optical transmission systems and sensor networks for several applications, including medical applications and structural health monitoring. He has authored or co-authored more than 180 international journal papers, book chapters and conference technical papers with an emphasis on optical sensing devices and applications.

Humberto Varum (Ph.D.) is a Full Professor at the Civil Engineering Department of the Faculty of Engineering of the University of Porto, Portugal. He is an Integrated Member of the CONSTRUCT research unit at the Institute of R&D in Structures and Construction. He was a Seconded National Expert at the ELSA laboratory, Joint Research Centre, European Commission, Italy, from July 2009 to August 2010. Since May 2015, he has been a member of the directorate body of the Construction Institute at the University of Porto. He has been a Member of the National Committee of the International Council on Monuments and Sites (ICOMOS) since 2009 and is an Expert Member of the ICOMOS's International Scientific Committee of Earthen Architectural Heritage (ISCEAH). He has participated in post-earthquake field reconnaissance missions, in particular to L'Aquila (Italy, 2009), Lorca (Spain, 2011), Emilia-Romagna (Italy, 2012), Gorkha (Nepal, 2015) and Puebla (Mexico, 2017). His main research interests include the assessment, strengthening and repair of structures; earthquake engineering; and historic constructions' conservation and strengthening. He has coordinated and participated in several national and international research projects. He has co-authored over 400 publications in international peer-reviewed scientific journals, books and conference proceedings, in the fields of earthquake engineering, the assessment and strengthening of structures and earth construction rehabilitation.

Preface to "Optical Sensors for Structural Health Monitoring"

The intention to prevent the severe damage and eventual collapse of structures, and consequent human, material and economic losses, fostered the need for the preservation, maintenance and retrofitting of existing structures. Our built heritage comprises multiple structural systems with different materials, designed and constructed in different periods, and thus with different design criteria, detailing and construction techniques. Effective monitoring enables the reduction of maintenance costs (optimized maintenance), or even reducing the costs of eventual interventions needed to upgrade the level of structural safety, supporting retrofitting strategies based on rich information about the structural behavior. For this, accurate assessment of structural safety is needed, which may benefit from innovative monitoring approaches, equipment and sensors. Photonic technologies have become vitally important in developing monitoring solutions. In recent decades, there has been a growing interest in monitoring research, especially sensors, not only in electronics but also in fiber optics. These have proved to be a promising technology because they are durable, stable, insensitive to electromagnetic interference and generally have minimal aesthetic impact; as such, they are particularly interesting for long-term structural health surveillance. The fact that no power supply at the measurement site is required, their multiplexing capability (allowing dozens of sensors on the same fiber optic cable) and low attenuation of the optical fiber are additional advantages. Their application may also allow the implementation of a safer monitoring system (without risk of short circuiting). The results of the information captured with the monitoring strategy may be used for the calibration of structural numerical models, based on data obtained in real-time, and eventually, the implementation of early warning systems in the case of danger.

This volume is a collection of papers that originated as a Special Issue, focused on some recent advances related to optical sensors and structural health monitoring. Applications including corrosion, load, crack opening and strain or deflection monitoring in different types of structural systems, such as pipelines, aircrafts, concrete beams and bridges, are presented.

The authors of this book are grateful to all the contributing authors, journal editors, reviewers and the production team.

Paulo Antunes, Humberto Varum

Editors

Article

External Corrosion Detection of Oil Pipelines Using Fiber Optics

Nader Vahdati [1,*], Xueting Wang [1], Oleg Shiryayev [2], Paul Rostron [1] and Fook Fah Yap [3]

1 Department of Mechanical Engineering, Khalifa University of Science and Technology, Sas Al Nakhl Campus, Abu Dhabi 999041, UAE; xuwang@pi.ac.ae (X.W.); paul.rostron@ku.ac.ae (P.R.)
2 Department of Mechanical Engineering, University of Alaska, Anchorage, AK 99508, USA; oshiryayev@alaska.edu
3 Department of Mechanical and Aerospace Engineering, Nanyang Technological University, Singapore 639798, Singapore; mffyap@ntu.edu.sg
* Correspondence: nader.vahdati@ku.ac.ae; Tel.: +971-26075787

Received: 12 December 2019; Accepted: 21 January 2020; Published: 26 January 2020

Abstract: Oil flowlines, the first "pipeline" system connected to the wellhead, are pipelines that are 5 to 30.5 cm (two to twelve inches) in diameter, most susceptible to corrosion, and very difficult to inspect. Herein, an external corrosion detection sensor for oil and gas pipelines, consisting of a semicircular plastic strip, a flat dog-bone-shaped sacrificial metal plate made out of the same pipeline material, and an optical fiber with Fiber Bragg Grating (FBG) sensors, is described. In the actual application, multiple FBG optical fibers are attached to an oil and gas pipeline using straps or strips or very large hose clamps, and, every few meters, our proposed corrosion detection sensor will be glued to the FBG sensors. When the plastic parts are attached to the sacrificial metals, the plastic parts will be deformed and stressed; thus, placing the FBG sensors in tension. When corrosion is severe at any given pipeline location, the sacrificial metal at that location will corrode till failure and the tension strain is relieved at that FBG Sensor location, and therefore, a signal is detected at the interrogator. Herein, the external corrosion detection sensor and its design equations are described, and experimental results, verifying our theory, are presented.

Keywords: corrosion sensor; oil and gas pipelines; optical fibers; Fiber Bragg Grating (FBG)

---

1. Introduction

Pipelines are the most practical, economical, and safest way of transporting crude or refined oil and gas (O&G) around the world. A study done by the Fraser Institute [1], comparing safety of transporting oil and gas (O&G) by rail versus pipelines, revealed that both ways are safe but pipelines are the safest transportation mode. Even all living creatures except flatworms, nematodes, and cnidarians have circulatory systems and use veins and arteries (pipelines) to transport blood, oxygen, and nutrients to their bodies. However, from time to time, rupture of the pipelines or veins and arteries can occur. In the case of O&G pipelines, as they are transporting flammable and very hazardous materials, any rupture or defect of the pipeline can potentially result in explosions, fires, release of toxic gases, loss of human lives, property damage, and environmental disasters. Many living creatures and humans have a nervous system, which can detect rupture of veins and arteries when it occurs (acting as a health monitoring system) to warn humans of occurrence of such events, but such health monitoring systems are nonexistent in most O&G fields.

The purpose of this research is to develop a health monitoring system (a corrosion detection sensor) using fiber optics to facilitate detection of external corrosion and help prevent leaks in exposed O&G pipelines. In this paper, an external corrosion detection sensor for O&G pipelines, consisting of a semicircular plastic strip, a flat dog-bone-shaped sacrificial metal plate made out of the same pipeline material, and an optical fiber with FBG sensors, is described. The external corrosion detection sensor and its design equations are described, and experimental results, verifying our theory, are also presented.

1.1. Literature Review—Corrosion Prevention

Protective coatings (zinc, epoxy, paint, and other polymers) applied to the outside of O&G pipelines, and cathodic protection are two most commonly used methods [2] by the oil industry to control and inhibit external corrosion of buried pipelines. For above the ground pipelines, which is the focus of this paper, cathodic protection is not an option since the electrolyte (water or soil) is not present and protective coating is the only external protection. The most common types of coatings used in the O&G industry are zinc coating, created during manufacturing of the pipelines and epoxy coating, which is a paint-like substance that seals the surface of the pipeline. Pipeline coating prevents the metal pipe from being in direct contact with the environment, thus extending its life.

Despite the use of coatings, due to mechanical and environmental damages to the coating, external corrosion still takes place on above the ground pipelines. As most oil fields lack a corrosion monitoring system, corrosion can occur undetected.

Another indirect method used to protect pipelines from external corrosion has been pipeline insulation, acting like coatings. Pipeline insulation is used to reduce energy loss, maintain temperature in O&G pipelines, control paraffin waxes from precipitation, and prevent pipelines from freezing and cracking, and is designed to be water tight to prevent infiltration of water from the outside environment onto the pipeline surface; thus, protecting the pipelines from external corrosion but due to mechanical and environmental damages to the insulation, water invariably seeps into the insulation, and pipeline corrosion occurs. Corrosion under insulation (CUI) or corrosion under fire-proofing (CUF) is reported by most O&G and petrochemical companies to be one of their worst nightmares. The cost associated with controlling corrosion is astronomical. On 21 June 2016, PHMSA (US Pipeline and Hazardous Materials Safety Administration) issued an advisory bulletin [3] warning the pipeline industry about Corrosion Under Insulation (CUI). Thermal insulation not only has failed to shield O&G pipelines from external corrosion, but has actually exacerbated the corrosion problem.

Corrosion of low carbon steel pipelines cannot be entirely eliminated but can only be controlled; meaning occurrence of corrosion is a certainty. Thus, there is a definite need for corrosion detection, and inspection. As corrosion is a major threat to O&G pipelines, its inhibition and timely detection are the two key parts of pipeline integrity practice.

1.2. Literature Review—Corrosion/Leak Detection Sensors

In the past four years, numerous in-depth review papers have been published [4–11] that show inspection techniques commonly used by the O&G industry for external and internal corrosion detection of O&G pipelines.

For above the ground O&G pipelines without an insulation or with insulations removed, visual inspection, ultrasonic thickness measurement method, Pipeline Inspection Gage (PIG) or Inline Inspection (ILI) tool, and Hydrostatic Pressure Testing (HT) are four main pipeline integrity inspection techniques used by most O&G companies to detect external corrosion and assess if a corroded pipeline is safe to be in operation or not. These techniques have their advantages and disadvantages, and each reflects a different, unique aspect of the overall pipeline integrity management.

The first and the simplest method of inspection is, of course, visual inspection, but it is also the most expensive and time-consuming method. It involves a person walking along the pipeline and checking the surface condition of the pipe, looking for dents, pitting corrosion, metal loss, cracks, and other defects. Visual inspections are usually performed with portable visual scanners (laser scanners), which allow for precise, traceable sizing of surface corrosion at the outer diameter of the pipeline. The task of visual inspection of O&G pipelines becomes harder when insulations are involved. Insulation has to be first removed before the visual inspection and later replaced when visual inspection is complete. This is why this method is labeled as expensive and time-consuming. Corrosion under insulation (CUI), as explained earlier, is one of the most difficult corrosion processes to detect and prevent, as the insulation covers the corrosion problem until it is too late.

When corrosion is found on the surface of the pipeline, ultrasonic thickness measurement method can be used to detect the depth of the corrosion and, at the same time, detect if internal corrosion is also occurring at the same location where external corrosion has been found. Ultrasonic thickness measurement method is a very effective tool in determining local wall thickness of a pipe, but this method is limited to small areas and it takes a long time to use this method to inspect a large area of a pipeline. This method is mostly used to see the severity of corrosion defects when visual inspection shows occurrence of external corrosion.

In-line inspection (ILI) tools, or also called smart pipeline inspection gauges (pigs) travel through a pipeline scanning, measuring, and recording wall thickness, and looking for metal loss, dents, corrosion, deformations, cracking, or other defects [12]. Smart pigs use magnetic flux leakage (MFL) [13] or ultrasonic waves [14] to identify potential problems. The resulting data is then analyzed to diagnose issues and schedule maintenance. Most O&G companies use ILI technology every 3 to 5 years, thus, due to this long inspection interval, the ILI method cannot be considered as a health monitoring system, but instead as an inspection method. ILI testing involves production shutdown which is very costly, even when no corrosion is found during the ILI inspection.

Another technique used by the O&G industry to test pipelines for strength and leaks is hydrostatic testing [15]. This technique is particularly used to test newly laid pipelines for leaks. However, the same technique can also be applied to existing pipelines with defects and corrosion damage. The test involves filling a segment of a corroded pipeline with a liquid, usually water, which may be dyed to aid in visual leak detection, and pressurization of the pipeline to the specified test pressure. The test pressure is normally chosen higher than the working pressure to create a factor of safely. After shutting off the supply valve, the pressure tightness can be tested by observing whether there is pressure loss. The location of a leak can be visually identified since the water contains a dye and repairs can be performed if a leak or severe corrosion is found. Hydrostatic testing involves production shutdown which is again very costly.

To inspect insulated O&G pipelines for external corrosion, in addition to ILIs and hydrostatic pressure testing, the following three techniques are also used; Neutron Backscatter, X-rays or Radiography [16], and Pulsed Eddy Current [17].

In the case of the Neutron Backscatter method, the technique is not used to directly detect external corrosion of O&G pipelines but to detect presence of water underneath the thermal insulation. A radioactive source emits high-energy neutrons into the insulation. If there is moisture in the insulation, the hydrogen nuclei attenuate the energy of the neutrons. If presence of water under the thermal insulation is detected using Neutron Backscatter method, then, most likely, external corrosion under insulation (CUI) is occurring.

When thermal insulation is present, X-rays or Radiography method can be used to detect change in pipe wall thickness due to corrosion. Sections of the pipe wall, suspected of having corrosion, can be exposed to Iridium 192 or Cobalt 60 gamma rays, and the radiation transmitted through the pipe is captured using sensitive films. The sensitive film carries the image of the pipe section and the image can be used to calculate the remaining wall thickness of the pipe. This method is effective in detecting CUI, but it is limited to small area coverage. The radiation hazard to radiography personnel who perform the inspection is also of concern.

The Pulsed Eddy Current method is another inspection technique used to detect corrosion under insulation. Eddy currents are generated in the pipeline wall due to magnetic field produced by a coil. The coil-induced magnetic field is created by applying and controlling the electrical current to the coil. The thicker the pipeline wall, the longer it takes for the eddy currents to decay to zero. This property and technique are used to detect remaining wall thickness of pipelines.

All the techniques mentioned above, are inspection techniques. They are not able to provide real-time or on-demand corrosion monitoring for the O&G pipelines. As the above-mentioned inspection techniques are typically used once every few years, aggressive pipeline corrosion can occur in between inspection intervals without the knowledge of pipeline operators.

The demand for developing a corrosion detection sensor for O&G pipelines is ever increasing due to industry regulations and an aging pipeline network. The fact that most O&G flowlines cannot be pigged or are very difficult to pig, also plays a role in the demand for development of corrosion detection sensors that can be permanently deployed in the field. The industry is still in the search of a sensing solution that could be permanently deployed in the field, does not affect oil production, will be safe in volatile environments, cost-effective, require no or little power, and will not require any alteration or intrusion in the pipeline wall. Many of the existing inspection or health monitoring technologies violate the above-mentioned requirements, but our proposed sensor meets all the mentioned requirements as will be seen later.

Several sensors have been proposed for monitoring of pipelines based on optical fibers. Ren et al. [18] proposed to monitor hoop strain in the pressurized pipe, which will change as the pipe wall gets thinner due to degradation from corrosion or erosion. This solution is suitable for determining both external and internal corrosion [19,20], but it also involves removal of protective coatings and is sensitive to pressure fluctuations during pipeline operations. Lawand et al. [21] proposed a corrosivity sensor that can be placed in the vicinity of the exposed pipeline. This solution was based on Radio-Frequency Identification (RFID) technology and required an inspection crew to walk along the pipeline in order to interrogate each sensor.

In this paper, an external corrosion detection sensor, based on fiber optics and strain change, is proposed. It can be placed on the exposed O&G pipelines and interrogated remotely at any time. The size of the sensor is determined using Castigliano's second theorem and the sensor design equations are verified using the Finite Element Analysis (FEA) method. The sensor prototype was manufactured and tested in an accelerated corrosion test. The OBR 4600 Optical Back-scatter Reflectometer (OBR) was used as the fiber optic interrogator in the experimental apparatus.

The results obtained from the FEA, closed form equations, and the experiment show excellent correlations. Experimental results prove the feasibility of the proposed sensor. This sensor is able to provide corrosivity environment near the O&G pipeline and help prevent leaks by providing early warning for the operators to perform detailed inspection of a specific location on the pipeline.

The sensor is very safe as it involves only light traveling through the optical fiber. The only challenge is that the proposed corrosion sensor is unable to measure the corrosion rate in real-time, but it is able provide an average corrosion rate when the sacrificial metal element in the sensor fails.

2. External Corrosion Detection Sensor

The corrosion detection sensor consists of a semicircular shaped plastic curved beam (shown in Figure 1a), attached to a dog-bone-shaped metal component (see Figure 1b), with exact material property as of the pipeline, and an optical fiber with FBG sensors. Each FBG sensor will be glued to a semicircular shaped plastic curved beam at point A, as shown in Figure 1a.

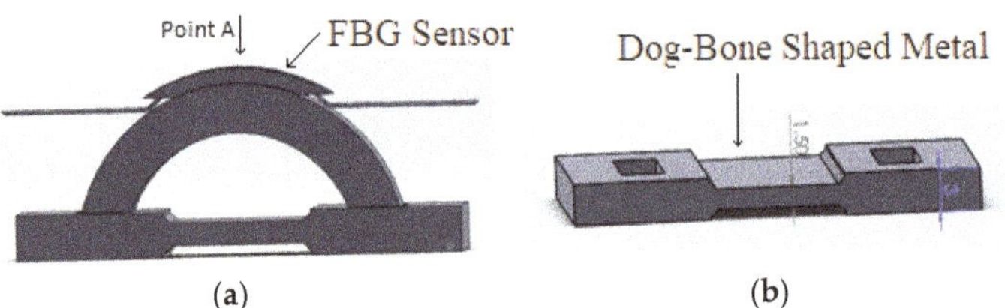

Figure 1. (**a**) External corrosion detection sensor. (**b**) Dog-bone-shaped metal.

As shown in Figure 2, the outer diameter of the semicircular shaped plastic curved beam, d_o, is chosen to be larger than the length L_o by design. When the semicircular shaped plastic curved beam is radially compressed inwards and inserted to the two holes of the dog-bone-shaped metal, the metal and the plastic curved beam at point A are placed in tension.

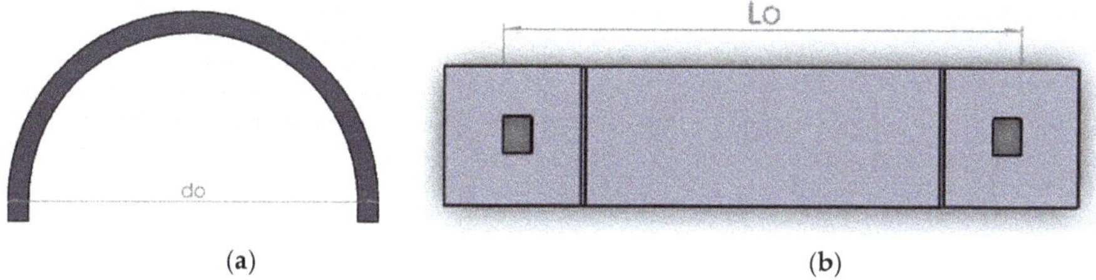

Figure 2. (**a**) Semicircular plastic component. (**b**) Metal similar to pipeline.

This tension is also felt by the FBG sensor as it is glued to the semicircular shaped plastic curved beam at point A. Once tension is observed at point A by the interrogator, the interrogator is zeroed. As the dog-bone-shaped metal corrodes, resulting in failure of the dog-bone-shaped metal, the tension at point A is relieved; thus, a signal is picked up by the interrogator. Figure 3 shows how the sensor of Figure 1 will be implemented in the field.

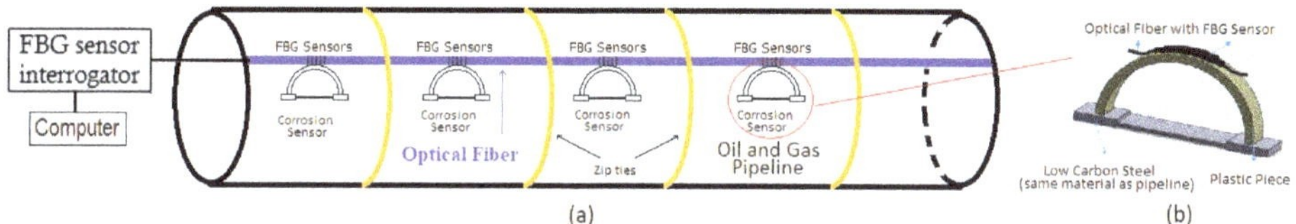

Figure 3. An O&G pipeline with external corrosion detection sensors, spaced every few meters. (**a**) Pipeline with an optical fiber and corrosion detection sensors, (**b**) Corrosion detection sensor.

In the field, multiple optical fibers with FBG sensors will be placed on each O&G pipeline. The fiber optic cables with corrosion sensors are attached to the pipeline by simple zip ties, straps, or large hose clamps, see Figure 3. Neither the sacrificial steel material nor the semicircular plastic part shall be rigidly attached to the pipe. Our proposed sensor sits very near the pipeline but not rigidly attached to the pipeline, basically acting as an environmental corrosivity sensor near the pipeline. When the sensor fails due to corrosion, most likely the pipeline may be already corroded since the corrosion sensor is very near to the pipeline. When corrosion is severe at any pipeline location, the dog-bone-shaped metal corrodes at that location and the prestressed semicircular plastic curved beam will break the dog-bone-shaped metal in two pieces, thus a signal is detected by the interrogator at the control room. When a sensor fails at any particular location, the pipeline inspector will visit the pipeline at that location. He or she will conduct a visual inspection first. If corrosion is observed on the pipeline, he or she will use the inspection techniques such as ultrasound technology or eddy current probe and other techniques to further assess the severity of the pipeline corrosion. If repairs are needed, the pipe will be repaired at that location. As the dimension of the dog-bone-shaped metal is known, and, following the ultrasound inspection, the thickness of pipeline corrosion damage is also known, the time to failure is also known, and an average corrosion rate can be calculated. If no corrosion is observed on the pipeline, only the dog-bone-shaped metal is replaced till the next sensor failure.

Note that as the plastic curved beam can be in the deformed state for a long periods of time, seeing extreme ambient temperatures (high and low temperatures) and humidity conditions, the plastic material needs to exhibit no permanent set or creep and should be able to withstand the temperature and humidity condition of the ambient environment. Also, the plastic material needs to have UV resistance, and its mechanical properties should remain constant with aging.

The number of FBG sensors that one can incorporate within a single fiber depends on the wavelength range of operation of each sensor and the total available wavelength range of the interrogator. FBG strain sensors are often given a 4 nm range. Most commercially available interrogators provide a measurement range of 60 to 80 nm. At 80 nm wavelength range, one can only incorporate 20 sensors per fiber. A new interrogator with 160 nm wavelength range and 16 channels has been recently made available to the public. Assuming 4 nm wavelength range of operation for each FBG sensor, and using this new interrogator, at least 40 FBG sensors can be implemented in a single fiber. With a sixteen-channel interrogator, at least 640 FBG sensors can be implemented per pipeline without using any optical switches. Most flowlines are 1 to 4 km long. Assuming a 4 km long oil or gas pipeline (flowline), without using any optical switches, one can monitor corrosion of O&G pipelines every 6.25 m. With the use of optical switches, we can further reduce the distance between the corrosion sensors. If a 3 nm range is used for each FBG sensor, then the total number of sensors that can be used on each pipeline will be 848, meaning every 5 m we can place an external corrosion detection sensor.

Crevice corrosion refers to the localized attack on a metal surface at, or immediately adjacent to, the gap or crevice between two joining surfaces. To eliminate the crevice corrosion between the dog-bone-shaped metal and the plastic curved beam, the ends of the dog-bone-shaped metal can be coated with a thin film of plastic or some anticorrosion coatings. By coating the ends, we force the corrosion to only occur at the center of the dog-bone-shaped metal; thus, eliminating crevice corrosion. Crevice corrosion between the pipeline and the sensor needs to be also avoided. That is why our sensor is not attached to the pipeline but only attached to the optical fibers.

As in most oil fields 3000 to 4000 flowlines maybe present and each pipeline may have anywhere from 600 to 800 sensors, there is a need to keep the cost of the corrosion detection sensor low. To keep the cost of the sensor low, 3D printers can be used to manufacture the plastic semicircular shaped component. The 3D printing technology to print the plastic component is very mature. As for metals, in recent years, major research has taken place with printing metal components and some 3D printing companies claim that they are able to print steel with 0.02% to 2% carbon content. The authors believe in few years' time, the technology to print all sensor components using 3D printers will be there, and the cost of the corrosion detection sensor will go down as time goes on.

Sensor Design Equations

To design the corrosion detection sensor of Figure 1 at different sizes, developing a closed form design equation for the above proposed sensor is required. Most optical fibers can only handle strains up to a limit and design equations will thus be necessary to make sure the strain of the optical fiber at the FBG sensor locations does not exceed the manufacturer specified strain limit. To verify the developed closed form design equations, ANSYS finite element software will be used to compare the analytical results with ANSYS results.

In Figure 2, as we mentioned earlier, the distance d_o is larger than L_o, meaning to insert the semicircular plastic curved beam to the holes of the dog-bone-shaped metal (see Figure 2b), one needs to compress the semicircular plastic curved beam radially inwards. When the semicircular plastic curved beam is compressed radially inwards and then inserted to the dog bone shaped metal piece, we get the picture of Figure 4a. The boundary conditions applied to the plastic curved beam will be as follows, see Figure 4b.

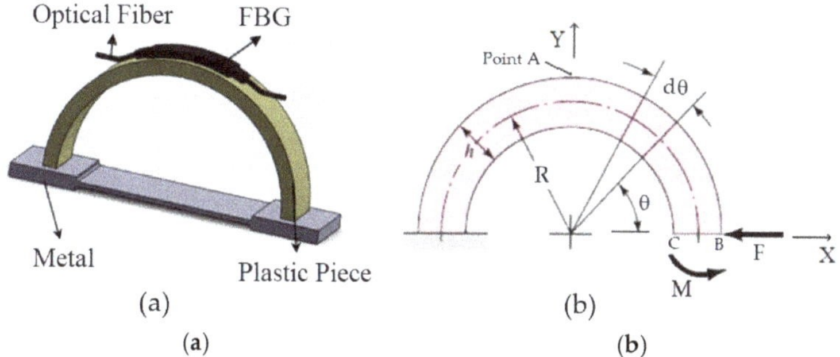

(**a**) (**b**)

Figure 4. (**a**) Assembled sensor. (**b**) Boundary conditions.

In actual practice, to insert the semicircular plastic curved beam to the dog-bone-shaped metal, a radial displacement in the negative X-direction is given to point B, but to apply that displacement, a force F is required to be applied to move point B in the negative x-direction. The force F causes not only to move point B and surface BC in the negative X-direction but also in the negative Y-direction.

Force F rotates surface BC in the clock-wise direction. A counterclockwise moment M is needed to be applied to the right end of the curved beam in order to bring the slope of surface BC to zero and keep the motion of the BC-surface in the Y-direction to zero.

At any angle θ, the internal forces and moments will be as follows, see Figure 5.

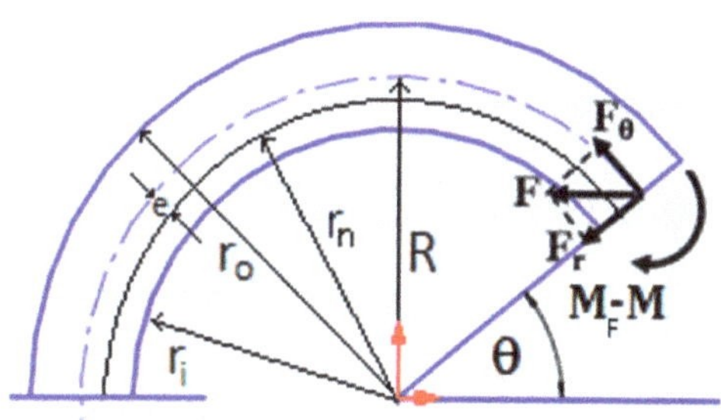

Figure 5. Plastic semicircular curved beam internal forces and moments.

The internal forces and moments (shown in Figure 5), at any angle θ, are equal to

$$F_\theta = F\sin\theta \quad (1)$$

$$F_r = F\cos\theta \quad (2)$$

$$M_F = FR\sin\theta \quad (3)$$

The curved beam stress, at any angle θ, and at any radius r is equal to [22]

$$\sigma = \frac{M_t(r - r_n)}{Aer} - \frac{F_\theta}{A} \quad (4)$$

where $M_t = (M_F - M)$. The strain, at any angle θ, and at any radius r will be equal to

$$\varepsilon = \frac{M_t(r - r_n)}{AeEr} - \frac{F_\theta}{AE} \quad (5)$$

The strain limit of the FBG sensor determines the maximum deflection that can be given to point B. The Castigliano's second theorem is used to develop the closed form equations.

In Figure 5, for a curved beam with a rectangular cross section, the width is assumed to be W, and the outer and the inner radii are r_o and r_i, respectively. The location of neutral axis is given by Equation (6).

$$r_n = \frac{A}{\int_A \frac{dA}{r}} = \frac{W(r_o - r_i)}{\int_r \frac{W(dr)}{r}} = \frac{W(r_o - r_i)}{W\int_{r_i}^{r_o} \frac{1}{r}dr} = \frac{(r_o - r_i)}{\ln r|_{r_i}^{r_o}} = \frac{(r_o - r_i)}{\ln r_o - \ln r_i} \quad (6)$$

If the curved beam is sectioned at an angle, θ, see Figure 5, as explained earlier, there will be two forces, F_r, and F_θ and a moment, M_t, at that section. The total strain energy of the semicircular curved beam from $0 < \theta < \pi$, can be calculated by adding four terms, shown below in Equation (7).

$$U = \int \frac{{M_t}^2}{2EAe}d\theta + \int \frac{F_\theta^2 r_c}{2EA}d\theta + \int \frac{M_t F_\theta}{EA}d\theta + \int \frac{CF_r^2 r_c}{2AG}d\theta \quad (7)$$

The first strain energy term in Equation (7) is generated by the moment M_t, the second term is due to axial force F_θ, the third term accounts for coupling energy due to M_t and axial force F_θ, and the fourth term is due to transverse shear energy due to radial force F_r [23]. The parameter C in the fourth term is the strain-energy correction factor for transverse shear, equal to 1.2 when the cross section is rectangular [24].

Using Equation (7) and conducting a lengthy mathematics, please see reference [22] for more details, we arrive to the following two important equations:

$$u_x = u_{xF} - u_{xM} = F\left(\frac{\pi r_c^2}{2EAe} - \frac{\pi r_c}{2EA} + \frac{\pi C r_c}{2AG}\right) - M\left(\frac{2r_c}{EAe} + \frac{2}{EA}\right) \quad (8)$$

$$u_y = u_{yF} - u_{yM} = F\left(\frac{2r_c^2}{EAe} - \frac{2r_c}{EA}\right) - M\frac{\pi r_c}{EAe} \quad (9)$$

For the design of Figure 4, $u_y = 0$, and u_x is known. Knowing u_x and u_y, force F and moment M can be calculated from Equations (8) and (9). Knowing F and M, we can now calculate the strain ε using Equation (5). The maximum strain occurs at θ = 90 degrees and $r = r_o$.

3. Equation Validation Using Finite Elements

To validate Equations (5), (8), and (9), ANSYS finite element (FE) software was used to model a semicircular plastic beam, made from PVC material, with following dimensions (see Table 1).

Table 1. Mechanical property and model geometry.

Parameters	Values, SI Units
Inner radius, r_i	14.22 mm
Outer radius, r_o	15.37 mm
Thickness, t	1.15 mm
Width, W	10 mm
Young's Modulus of PVC, E	3.4 GPa
Poisson's ratio, ν	0.4
End deflection given to the right end, δ	1.5 mm
Strain, ε_x, obtained from ANSYS, με	4487
Strain, ε_x, calculated from Equation (5), με	4492
Error	0.1%

As was explained earlier in Section 2, the semicircular shaped plastic curved beam is chosen to be larger than the dog-bone-shaped metal. To insert the plastic curved beam to the metal, the beam is radially compressed inwards and inserted to the two holes of the dog-bone-shaped metal. The radial inward motion here is 1.5 mm, as shown in Table 1. The ANSYS 2D FE model is shown in Figure 6. Plane183, 2D 8-node element with "plane stress with thickness" option was used to mesh the curved beam. The thickness of the curved beam (thickness is in to the paper) is set as W = 10 mm. Both ends of the curved beam are held fixed in the "u_y" direction. The left end is given u_x = +0.75 mm and the right end is given u_x = −0.75 mm motion, simulating 1.5 mm inward radial motion. The FE model of Figure 6 shows eight elements through the radial thickness.

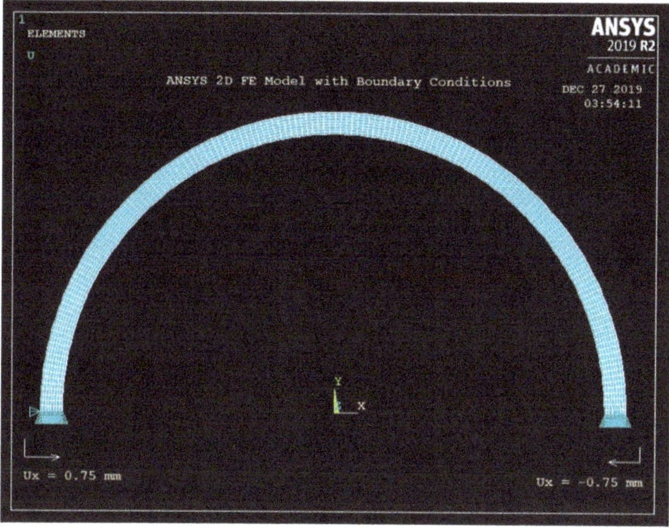

Figure 6. ANSYS Finite Element (FE) Model.

Figure 7 shows the deformed and the undeformed shape of the semicircular curved beam due to 1.5 mm radial displacement.

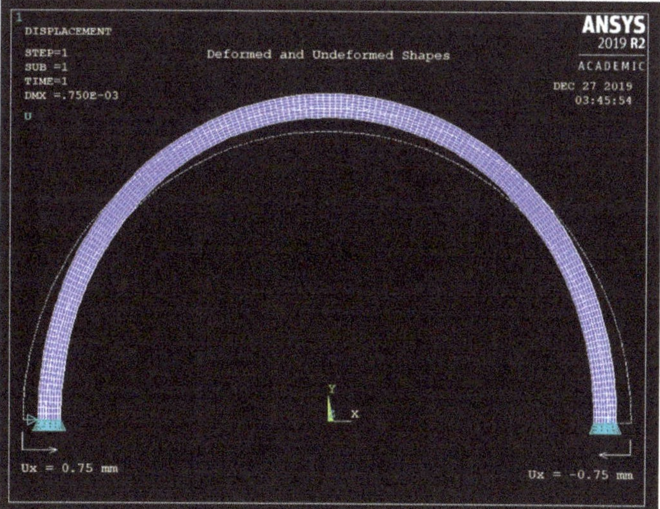

Figure 7. The deformed and the undeformed shapes with boundary conditions.

As can be seen from Figure 8, the maximum strain occurs at $r = r_o$, at $\theta = 90$ degrees, and at point A (see Figure 4), and it is equal to 4487 $\mu\varepsilon$. The maximum strain was calculated using Equations (8), (9) and finally (5) and was found to be 4492 $\mu\varepsilon$. The error is only 0.1% (see Table 1). ANSYS FE stress analysis validates the derived equations. Equations (1)–(9) can now be used to design the corrosion sensor of Figure 4.

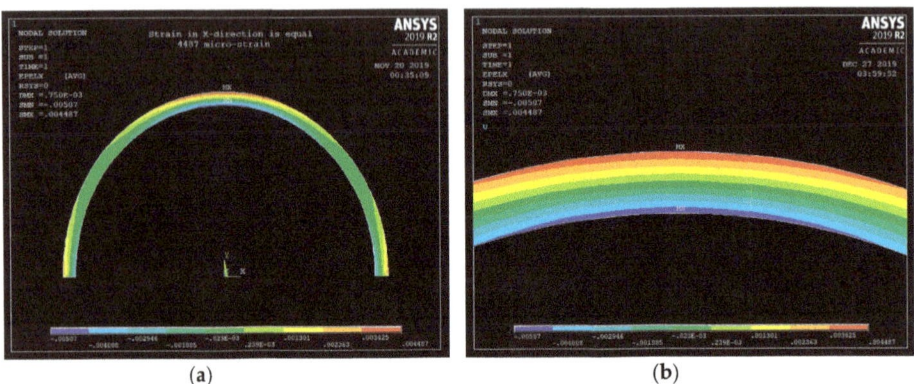

(a) (b)

Figure 8. (**a**) Curved beam strain in the x-direction, (**b**) Strain in the x-direction zoomed to point A.

4. Experimental Validation

Figure 9 shows the actual corrosion detection sensor and the dog-bone-shaped metals (low carbon steel), made from the same O&G pipeline material. The semicircular curved beam is made of PVC for the experiment, but long-term, it will be constructed using a 3D printer. Three corrosion detection sensors were constructed using the three dog-bone-shaped metals shown in Figure 9. The dimensions of the dog-bone-shaped metals are shown in Figure 10. The three corrosion sensors were placed in series as shown in Figure 11; Figure 12. Figure 11; Figure 12 show the experimental apparatus used for validating the performance of the proposed external corrosion detection sensor.

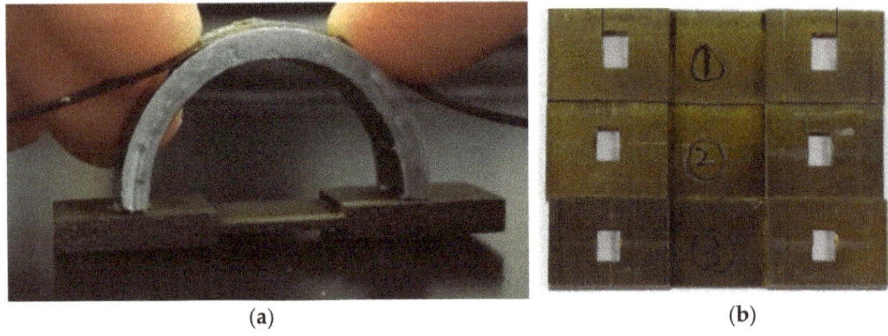

(a) (b)

Figure 9. (**a**) Actual corrosion detection sensor. (**b**) The dog-bone-shaped metals.

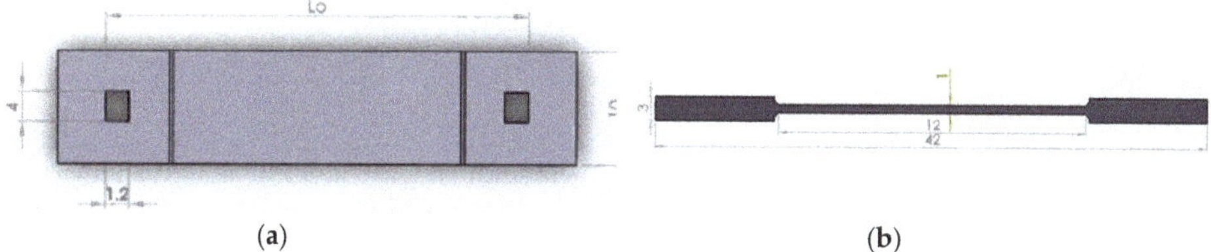

Figure 10. Dimensions of the dog-bone-shaped metal (Lo = 29.24 mm). (**a**) Top view; (**b**) Side view.

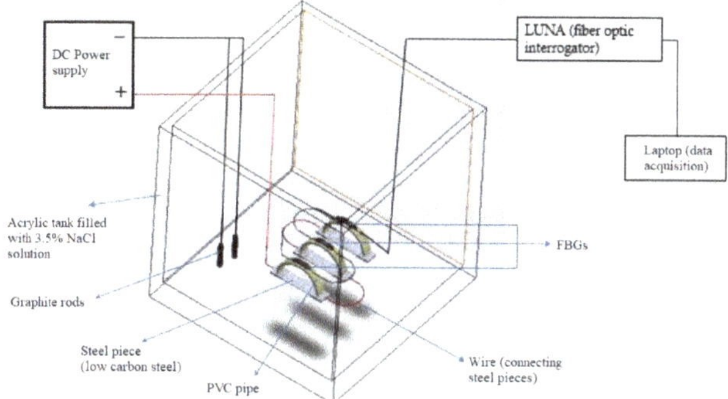

Figure 11. Schematic of the experimental test setup to test the proposed corrosion detection sensor.

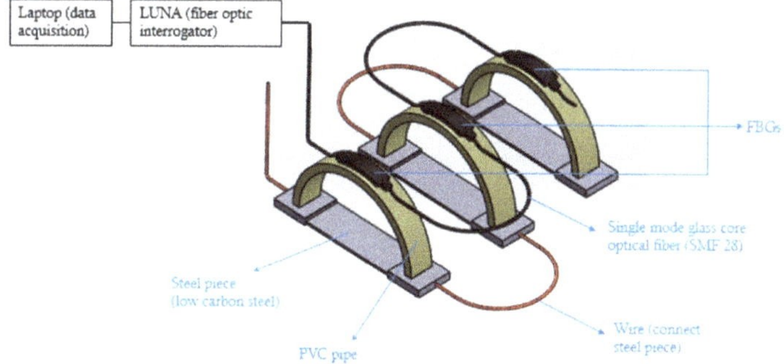

Figure 12. Schematic of details of proposed sensor.

The optical fiber used in our experiment is a single mode fiber (SMF 28) with 5 FBG sensors, with FBG sensors 1 m apart. The first FBG Sensor is located at 0.5 m from the end of the optical fiber that connects to the interrogator. Glue was used to attach the outer surface of the PVC rings to the bottom surface of the FBG sensors (see Figure 9; Figure 12). To make sure the bond between the FBG sensor and the semicircular plastic curved beam remains intact at high temperature of the pipeline (≤150 °C) and high humidity of the desert for prolonged periods of time, the authors of [25] recommend a compound, based on a combination of ceramic fillers with an epoxy binder that is applied with a brush technique, and this compound can withstand temperatures in the region of 260 °C and humidity of 75%.

The optical fiber was connected to an interrogator with wavelength range of 1270–1340 nm. The connection termination type between the optical fiber and the interrogator was FC-APC.

Most O&G pipelines are made of API 5L X42 to X70 material with carbon content from 0.16% to 0.28% (mild or low carbon steels). Other mild or low carbon steel materials are also used in the O&G industry. The dog-bone-shaped metal pieces are required to be made from the same pipeline material. For our experiment, API 5L X65 material was used. As can be seen from Figure 11, the dog-bone-shaped metals are linked to one another using wires. Wires are soldered to the dog-bone-shaped metals and the solder joints were coated with an anticorrosion coating. Figure 13 shows the entire experimental test set-up consisting of a laptop, an interrogator, corrosion cell, and a power supply.

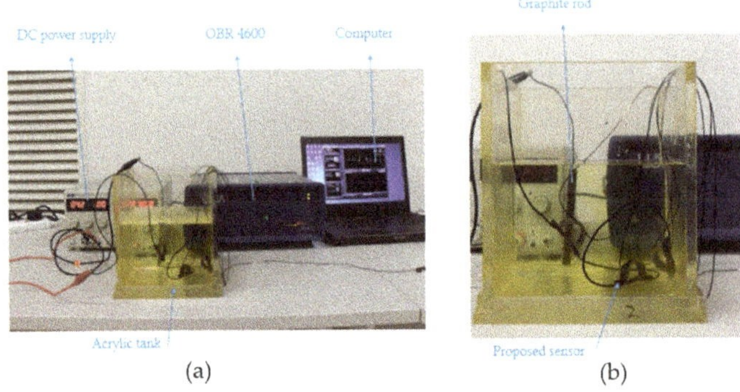

Figure 13. (**a**) Experimental test set-up. (**b**) and details of the apparatus inside acrylic tank.

A DC power supply (4 V and 1 A) was used to accelerate the corrosion reaction. Graphite rods were the cathode and the low carbon steel pieces were the anode, all of them are placed into 3.5% by weight NaCl solution which is regarded as the electrolyte. The thinnest section of the dog-bone-shaped metals is in the middle, with thickness of 1 mm. It was at this location where the corrosion failure first occurred. The experiment kept running until complete failure of one of the sensors. It took almost 14 h to corrode one of the sensors. The three corroded sensors are shown in Figure 14.

Figure 14. The three corrosion sensors after accelerated corrosion failure.

As was explained on page 10, the optical fiber used in our experiment is a single mode fiber (SMF 28) with five FBG sensors, with FBG sensors 1 m apart. The first FBG Sensor is located at 0.5 m from the end of the optical fiber that connects to the interrogator. Thus, the first FBG Sensor is located at 0.5 m, the 2nd FBG sensor at 1.5 m, the third at 2.5 m, and so on. In our experiment, only three FBG sensors out of five were used. Figure 15 shows the strain observed by all the FBG sensors before and after the failure of the first corrosion sensor. The yellow color trace is the strain observed by all the FBG sensors before the failure of the 1st corrosion sensor, which is about zero and the blue color trace is the strain observed by all the FBG sensors after the failure of the 1st corrosion sensor. The y-axis shows the strain (in micro strain) of all the FBG sensors and the x-axis indicates the location (in meters) of the FBG sensors. In the upper right corner of the Figure 15, the graph indicates "time domain". This implies that the strain at each FBG sensor location can change with time. If we would have continued the accelerated corrosion, we would have had the failure of the 2nd or 3rd corrosion sensors and in Figure 15, we would have seen additional peaks, as time goes by.

The peak value of the strain, at 0.5 m, where the first FBG sensor is located, is −3100 $\mu\varepsilon$. When the semicircular plastic curved beam is inserted to the dog-bone-shaped metal, point A (where the FBG sensor is attached to) will be strained. The strain at point A will be in tension. We tared (zeroed out) that strain in the interrogator. However, when the dog-bone-shaped metal corrodes, the semicircular plastic curved beam is released thus negative strain is observed on the interrogator display monitor.

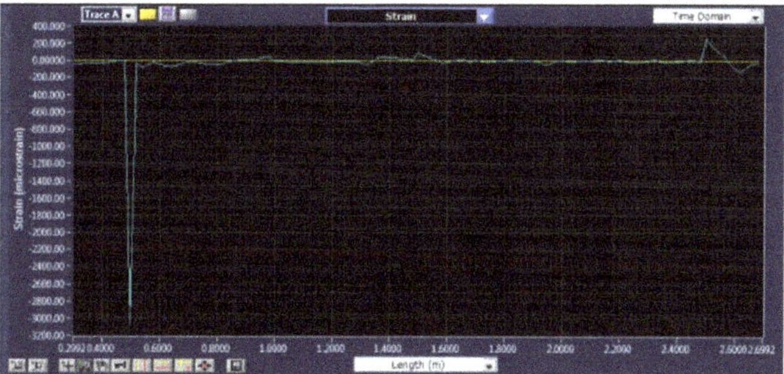

Figure 15. (**a**) Yellow trace displays strain before sensor failure. (**b**) Blue trace displays strain after sensor failure due to corrosion.

It was not necessary to tare the strain. We could have done the opposite, meaning not zero out the strain and leave the strain at point A as it is and when the dog-bone-shaped metal is corroded, the strain at point A would have gone back to zero.

5. Discussion

The external corrosion rate of O&G pipelines, based on NACE Standard (NACE RP0-502: Pipeline External Corrosion Direct Assessment Methodology), is ~0.4 mm/year [26]. Due to this low corrosion rate, larger diameter O&G pipelines are inspected every 3 to 5 years. Pipeline pigging and data analysis is very costly, and due to this high cost, most O&G companies have opted for 5-year inspection interval rather than the shorter ones. The smaller diameter pipelines (flowlines) are not even pigged due to diameter of the pipeline being small and no launch and retrieval stations being available to launch and retrieve the PIG. Even if the flowlines were piggable, since pigging occurs every 5 years, if there is an aggressive corrosion occurring due to the change in the pipeline environment or surrounding, O&G industry is vulnerable to possible field accidents in between inspection interval.

As, in normal conditions, the external corrosion rate of O&G pipelines is low, real-time corrosion monitoring is not really necessary, but as long as corrosion state of the pipelines are monitored monthly or every few months, this monitoring protects the O&G industry against aggressive pipeline corrosion. Our proposed sensor can provide corrosion state of the O&G pipelines, as and when needed. Our sensor is basically monitoring the corrosivity environment near the O&G pipelines.

In this paper, we embarked upon designing a very low-cost corrosion detection sensor for O&G pipelines, requiring no power. The sensors are passive, very simple, and they are permanently deployed in the field, unlike ILIs, ultrasound probes, X-ray, or radiography tools. The sensor involves a semicircular plastic component, a sacrificial dog-bone-shaped metal made from the same pipeline material, and optical fibers. The optical fibers have up to 20 to 40 FBG sensors per optical fiber, depending on which type of interrogator used. The optical fibers are attached to the gas and oil pipelines using zip ties, straps, or large hose clamps and our sensors are attached to the FBG sensors. When corrosion is severe at any pipeline location, the dog-bone-shaped metal corrodes at that location and eventually fails and thus a signal is detected by the interrogator at the control room. Once a signal is picked up at the control room, inspection personnel will visit the pipeline at that location and conduct visual inspection first, and possibly ultrasound, X-ray or radiography inspection. If the corrosion is not severe, the dog-bone-shaped metal is replaced until the next failure. If severe corrosion is observed at any pipeline location, the pipeline is inspected using ultrasound probes, or X-rays, or eddy current probes or other inspection methods. Depending on the severity of the corrosion, pipeline may be repaired and the metal dog-bone-shaped metal is replaced till the next sensor failure. In the proposed corrosion monitoring system of this paper, all the communication between the corrosion detection sensors and control room is through the optical fibers. As only light is involved, and there are no batteries or electricity of any kind (since interrogator sits inside the control room), this corrosion monitoring system is very safe.

The thickness of the sacrificial dog-bone-shaped metal was chosen to be 1 mm. With corrosion rate of 0.4 mm/yr, it would take 2.5 years for the sacrificial dog-bone-shaped metal to corrode completely through the 1 mm thickness. Two and a half years is the mid-span of the 5-year inspection interval. If after 2.5 years, no signals are observed on the remote interrogator for any of the corrosion sensors, then one can conclude the corrosion rate is less than 0.4 mm per year for that pipeline. When few sensors fail earlier than 2.5 years, then it implies we are having aggressive corrosion occurring at some specific pipeline locations. Our proposed sensor is very helpful to the pipeline operators as they can now go to those specific locations and find out why there is a higher rate of corrosion at those locations. Steps can be taken at those high corrosion rate locations to lower the corrosion rate back to 0.4 mm/yr or lower.

If there is a leakage of the crude oil, as crude oil is hot, the heat can create tension strain on the FBG sensors thus a signal can be picked up by the interrogator in the control room. The proposed sensor of this paper can not only detect occurrence of pipeline corrosion but also the pipeline fluid leakage.

Ground settlement is a geological phenomenon of ground elevation changing (vertical movement of the ground) caused by the compression of earth's crust surface soil due to natural and unnatural events. Caving in or sinking of the ground is one form of ground settlement. If the ground caves in, naturally, the O&G pipelines foundations will also cave in. Our fiber optic-based corrosion monitoring system proposed in this paper, most likely, does not capture ground settlement since neither the optical fiber nor the corrosion sensors are rigidly attached to the pipeline unless the ground cave-in is deep. There are few O&G companies already using optical fibers on their O&G pipelines to monitor the potential security risks, detect pipeline temperature, and detect ground settlement. For those companies, presently having optical fibers on their pipelines, it makes sense to bundle our corrosion detection system to their existing fiber optics monitoring system to also detect pipeline corrosion.

6. Conclusions

The grating length of most FBG sensors is anywhere from 5 to 20 mm. If the grating length is too large, the strain experienced by the FBG sensor won't be the strain of point A but the average strain near point A (See Figure 4b). The FBG sensor, used in our experiment, had a grating length of 10 mm. The FBG sensor was picking up the average strain underneath the FBG sensor and not strain at point A only, and that is why the experimental tension strains are below the predicted ANSYS strain results.

Due to the differences in physical–mechanical properties of the matrix material and the FBG sensor, and the adhesive, the strains measured by the FBG sensor may not be equal to the actual strains experienced by the matrix material. When the FBG sensor is glued on to the PVC semicircular curved beam, the area of moment of inertia is changed and the FBG sensor thickness, material and adhesive used can have some impact on the actual measured strain.

The exact value of strain obtained at point A is not important as long as the strain is below the strain limit of the optical fiber and the strain is large enough to be detected by the interrogator; thus, the proposed sensor of this paper can be used to detect occurrence of pipeline corrosion and pipeline leakage, and provide average pipeline corrosion rate.

Author Contributions: N.V. and X.W. conceived the sensor design, P.R. conceived and designed the experiments; X.W. and O.S. performed the experiments; N.V., X.W. and F.F.Y. conducted the FE analysis, F.F.Y. and N.V. looked into wireless communications, X.W. and N.V. developed the closed form equations and analyzed the data; and N.V., X.W. and O.S. wrote the paper. All authors have read and agreed to the published version of the manuscript.

Acknowledgments: Funding for this research project came from the Khalifa University of Science and Technology.

Conflicts of Interest: The authors declare no conflicts of interest. The funding sponsors had no role in the design of the study; in the collection, analyses, or interpretation of data; in the writing of the manuscript; and in the decision to publish the results.

References

1. Green, K.P.; Jackson, T. Safety in the Transportation of Oil and Gas: Pipelines or Rail? In *Fraser Research Bulletin*; Fraser Institute: Vancouver, BC, Canada, 2015; ISSN 2291-8620. pp. 1–14.
2. Peabody, A.W. *Peabody's Control of Pipeline Corrosion*, 3rd ed.; Bianchetti, R., Ed.; NACE International: Houston, TX, USA, 2018; chapters 2 and 3.
3. Pipeline Safety: Ineffective Protection, Detection, and Mitigation of Corrosion Resulting from Insulated Coatings on Buried Pipelines. PHMSA (US Pipeline and Hazardous Materials Safety Administration) Advisory Bulletin, Docket No. PHMSA–2016–0071. 21 June 2016. Available online: https://www.phmsa.dot.gov/regulations-fr/notices/2016-14651 (accessed on 10 December 2019).
4. Ho, M.; El-Borgi, S.; Patil, D.; Song, G. Inspection and monitoring systems subsea pipelines: A review paper. *Struct Health Monit* **2019**. [CrossRef]
5. Wright, R.F.; Lu, P.; Devkota, J.; Lu, F.; Ziomek-Moroz, M.; Ohodnicki, P. Corrosion Sensors for Structural Health Monitoring of Oil and Natural Gas Infrastructure: A Review. *Sensors* **2019**, *19*, 3964. [CrossRef] [PubMed]
6. Xie, M.; Tian, Z. A review on pipeline integrity management utilizing in-line inspection data. *Eng. Fail Anal.* **2018**, *92*, 222–239. [CrossRef]
7. Sundaram, B.A.; Kesavan, K.; Parivallal, S. Recent Advances in Health Monitoring and Assessment of Inservice O&G Buried Pipelines. *J. Inst. Eng India Ser. A* **2018**, *99*, 729–740.
8. Vanaei, H.R.; Eslami, A.; Egbewande, A. A review on pipeline corrosion, in-line inspection (ILI), and corrosion growth rate models. *Int. J. Pres Ves. Pip.* **2017**, *149*, 43–54. [CrossRef]
9. Ameh, E.S.; Ikpeseni, S.C.; Lawal, L.S. A Review of Field Corrosion Control and Monitoring Techniques of the Upstream Oil and Gas Pipelines. *NJTD* **2017**, *14*, 67–73. [CrossRef]
10. Hedges, B.; Papavinasam, S.; Knox, T.; Sprague, K. Monitoring and Inspection Techniques for Corrosion in Oil and Gas Production. In Proceedings of the NACE International. Corrosion 2015 Conference and Expo, Houston, TX, USA, 15–18 March 2015.
11. Varela, F.; Tan, M.Y.; Forsyth, M. An overview of major methods for inspecting and monitoring external corrosion of on-shore transportation pipelines. *Corros. Eng. Sci. Technol.* **2015**, *50*, 226–235. [CrossRef]
12. Tiratsoo, J. *Pipeline Pigging & Integrity Technology*, 4th ed.; Tiratsoo, J., Ed.; Clarion Technical Publishers: Houston, TX, USA, 2013.
13. Shi, Y.; Zhang, C.; Li, R.; Cai, M.; Jia, G. Theory and application of magnetic flux leakage pipeline detection. *Sensors* **2015**, *15*, 31036–31055. [CrossRef] [PubMed]
14. Reber, K.; Beller, M.; Willems, H.; Barbian, O.A. A new generation of ultrasonic in-line inspection tools for detecting, sizing and locating metal loss and cracks in transmission pipelines. *Proc. IEEE Ultrason. Symp.* **2002**, *1*, 665–671.
15. McAllister, E.W. *Pipeline Rules of Thumb Handbook: A Manual of Quick, Accurate Solutions to Everyday Pipeline Engineering Problems*; Elsevier Science & Technology: Waltham, MA, USA, 2013; pp. 150–172.
16. Bossi, R.H.; Iddings, F.A.; Wheeler, G.C. *Nondestructive Testing Handbook: Radiographic Testing*, 3rd ed.; American Society for Nondestructive Testing: Columbus, OH, USA, 2002; Volume 4, Chapter 19.
17. Udpa, S.S.; Moore, P.O. *Nondestructive Testing Handbook: Vol 5. Electromagnetic Testing (ET)*, 3rd ed.; Chapter 15: Chemical and Petroleum Applications of Electromagnetic Testing; American Society for Nondestructive Testing: Columbus, OH, USA, 2004; pp. 381–398.
18. Ren, L.; Jia, Z.G.; Li, H.N.; Song, G. Design and experimental study on FBG hoop-strain sensor in pipeline monitoring. *Opt. Fiber Technol.* **2014**, *20*, 15–23. [CrossRef]
19. Jiang, T.; Ren, L.; Jia, Z.; Li, D.; Li, H. Pipeline internal corrosion monitoring based on distributed strain measurement technique. *Struct Control Health Monit.* **2017**, *24*, 1–11. [CrossRef]
20. Ren, L.; Jiang, T.; Jia, Z.; Li, D.; Yuan, C.; Li, H. Pipeline corrosion and leakage monitoring based on the distributed optical fiber sensing technology. *Meas. J. Int. Meas. Confed.* **2018**, *122*, 57–65. [CrossRef]
21. Lawand, L.; Shiryayev, O.; Al Handawi, K.; Vahdati, N.; Rostron, P. Corrosivity sensor for exposed pipelines based on wireless energy transfer. *Sensors* **2017**, *17*, 1238. [CrossRef] [PubMed]
22. Wang, X. External Corrosion Detection of Oil and Gas Pipelines Using Fiber Optics. Master's Thesis, Khalifa University, Abu Dhabi, UAE, December 2017.

23. Budynas, R.G.; Nisbett, J.K. *Shigley's Mechanical Engineering Design*, 9th ed.; McGraw Hill: New York, NY, USA, 2011; pp. 169–171.
24. Budynas, R.G. *Advanced Strength and Applied Stress Analysis*; McGraw-Hill Education: Boston, MA, USA, 1999.
25. Wnuk, V.P.; Mendez, A.; Ferguson, S.; Graver, T. Process for Mounting and Packaging of Fiber Bragg Grating Strain Sensors for use in Harsh Environment Applications. *Proc. SPIE* **2005**, *5758*. [CrossRef]
26. *NACE RP-0502: Pipeline External Corrosion Direct Assessment Methodology, Item No. 21097*; NACE International: Houston, TX, USA, 2002; p. 53.

Article

Realtime Localization and Estimation of Loads on Aircraft Wings from Depth Images

Diyar Khalis Bilal [1,2], Mustafa Unel [1,2,*], Mehmet Yildiz [1,2] and Bahattin Koc [1,2]

1 Faculty of Engineering and Natural Sciences, Sabanci University, Istanbul 34956, Turkey; diyarbilal@sabanciuniv.edu (D.K.B.); meyildiz@sabanciuniv.edu (M.Y.); bahattinkoc@sabanciuniv.edu (B.K.)
2 Integrated Manufacturing Technologies Research and Application Center, Sabanci University, Istanbul 34906, Turkey
* Correspondence: munel@sabanciuniv.edu

Received: 16 May 2020; Accepted: 12 June 2020; Published: 16 June 2020

Abstract: This paper deals with the development of a realtime structural health monitoring system for airframe structures to localize and estimate the magnitude of the loads causing deflections to the critical components, such as wings. To this end, a framework that is based on artificial neural networks is developed where features that are extracted from a depth camera are utilized. The localization of the load is treated as a multinomial logistic classification problem and the load magnitude estimation as a logistic regression problem. The neural networks trained for classification and regression are preceded with an autoencoder, through which maximum informative data at a much smaller scale are extracted from the depth features. The effectiveness of the proposed method is validated by an experimental study performed on a composite unmanned aerial vehicle (UAV) wing subject to concentrated and distributed loads, and the results obtained by the proposed method are superior when compared with a method based on Castigliano's theorem.

Keywords: structural health monitoring; load localization; load estimation; depth sensor; artificial neural networks; castigliano's theorem

1. Introduction

Structural Health Monitoring (SHM) has been an increasingly important technology in monitoring the structural integrity of composite materials used in the aerospace industry. Because airframes operate under continuous external loads, they will be exposed to large deflections that may adversely affect their structural integrity. Critical components, such as fuselage and wings, should be monitored to ensure long service life. Although these components are designed to withstand different types of loading conditions, such as bending, torsion, tension, and compression, among others, a robust SHM system will be extremely valuable for the aerospace industry for realtime monitoring of loads.

Current aircraft maintenance and repair systems used for structural monitoring rely on load monitoring systems while using strain gauges [1,2], optical measurement systems [3–7], and fiber brag grating (FBG) [8–11] sensors.

The strain gauge based measurements are widely used both in literature and the industry for aircraft wing deflection measurements due to their ability to fit into almost any space and proven high accuracy measurements [1,2]. However, strain gauges have many limitations, such that they cannot be attached to every kind of material, they are easily affected by external temperature variations, and physical scratches or cuts can easily damage them. More importantly, a large number of them need to be installed if one needs to monitor the whole wing due to their small size.

Besides strain gauges, various approaches that were based on optical methods were investigated in the literature for measurement of wing deflections and loads acting on them. Burner et al. [3] presented the theoretical foundations of video grammetric model deflection (VMD) measurement technique, which was implemented by National Aeronautics and Space Administration (NASA) for wind tunnel testing [4]. Afterwards, many research on wing deflection measurement and analysis were motivated by the catastrophic failure of the unmanned aerial vehicle (UAV) Helios [5,6]. Lizotte et al. [7] proposed estimation of aircraft structural loads based on wing deflection measurements. Their approach is based on the installation of infrared lightemitting diodes (LEDs) on the wings; however, the deflection measurements are local, which do not cover the whole wing structure unless a large number of LEDs are installed.

Motivated by the catastrophic failure of Helios, a realtime in flight wing deflection monitoring of Ikhana and Global Observer UAVs were performed by Richards et al. [8] by utilizing the spatial resolution and equal spacing of FBGs. Moreover, FBGs have also been used by Alvarenga et al. [9] for realtime wing deflection measurements on lightweight UAVs. Additionally, chord wise strain distribution measurements that were obtained from a network of FBG sensors were also used by Ciminello et al. [10] for development of an in flight shape monitoring system as a part of the European Smart Intelligent Aircraft project. More recently, Nicolas et al. [11] proposed the usage of FBG sensors for determination of wing deflection shape as well as the associated out of plane load magnitudes causing such deflections. To simulate in flight loading conditions, both concentrated and distributed loads were applied on the wings each with incrementally increasing loads. They reported that their calculated out of plane displacements and load magnitudes were within 4.2% of the actual measured data by strain gauges. As seen from these works in literature, even though FBGs used for SHM purposes have advantages over conventional sensors, they are still highly affected by temperature changes. Moreover, their installation is not an easy task due to their fragile nature, and special attention must be given to the problems of ingress and egress of the optical fibers [12].

Although the aforementioned sensors used in the literature can be used for load monitoring in aerospace vehicles, better technologies are needed to achieve usable sensitivity, robustness, and high resolution requirements. The need for the use of a large quantity of sensors to cover the whole structure is one of the major drawbacks of these sensor technologies. Therefore, a sensor that is capable of full field load measurement from a single unit with high accuracy and precision can become an important alternative. This will also result in a considerable reduction in costs, especially when a fleet of airframes need to be inspected and monitored.

From these works in the literature, it is observed that, in general, a mathematical model for describing the deflections of an aircraft wing is used to study its behavior under different types of loads. However, obtaining physics based models of systems can easily become a difficult problem due to system complexity and uncertainties; thus, effectively decreasing their usefulness. This is especially the case in systems, where lots of data are obtained using different types of sensors, which, in turn, adds more complexity to the system due to inherent sensor noise. In such cases data driven modeling techniques have been found to be more effective since the acquired data already contains all kinds of uncertainties, sensor errors and sensor noise [13]. One of the most effective data driven modeling techniques has been proven to be artificial neural networks (ANN)s [14]. In this regard, many recent applications of neural networks have emerged in literature for monitoring of strains and stresses during load cycles using strain gauges [15], pavement defect segmentation using a deep autoencoder [16], and machine learning based continuous deflection detection for bridges using fiber optic gyroscope [17].

In this work, an ANN based approach for realtime localization and the estimation of loads acting on aircraft wings from full field depth measurements is proposed. The proposed methodology can work with a single external depth image sensor with full field measurement capability for a single wing; thus, one sensor is enough for inspection of the whole wing. Moreover, depth cameras do not require any calibration and can be directly used on any kind of wing regardless if it was made of composites or not due to optical measurement. The proposed framework is able to estimate the magnitude of the load causing wing deflections under both bending and twisting loading conditions; therefore, it is not limited to pure bending case, as was in the work of Nicolas et al. [11]. Moreover, the proposed method is not just limited to the estimation of load magnitudes, but it will also be able to estimate the location of the load causing bending and twisting deflections; therefore, making the localization of the loads possible. The localization of loads can become a very useful tool, especially in the case when one needs to know the nature of the external loads occurring in flight. Using this information, one can estimate the exact flight conditions and, as such, can improve the design of the aircraft based on this new data. More importantly, the proposed framework can operate in real time. To the best of the authors' knowledge, this is the first work in the literature to address the problems of real time localization and estimation of bending and twisting loads causing deflections to structures based on depth imaging and ANN.

The rest of the paper is structured, as follows; in Section 2, the proposed method for real time load monitoring from depth measurements using neural networks is presented. In Section 3, the experimental setup, data collection procedure, and evaluation of the used depth sensor for load monitoring are described. The effectiveness of the proposed approach is validated by an experimental study in Section 4, followed by the conclusion in Section 5.

2. A Data Driven Methodology for Realtime Monitoring of Loads from Depth Images

This work proposes the development of a data driven modeling approach for localization and estimation of loads acting on aircraft wings from full field depth measurements. These measurements can be provided by a multitude of sensors, such as depth cameras. An autoencoder coupled with two different supervised ANN architectures are proposed for the localization and estimation of loads in order to develop these models and ensure realtime performance. The localization part is treated as a multinomial logistic classification problem and the load magnitude estimation as a logistic regression problem, which are explained in detail in the following subsections.

2.1. Dimensionality Reduction Using Autoencoders

To develop data driven models for localization and estimation of loads from depth measurements while providing realtime performance, an autoencoder [18] framework is proposed to be utilized. This is because the full field measurements that are acquired from the depth sensors are inherently rich, but can be very large in size, thus working with them becomes computationally expensive and can hinder realtime performance. Autoencoders can effectively reduce the large number of features obtained from depth sensors while retaining the critical information, thus encoding the original input at a much smaller dimension. Furthermore, to ensure that maximum informative data is obtained, Kullback-Leibler divergence (KL_{Div}) [19] was used to avoid obtaining binary encoded data by enforcing the mean and standard deviation of the encoded data to be some desired values. In this work, logarithmic normalization was utilized to minimize this large range of data due to the possible presence of a large gap between the values of the input depth measurements. The overall algorithm for the utilized autoencoder is given, as follows:

$$Y = \Gamma(< \log(X), W_1 > +B_1) \tag{1}$$

$$Z = \Gamma(< Y, W_2 > +B_2) \tag{2}$$

$$KL_{Div} = \alpha_d \log \frac{\alpha_d}{\alpha} + (1-\alpha_d) \log \frac{1-\alpha_d}{1-\alpha} \tag{3}$$

$$CF_{AE} = \frac{1}{N} \sum_{i=1}^{N} (X_i - Z_i)^2 + \beta KL_{Div} \tag{4}$$

where X is the input depth vector, Y is the output of the encoder, Z is the output of the decoder, Γ is the activation function, α_d, and α are the desired and actual mean and/or standard deviation of the encoded data, respectively, W_1 and W_2 are the weight matrices, B_1 and B_2 are the bias vectors, CF_{AE} is the autoencoder cost function to be minimized, and $< \cdot, \cdot >$ is the dot product.

After the critical information is extracted from the input depth features and is encoded at a smaller dimension using the proposed autoencoder, two different supervised ANNs for realtime localization and estimation of loads can then be utilized, as illustrated in Figure 1.

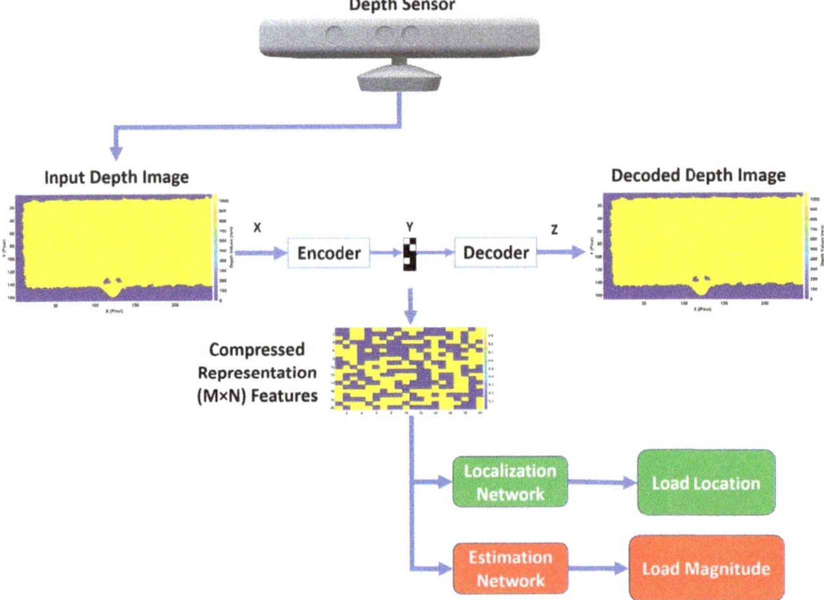

Figure 1. Proposed realtime load localization and estimation framework for SHM.

2.2. Load Localization from Depth Images Using ANN

A supervised feed forward ANN with a classification structure is proposed to localize where the load is acting on the wing. The proposed ANN classifier, as depicted in Figure 2, takes the encoded depth images as inputs and provides output based on the location of the loads.

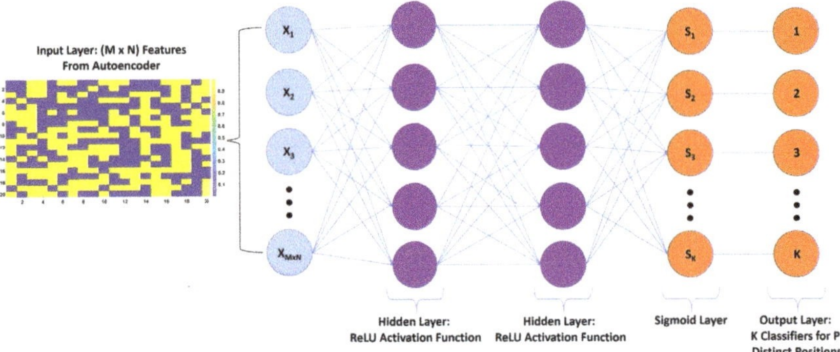

Figure 2. The proposed load localization ANN with 2 hidden layers and ReLU activation functions.

The encoded depth features for each sample in the training set were normalized across the samples by making use of the standardization technique so that the input features had zero mean and unit standard deviation. The formula used for standardization is given by Equation (5). As for the test set, they were standardized using the mean and standard deviation of the training set. Standardization was performed due to inevitable sensor noise, which can hinder the generalization capabilities of neural networks. Afterwards, the standardized inputs were fed into the localization ANN consisting of two hidden layers. The activation functions in both layers were chosen to be ReLU (Rectified Linear Unit), among other functions, such as sigmoid and tanh, due to ReLU's fast convergence.

$$\hat{X}_i = \frac{X_i - mean(X_j)}{\sigma(X_j)} \tag{5}$$

where X_i vector contains the encoded features in each sample, X_j vector contains the features across the samples, $\hat{X}_i$ vector contains the standardized features for each sample, and σ is the standard deviation.

Typically, in classification problems, the output labels are one hot encoded, through which, categorical data, in this case, the load positions, are converted into numerical data. The output of the last layer of the neural network, which is now a one hot encoded vector is passed through a sigmoid function. The Sigmoid function changes the arbitrary output scores to a range of probabilities that range between zero and one. Sigmoid, instead of other activation functions, was chosen to be in the output layer. This is because the load localization in this work is a multi label classification problem where more than one correct label exists in the output. Therefore, the output labels are not mutually exclusive i.e. the output labels are independent. The closeness between the output of the sigmoid function and the true labels (T) is defined as loss or cost function. The cost function of the classification (CF_{CL}) is defined as the average of Cross Entropy Error Function (CEEF) over a batch of multiple samples of size N and labels of size K, as follows:

$$CF_{CL} = \frac{1}{N}\sum_{i=1}^{N}\sum_{j=1}^{K} T_{ij}\log(S(x_{ij})) \tag{6}$$

The optimizer in the backpropagation algorithm updates the weights and biases, so as to minimize this loss and, as such, the loss decreases and the accuracy of the neural network increases. A classification ANN with two hidden layers of ReLU activation functions was determined to be sufficient to successfully localize the loads causing bending and/or twisting deflections. The proposed ANN was trained using Adam [20] optimizer. Both L2 regularization and dropout [21] techniques were utilized in order to increase the generalization performance of the network and prevent overfitting. This resulted in obtaining a new cost function, which consists of both the cost function defined by Equation (6) and the new scalar regularization value β due to L2 regularization. The final cost function FCF_{CL} is given by Equation (7) and the metric used for calculating the accuracy of predictions in load localization is given by Equation (8). The localization results obtained for both concentrated and distributed loading scenarios are presented and evaluated in detail in the results section.

$$FCF_{CL} = CF_{CL} + \beta\sum||Weights||_2 \tag{7}$$

$$Accuracy_{CL} = \frac{\sum(Y = \hat{Y})}{N} \times 100 \tag{8}$$

where Y is the ground truth, $\hat{Y}$ is the prediction, and N is the number of samples.

2.3. Load Estimation from Depth Images Using ANN

In this section, a logistic regression ANN for the estimation of the magnitude of loads acting on the wing is proposed. The input to this network is again the encoded depth images, and the output is the magnitude of the load. Unlike the ANN classifier, the output layer here consists of only a single node which provides continuous type numeric outputs in terms of loads. Because the output, in this case, is a single numeric value, there is no need to use sigmoid function in the output, as was the case in logistic classification. Moreover, the cost function for the estimation of load magnitudes (CF_E) is simply defined as the sum of the squared difference between the predicted value and the ground truth as given by Equation (9). Similar to the localization part, the estimation of load magnitudes was performed using two hidden layers, but the activation functions used in the first and second hidden layers were chosen to be tanh and sigmoid, respectively. The proposed load estimation ANN is illustrated in Figure 3.

$$CF_E = \frac{1}{N}\sum_{i=1}^{N}(Y - \hat{Y})^2 \tag{9}$$

where Y is the ground truth, $\hat{Y}$ is the prediction, and N is the number of samples.

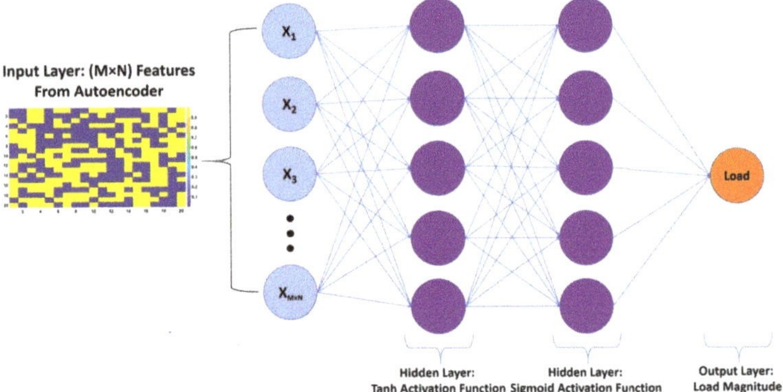

Figure 3. The proposed artificial neural networks (ANN) for estimation of loads with two hidden layers of tanh and sigmoid activation functions, respectively.

Both tanh and sigmoid functions belong to the family of sigmoid functions. The difference between these two is that the output of sigmoid function ranges from zero to one while the output of tanh ranges from −1 to +1. Moreover, the tanh function often converges faster than sigmoid due to tanh's symmetric nature [22]. The formula used for calculating the accuracy of predictions [23,24] for load estimation is given by Equation (10). The results obtained for both concentrated and distributed loading scenarios are presented and evaluated in detail in the experimental results section.

$$Accuracy_E = (1 - \frac{||Y - \hat{Y}||}{||Y - \bar{Y}||}) \times 100 \tag{10}$$

where Y is the ground truth, $\hat{Y}$ is the prediction, and $\bar{Y}$ is the mean of the ground truth.

3. Experimental Setup and Evaluation of the Depth Sensor for Load Monitoring

3.1. Experimental Setup

In order to validate the effectiveness of the proposed framework, an experimental setup that consists of a composite wing of a quad tilt-wing aircraft called SUAVI [25] was used. The wing has a size of 50 × 25 cm in length and width, respectively. The root side of the wing was fixed, so that no tilting was induced under applied loads. Similar to the works in the literature, ground tests were performed to experimentally mimic the deflections that may occur on a wing due to some external loads during flight. In the experiment, different types of loads that cause bending and twisting deflections on the wing were applied in two different loading scenarios. First, the load was designed to be acting on one of the eight positions depicted in the left image of Figure 4 and it is called concentrated loading case in this work. Six calibrated loads with magnitudes of [2.45, 4.9, 7.35, 9.81, 12.26, 14.71] N were chosen to be acting on these positions. Therefore, in total, eight different positions exist with each one containing six distinct loads, resulting in forty eight configurations for the first case. In the second scenario, the loads were chosen to be distributed loads placed in between each of the eight locations. This way the loads were made to be acting on multiple locations of the wing at the same time, as indicated in the right image of Figure 4. In loading positions 9 to 21, except for positions 13, 17, and 21, the loads were made to act at two positions simultaneously, for example position 9 represents two loads acting at positions 1 and 5. As for positions 13, 17, and 21, the loads were made to act at four positions simultaneously, for example position 13 represents four loads acting at positions 1, 2, 5, and 6 at the same time. Therefore, the total number of positions corresponding to distributed loads are thirteen. The magnitude of loads used for distributed scenario were the same as the concentrated loading case, but their magnitudes were distributed among the multiple positions they were acting upon, for example at position 9 two loads of 1.225 N were acting at positions 1 and 5 simultaneously for a loading case of 2.45 N. Therefore, six distinct load magnitudes per location exist in the distributed loading case, thus resulting in seventy eight different configurations for the second case. Therefore, in total, 126 distinct loading cases were performed during the experiments. In order to measure the deflections occurring over the span of the wing, this work proposes the use of a single RGB-D camera. RGB-D cameras are sensors that are capable of providing pixel wise depth information from images, thus making them suitable for full field measurement purposes. The RGB-D sensor used for data collection in this work was chosen to be a Microsoft Kinect V1 [26] sensor. The reasons for choosing Microsoft Kinect V1 for this work are as follows:

- It has high resolution depth and visual (RGB) sensing and is offered at a very affordable price when compared to other three-dimensional (3D) cameras, such as SwissRanger [27] and other Time of Flight cameras [28].
- It works based on structured light thus making it suitable for measurement from an inclined angle.
- It works in realtime (@ 30 Hz) with a field of view (FOV) of 43° (vertical) × 57° (horizontal) and can measure an area of 1.5 × 1 m from a distance of 1 m.
- One of the advantages of using a depth camera like Microsoft Kinect is that it does not require the sophisticated and time consuming camera calibration procedures since it is already calibrated, and it directly provides X, Y, and Z information in the camera frame and without loss of generality this frame is also the world frame.
- Moreover, depth cameras have many advantages over conventional intensity sensors in that they can work in low light conditions and are invariant to texture and color changes [29].

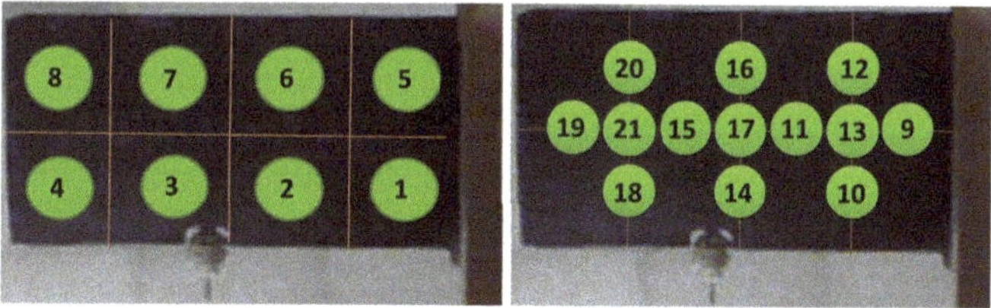

Figure 4. Positions of concentrated (**Left**) and distributed (**Right**) loads (Green) acting on the UAV wing.

Khoshelham et al. [30] theoretically and experimentally evaluated the geometric quality of the depth data that were collected by Kinect V1. They quantified the random error of depth measurement to be 4 cm at a ranging distance of 5 m, and concluded that the error increased quadratically by the increase in the ranging distance from Kinect. Based on their results, Khoshelham et al. made a recommendation for the use of Kinect sensor for general mapping applications, and they indicated that the data should be acquired at a distance of 1 to 3 m from the sensor due to the reduced quality of range data at further distances. Therefore, in this work, the Kinect was placed at a distance of one meter from the wing during the tests, and it was placed under the wing of the UAV, so as to capture the whole wing. Figure 5 shows the experimental setup used in this work. Although the methodology will be evaluated for a relatively small aircraft, the same technique can be utilized for structural health monitoring of much larger ones with the use of a depth camera with larger field of view, such as Carnegie Robotics® MultiSense™ S21B [31], Arcure Omega [32], and MYNT EYE [33]. Moreover, because the UAV used in this work was small in size, the depth sensor was not installed on it. Nonetheless, depth cameras can be installed on a larger aircraft with a minimum distance according to the depth sensor's operation range between the wing and the installation location. Depth cameras could be installed in place of RGB cameras that were fitted on the aircraft [3,4,7] but with the advantage of not requiring installation of additional marker's or LEDs on the wing as shown in Figure 6. The installation of the depth sensors at an angle, as shown in Figure 6, will not affect their operation, since, once a deflection occurs over the span of the wing, the wing's depth will change with respect to the pose of the installed depth sensor. For large aircrafts, the depth sensors can be connected to an onboard pc via wired or wireless connections. Moreover, dampers can be utilized for reducing the impact of vibrations on the depth sensors in order to take into account the vibrations that may occur in flight. Furthermore, if the proposed method is trained with the data obtained from in flight conditions then the proposed method can take into account all of the disturbances acting on the depth camera, since the disturbances will manifest themselves in the acquired data.

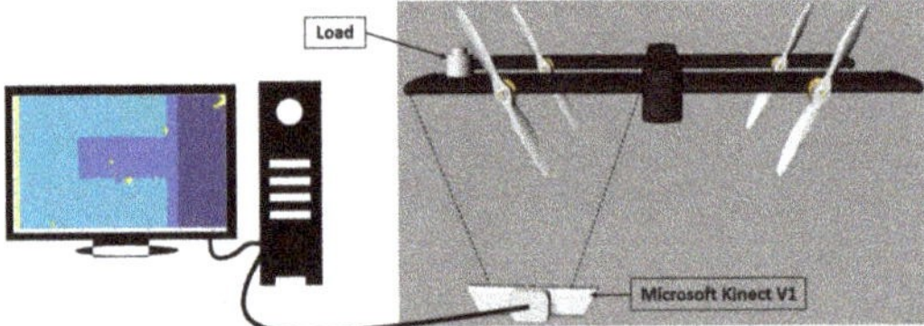

Figure 5. Experimental setup.

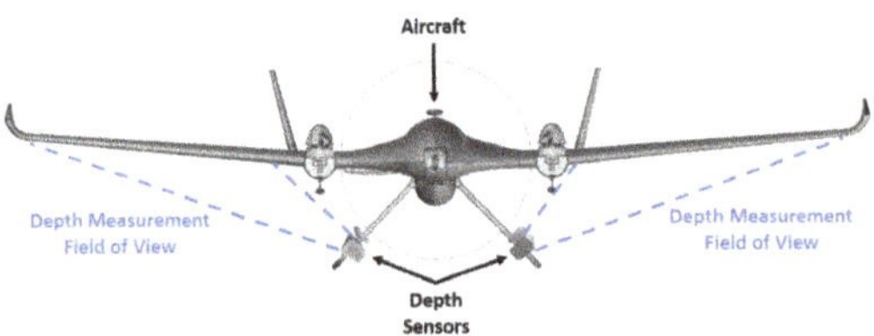

Figure 6. Possible installation locations of depth sensors on large aircraft.

3.2. Data Collection Procedure

The data collection procedure was performed similar to the works in the literature [1–3]. For both concentrated and distributed loading scenarios, first, loads were applied at each position as depicted in Figure 4 separately, and then depth images of the wing were acquired. For example, a load of 2.45 N was applied on position 1, while no other load was applied at any other location, and then the data was collected. Afterward, the next load of 4.9 N was applied at the same position, and the data were recorded. This procedure was repeated for all other locations in the same manner until data from all of the positions with all of the different load magnitudes were recorded. The acquired depth images by Kinect V1 are shown in Figure 7, in which the pixel values correspond to the actual measured distance in mm. Because the images were captured at one meter distance with a resolution of 640 $\times$ 480 pixels, the captured scene encompassed much more information than required; therefore, the images were cropped to include only the UAV wing, and the size of the acquired image was reduced to 247 $\times$ 166 pixels. It should be noted that this was only done in this case and, if the whole wing encompasses the image, then there is no need for cropping the image. Moreover, any depth values above 2000 mm and below 800 mm were changed to zero in order to get rid of redundancies, as shown in the right image of Figure 7. Moreover, since the gap between the depth values of the wing and its surrounding were very large, the color distribution resulted in a binarized representation.

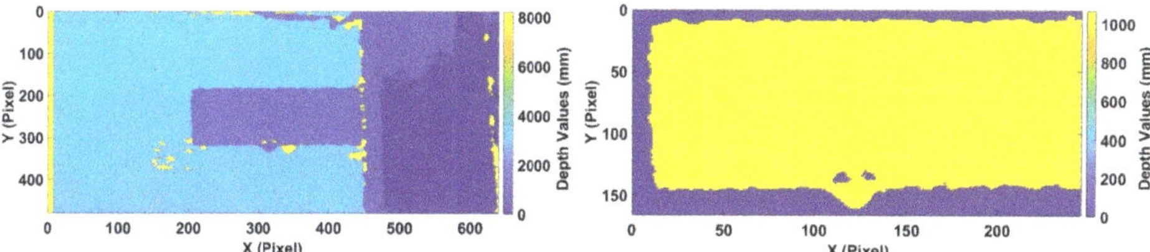

Figure 7. (**Left**) Acquired and (**Right**) Extracted depth image of the composite UAV wing.

3.3. *Evaluating Accuracy and Precision of Microsoft Kinect V1*

Even though Microsoft Kinect V1's overall accuracy and precision were previously evaluated by Khoshelham et al. [30]; in this work, its evaluation was performed for the specific working conditions required for structural health monitoring. To this end, the deflections at 32 reference points of the wing were measured using Kinect V1 sensor, and they were compared with the measurements from a laser ranger. To track the locations of reference points in the image plane, eight ArUco [34] markers, each having four corners were used. The patterns known as ArUco markers are small two-dimensional (2D) barcodes often used in augmented reality and robotics, as shown in Figure 8. ArUco was developed by Garrido-Jurado et al. [34], where they showed the superiority of their work to other known markers in literature such as ARTOOLKIT [35], ARToolKit Plus [36], and ARtag [37]. The locations of these 32 corners in image plane were detected in subpixel accuracy using the algorithm provided by Garrido-Jurado et al. [34] and the obtained results are shown in Figure 8.

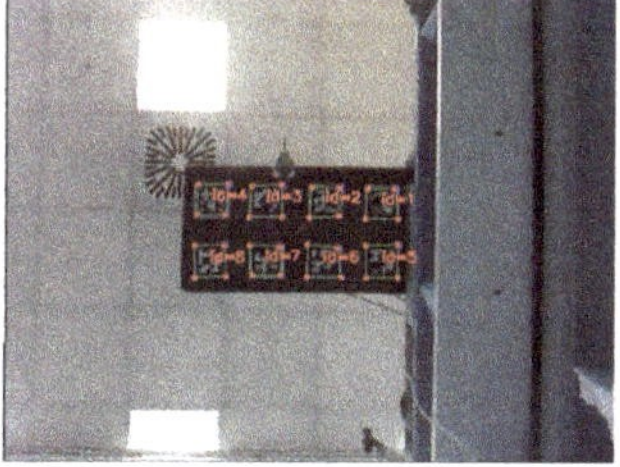

Figure 8. (**Left**) Sample ArUco Markers. (**Right**) Corner extraction in subpixel accuracy from ArUco markers.

3.3.1. Mapping Depth and RGB Images of Microsoft Kinect V1

To use ArUco markers for tracking the corners during deflection measurements, the RGB and depth images obtained from Kinect V1 were mapped using an affine transformation. Using n known points of the wing, such as its four corners, the unknown parameters of the transformation can be calculated, as follows:

$$C = AD \tag{11}$$

$$A = CD^{-1} \tag{12}$$

where $C \in \Re^{2\times n}$ contains the locations of a point in the colored image, $D \in \Re^{3\times n}$ contains the corresponding locations in the depth image, and $A \in \Re^{2\times 3}$ is the affine transformation matrix.

By following the above steps, depth images were successfully mapped to the RGB images that were acquired from the Kinect V1, as shown in Figure 9. As seen from this figure, the color images were changed to grayscale, so as not to work with unnecessary color channels. This way, the corners of the ArUco markers, which act as reference points, can be used for tracking the deflection changes at these points both by Kinect V1 and a laser meter, which will act as ground truth.

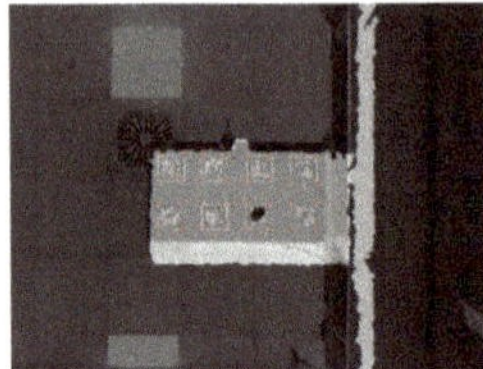
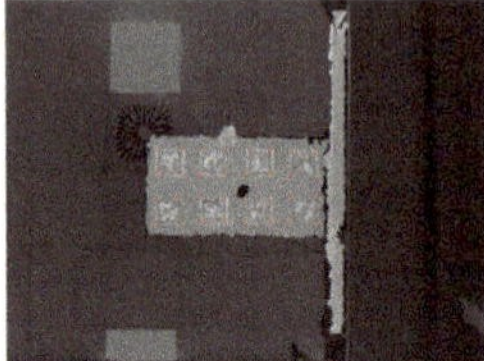

Figure 9. Kinect V1 depth and RGB images in grayscale (**Left**) Before mapping. (**Right**) After mapping.

3.3.2. Microsoft Kinect V1 vs Leica DISTO X310 Laser Meter

After the RGB and Depth images were mapped, the difference between the deflections that were measured by Kinect V1 and laser ranger at corner points were calculated. To account for errors in corner detection and mapping, for each detected corner pixel from RGB image, a 3×3 window of pixels was chosen from the corresponding depth image. A Leica DISTO X310 laser meter [38] with ± 1 mm accuracy was used for evaluating the accuracy of Kinect V1. Loads with magnitudes of 2.45, 4.9, 7.35, 9.81, 12.26, and 14.71 N were placed on position 4 of the wing and the deflections were measured using both the Kinect V1 and Leica DX310 sensors. The obtained results are shown in the left image of Figure 10 for 2.45 N load, in which the mean absolute error for the 32 corner points was measured to be 2.0556 mm with a standard deviation of 1.74 mm. This was repeated for 100 frames, and the obtained results are shown in the left image of Figure 11, in which the mean of absolute errors was measured to be 2.0642 mm with a standard deviation of 0.21 mm. Similarly, the results obtained for 14.71 N load are shown in right images of Figure 10 and Figure 11, in which the mean absolute error for 32 corner points was calculated to be 2.6875 mm and 2.1520 mm for a single and 100 frames, respectively, with standard deviations of 1.615 mm and 0.26 mm. The same procedure was repeated for all of the aforementioned loads that range from 2.45 to 14.71 N, and the obtained results are summarized in Table 1. These results show that for the required working conditions, the Kinect V1 has an accuracy of ± 2.25 mm with a standard deviation of 0.28 mm and a resolution of 1 mm. Therefore, since the deflections due to different loads range from 1 mm to 19 mm in this work, the Kinect V1 is proved to be suitable for acquiring the deflection measurements reliably. Moreover, neural networks already take the noise of the sensors into account when they are trained with the data acquired from them; therefore, the small noises in the acquired data are taken care of by the proposed method.

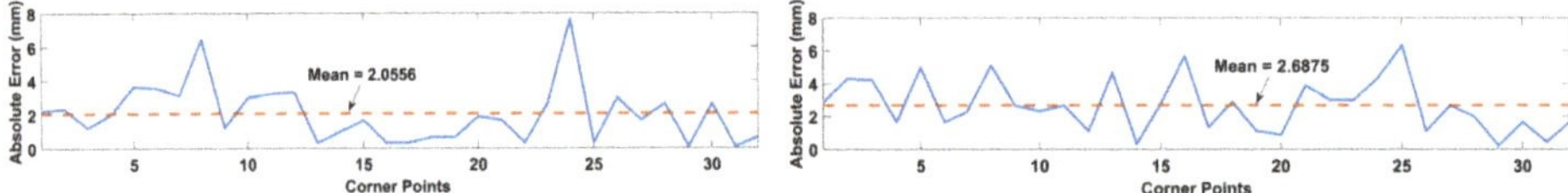

Figure 10. Absolute errors at 32 corner points of the wing for (**Left**) 2.45 N load. (**Right**) 14.71 N load.

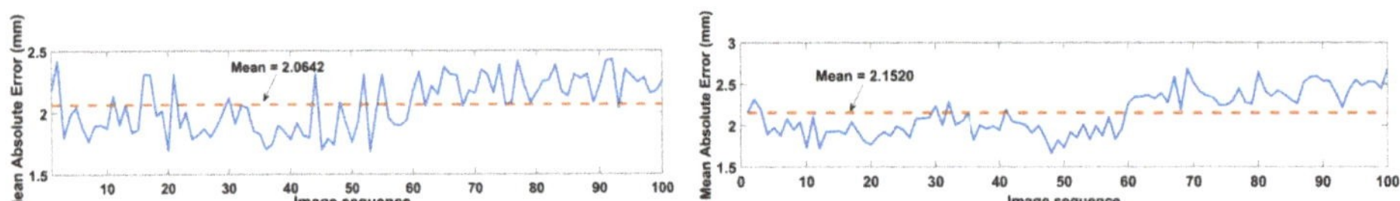

Figure 11. Mean of absolute errors in 100 frames for (**Left**) 2.45 N load. (**Right**) 14.71 N load.

Table 1. Accuracy and Precision of Microsoft Kinect V1.

Loads (N)	2.45	4.9	7.35	9.81	12.26	14.71	Average
Absolute mean error (mm)	2.06	2.18	2.3	2.59	2.21	2.15	2.25
Standard deviation (mm)	0.21	0.36	0.46	0.22	0.15	0.26	0.28

3.4. Visualization of Depth Features Acquired from Microsoft Kinect V1

Here, the visualization of depth features that were obtained from Kinect V1 for loads acting on the wing is shown. As mentioned, the spanwise deflections of the wing due to the application of the external loads were captured by the depth camera. Figure 12 illustrates these deflection measurements for the distinct loads acting on position 1 and 8. These illustrations were made by reconstructing the surface of the wing from the acquired depth images after the application of each load. These images show how the overall bending profile of the wing changes due to the magnitudes and locations of the external loads. The Kinect V1 sensor can measure deflections due to loads as small as 2.45 N acting on position 1 of the wing, as seen from these measurements. Deflection profile is especially visible for loads acting on Position 8, since it is the furthest point from the fixed part of the wing.

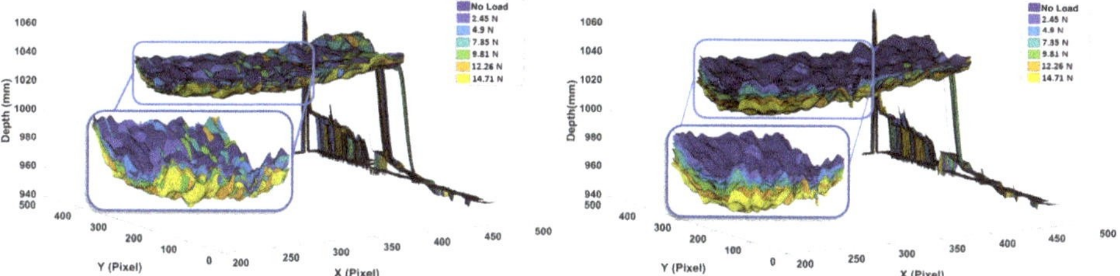

Figure 12. Wing deflections due to loads placed at (**Left**) Position 1 (**Right**) Position 8.

Moreover, in order to show the effect of the same load applied on different sections of the wing, loads of 2.45 and 14.71 N were applied on all eight distinct partitions of the wing, and the resulting depth images were plotted in Figure 13. These results clearly indicate that the wing deflection profile is different even when the same load is applied on different locations. Furthermore, by observing these results one can see a strong relationship between the behavior of the deflection of the wing for loads placed in parallel, namely loads placed at Positions 1 and 5, 2 and 6, 3 and 6, 4 and 8. These results prove the effectiveness of Microsoft Kinect V1 sensor for wing deflection measurement under various loading conditions acting on distinct sections of the wing.

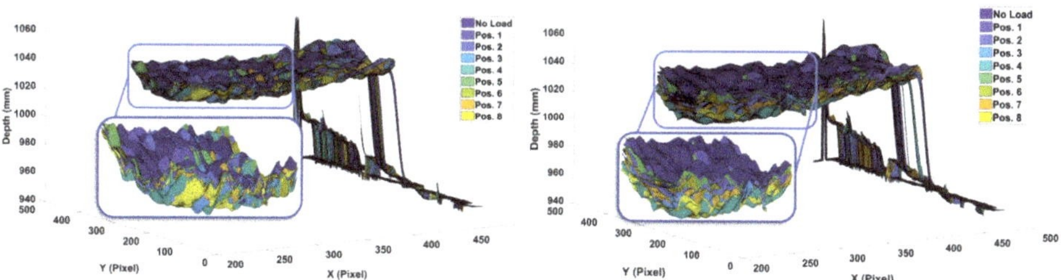

Figure 13. Wing deflections due to (**Left**) 2.45 N and (**Right**) 14.71 N loads applied on distinct positions of the wing.

4. Dataset Creation and Experimental Results

4.1. Dataset Creation

A dataset of wing deflections due to loads was constructed by extracting the depth features from the Microsoft Kinect V1 sensor operating at 30 Hz by following the procedure explained in Section 3.2 in order to train and test the proposed ANN models. To construct the training dataset, 100 samples for each loading case was acquired. Initially, depth images were acquired when no load was acting on the wing. Afterward, data were collected for six different load magnitudes acting on 21 distinct sections of the wing in sequence, which included the concentrated and distributed loading scenarios. Therefore, 12,700 samples were collected for the training dataset. As for the test dataset, the depth images were acquired for the duration of only one second, thus resulting in 30 samples for each loading case for this dataset. Unlike the training dataset, the six distinct loads on the 21 sections of the wing were applied at random in this case, so as to evaluate the robustness of the proposed approach. This resulted in the acquisition of 3810 samples for the test dataset. The details of the constructed training and test datasets are tabulated in Table 2.

Table 2. Training and test datasets.

	No. of Samples per Load per Position	No. of Distinct Loads per Position	No. of Distinct Positions	Total No. of Samples	No. of Features per Sample
No Load	100 (30)	1	1	100 (30)	247 × 166 (41,002)
Concentrated Loads	100 (30)	6	8	4800 (1440)	247 × 166 (41,002)
Distributed Loads	100 (30)	6	13	7800 (2340)	247 × 166 (41,002)
Training Dataset	100	—	22	12,700	247 × 166 (41,002)
Test Dataset	30	—	22	3810	247 × 166 (41,002)

The ( ) show the number of samples in the test dataset.

In this paper, TensorFlow [39] software was used to build and test the proposed ANN models. TensorFlow is an open source platform developed by Google Brain team, and it is widely used in literature to conduct machine learning based research due to its highly efficient computation framework. The computer used for developing the proposed ANN models had an Intel Xeon 3.6 GHz, twelve thread central processing unit (CPU) with 16 GB of RAM, and the whole network was trained on the CPU only, without the need of GPU.

4.2. Experimental Results and Discussions

In this section, the performance of the proposed framework for load localization and estimation are analyzed and discussed in detail. Initially, autoencoders were used to extract informative data from the depth images at a much smaller scale. subsequently, the proposed ANNs were trained using the training set described in Section 4.1, and their performance was evaluated and compared with a modified version of Castigliano's theorem (The readers can refer to the Appendix A for the details of this algorithm.) on the constructed test set.

4.2.1. Load Localization From Depth Images Using ANN

An autoencoder was used to obtain informative data at a much smaller scale from the acquired depth images to be fed into the localization and estimation networks, as stated in Section 2.1. The autoencoder was run in series with the classification network so as to use its accuracy as a measure of performance in order to determine the smallest size of the encoded features required for load localization. The encoded feature size was initially set to be 400 and then was increased with increments of 200 until satisfactory load localization performance was obtained. As described in Section 3.1, two different loading scenarios were considered in this work. In the concentrated loading scenario, the output locations were labeled positions 1 to 8 and, in the distributed loading scenario, they were labeled positions 9 to 21. Moreover, 0 label was chosen to represent the no load condition. Therefore, the total number of distinct positions in the output layer amounts to 22. By making use of one hot encoding technique, the aforementioned 22 distinct positions were encoded using only nine classifiers in the output of the localization network by setting K to 9 and P to 22. Labels 1 to 21 were one hot encoded using the first eight classifiers, and the no load case was encoded using the ninth classifier.

The proposed autoencoder and classification network was trained by varying the encoded data size. The activation function of the autoencoder was chosen to be a sigmoid and the number of neurons in the first and second hidden layers of the classification network were set to 60% and 30% of the encoded data size, respectively. The mean and standard deviation of the encoded data to be used in Equation (3) were chosen as 0.5 and 0.2, respectively. The β coefficient for regularization was chosen to be 0.1, and the dropout ratio of 0.8 was chosen in order to increase the network's generalization capabilities. Moreover, the starting learning rate was chosen to be 0.0005 and the network was trained using Adam optimizer. The training of the network was performed for 8000 iterations for each encoded data size, and the results are tabulated in Table 3. From these results, it is observed that, as the size of encoded data increases, the difference between the training and test accuracies decrease. This suggests that more distinctive data is being extracted as encoded data size increases. Moreover, an encoded data size of only 1200 was enough for obtaining very high accuracies of 96.4% and 94.3% when evaluated on the training and test datasets, respectively.

Table 3. Accuracy of the proposed localization ANN with varying encoded data size.

Encoded Data Size	CF_{AE}	Layer 1 Neurons	Layer 2 Neurons	Training $Accuracy_{CL}$ (%)	Test $Accuracy_{CL}$ (%)
400	31.76	240	120	66.1	60.3
600	31.58	360	180	93.9	88.8
800	31.05	480	240	95.1	91.3
1000	29.98	600	300	96.2	93.6
1200	29.44	720	360	96.4	94.3

As for load localization that is based on the modified Castigliano's theorem, its performance was also evaluated on the constructed test dataset. The proposed localization ANN was trained with the load locations shown in Figure 14. Outputs of the proposed localization network with an encoded data size of 1200 and the one based on Castigliano's theorem are plotted against the ground truth positions in Figure 15. In this figure, the label denoted as zero represents the no load case, labels 1 to 8 represent the concentrated load positions, labels 9 to 21 represent the distributed load positions, and label 22 represents misclassified outputs that do not belong to any of the aforementioned load positions. From these results, one can see that both frameworks are able to discern the locations of the loads causing different kinds of bending and twisting deflections on the wing rather successfully. However, the accuracy of the proposed ANN based framework is superior to the one that is based on Castigliano's theorem, as visible from the results obtained in Figure 15 and the obtained accuracies given in Table 4. More importantly, the proposed neural network based method exhibits invariance to the type of the applied load and is able to successfully localize both concentrated and distributed loads causing wing deflections, which the framework based on modified Castigliano's theorem fails to do so properly.

Table 4. Localization performance of the proposed ANN and modified Castigliano's theorem, evaluated on test dataset.

Localization Method	Accuracy (%)
Localization ANN	94.3
Modified Castigliano's Theorem	57.7

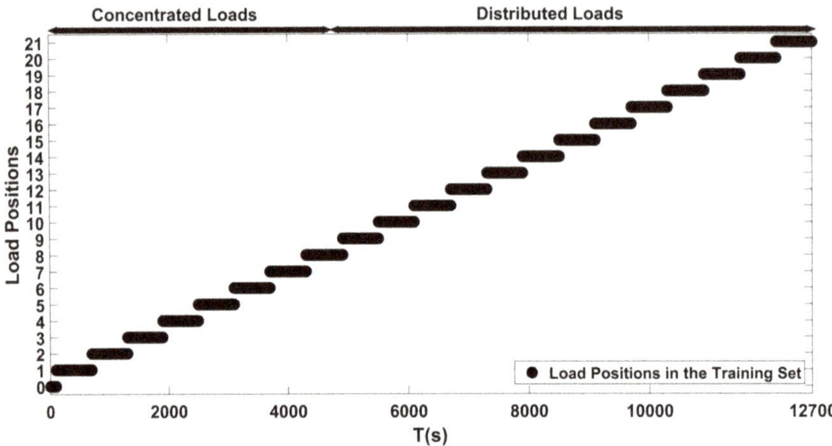

Figure 14. Load positions used for training the proposed localization ANN.

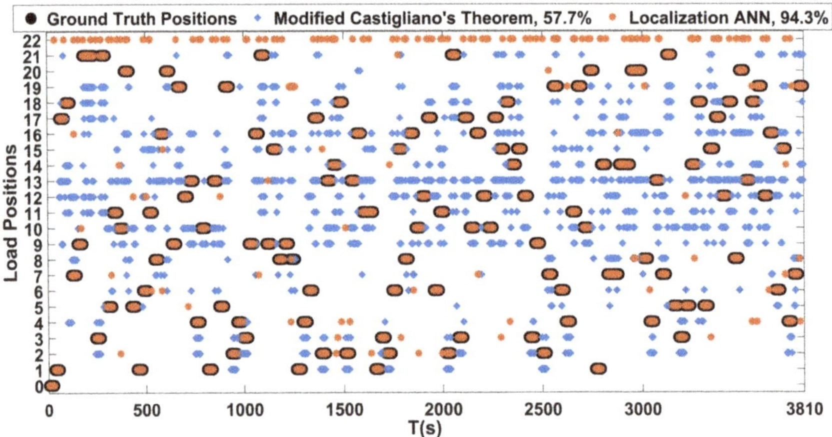

Figure 15. Position predictions based on the proposed localization ANN and modified Castigliano's theorem, evaluated on the test dataset.

The relationship between the ground truth and the estimations of both frameworks is illustrated as a confusion matrix in order to get more insight into the localization performance of both methods. Figure 16 shows the confusion matrix for the proposed localization ANN, and Figure 17 is for the modified Castigliano's theorem. From these plots, it can be seen that the correct estimations are on the diagonal and any points not located on this line represent misclassified outputs. For the ANN framework, all of the position estimations have accuracies of higher than 90%, except for Positions 8 and 14. Nonetheless, their accuracies are still high, being 76.7% and 86.1%, respectively. As observed from Figure 16, Position 8, which is a concentrated load, is misclassified 16.1% of the time as Position 19, which is a distributed load. This observation is due to the wing deflection profiles under these loading conditions having similar patterns. However, these observations seem to be worse for the results obtained through the modified Castigliano's theorem, especially in the case of distributed loads, as seen in Figure 17. This is due to the fact that measured deflections exhibit very similar patterns at different points and, therefore, can not be accurately captured unless a more robust model such as the proposed ANN one is used to classify them in an appropriate manner.

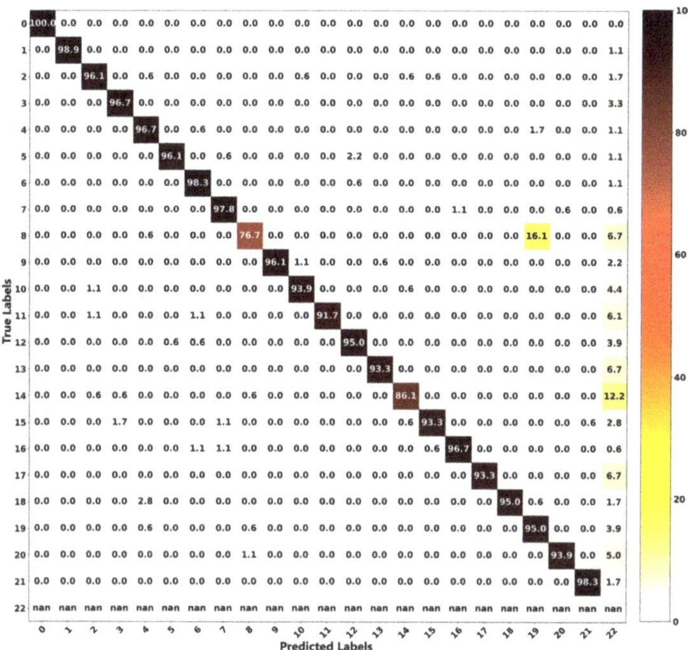

Figure 16. Confusion matrix for localization based on the proposed ANN, evaluated on test dataset.

Figure 17. Confusion matrix for localization based on modified Castigliano's theorem, evaluated on test dataset.

4.2.2. Load Estimation from Depth Images Using ANN

In this subsection, the proposed autoencoder and regression ANN's performance for load estimation is evaluated on the same dataset as the localization one. Unlike the localization networks' output, the output of this network is a single continuous variable that represents the magnitude of applied load under both concentrated and distributed loading conditions. Similar to the localization part, the proposed load estimation network was trained with varying encoded data size and the same parameters of the autoencoder network. As for the regularization coefficient β and dropout rates, they were fine tuned to be 0.1 and 0.35, respectively. This time the training was performed for 30,000 iterations for each encoded data size, and the obtained results are shown in Table 5. Based on the obtained results, it is seen that, again, an encoded data size of 1200 neurons was enough for obtaining very high accuracies of 97.3% and 92.7% when evaluated on the training and test datasets, respectively. Besides, it is observed that training such a network with good performance requires significantly more iterations when compared with the classification one, since the output in regression problems is a continuous variable.

Table 5. Accuracy of the proposed load estimation ANN with varying encoded data size.

Encoded Data Size	CF_{AE}	Layer 1 Neurons	Layer 2 Neurons	Training $Accuracy_E$ (%)	Test $Accuracy_E$ (%)
400	31.76	240	120	93.9	83.1
600	31.58	360	180	96.0	88.3
800	31.05	480	240	96.7	90.8
1000	29.98	600	300	96.4	91.5
1200	29.44	720	360	97.3	92.7

Moreover, the performance of load estimation that is based on modified Castigliano's theorem was also evaluated on the constructed test dataset. The proposed estimation ANN was trained with the loads that are shown in Figure 18. The outputs of the proposed load estimation ANN with 1200 encoded data size and the modified Castigliano's theorem are plotted against the ground truth loads presented in Figure 19. These results show that magnitudes of loads causing bending and twisting deflections can be estimated with very high confidence, regardless of where the load is acting on the wing. However, the accuracy of the proposed ANN based framework is far more superior when compared with the one that is based on the modified Castigliano's theorem, as visible from the results obtained in Figure 19 and the obtained accuracies tabulated in Table 6.

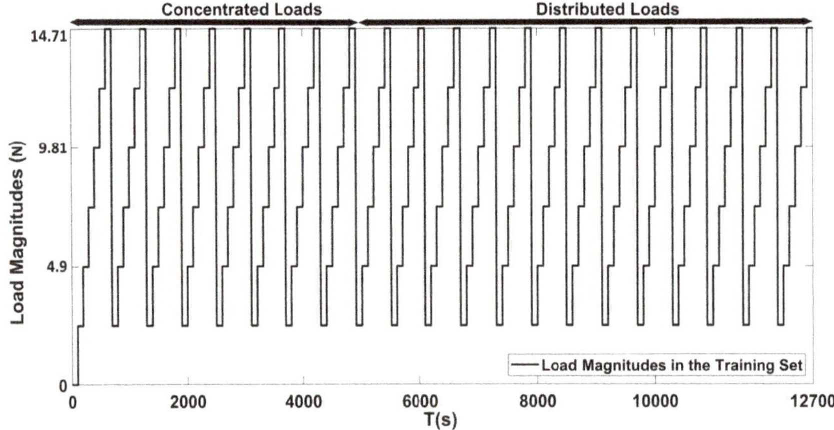

Figure 18. Load magnitudes used for training the proposed estimation ANN.

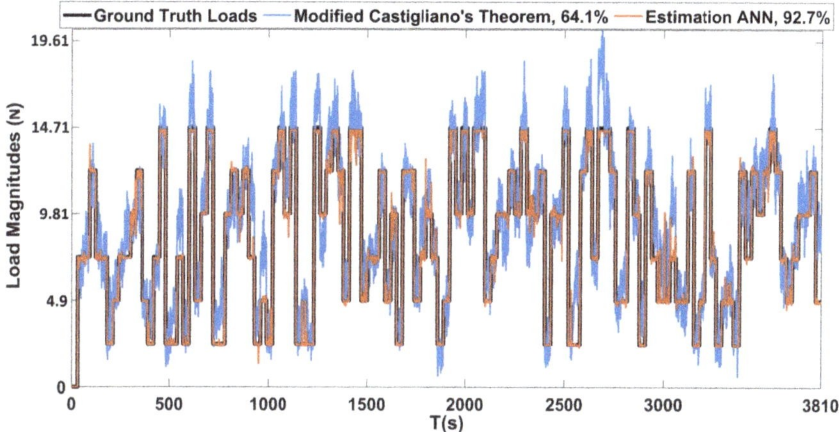

Figure 19. Load estimation based on the proposed ANN and modified Castigliano's theorem, evaluated on test dataset.

Table 6. Load estimation performance of the proposed ANN and modified Castigliano's theorem, evaluated on the test dataset.

Load Estimation Method	Accuracy (%)
Estimation ANN	92.7
Modified Castigliano's Theorem	64.1

These results show that the proposed method can localize and estimate the loads acting on an aircraft wing with very high accuracies under both concentrated and distributed loading conditions. It should be noted that, since the proposed method was trained with elastic loading cases, it will only work with elastic loads. If the wing is damaged, then the relationship between the deflections and the load will change; therefore, the proposed method will not work unless it is trained for the damaged cases as well. Even though the training of the proposed networks is performed offline, the proposed method requires only 0.02 s for data reduction from a full frame image with $640 \times 480 = 307{,}200$ pixel features using an autencoder with an encoded data size of 1200. As for data reduction from the cropped images used in this work that had $247 \times 166 = 41{,}002$ pixel features, an autoencoder with an encoded data size of 1200 requires only 0.008 s. Another 0.001 second is required for localization or estimation of loads from the encoded data. Therefore, the proposed method can operate at 0.021 s or around 47 Hz for a single full frame image, thus realizing real time performance.

5. Conclusions

In this work, a robust structural health monitoring system based on depth imaging and artificial neural networks for localization and estimation of bending and twisting loads acting on an aircraft wing in real time is proposed. The proposed framework is based on the usage of depth images obtained from a depth camera as input features to an autoencoder and load location or magnitude as output labels of supervisory neural networks placed in series with the autoencoder. Initially, the Microsoft Kinect V1 depth sensor's accuracy and precision were evaluated for monitoring of aircraft wings by making use of ArUco markers and a Leica DISTO X310 laser meter having an accuracy of ± 1 mm. The Kinect V1 proved to be reliable for SHM purposes, since it provided full field measurements with accuracy and standard deviation of 2.25 mm and 0.28 mm, respectively, when compared with the single point measurements provided by the laser meter.

As for the proposed method, first, an ANN consisting of an autoencoder and two hidden layer classification network with ReLU activation functions was proposed for estimating the location of loads. Second, an autoencoder and logistic regression network of two hidden layers with tanh and sigmoid activation functions was proposed for estimating the magnitude of these loads. Both of the proposed networks were trained and validated on an experimental setup, in which the application of concentrated and distributed loads were applied on a composite UAV wing.

In addition, a comparison with an approach based on Castigliano's theorem was performed, and the proposed method proved to have superior performance in terms of the localization and estimation of loads. The proposed localization and estimation ANNs achieved accuracies of 94.3% and 92.7%, while the framework based on Castigliano's theorem achieved average accuracies of 57.7% and 64.1%, respectively, when both of the methods were evaluated on a dataset containing randomly applied concentrated and distributed loads. As demonstrated, the proposed ANN based framework can localize and estimate the magnitudes of loads acting on aircraft wings with very high accuracies from a single depth sensor in realtime.

In the near future, it is planned to extend the current study for the localization and estimation of highly dynamic loads on larger aircraft wings.

Author Contributions: Conceptualization, D.K.B., M.U., M.Y., and B.K.; methodology, D.K.B. and M.U.; resources, M.U.; data curation, D.K.B. and M.U.; visualization, D.K.B. and M.U., writing—original draft preparation, D.K.B.; writing—review and editing, M.U., M.Y., and B.K.; supervision, M.U. All authors have read and agreed to the published version of the manuscript.

Funding: This research received no funding.

Conflicts of Interest: The authors declare no conflict of interest.

Appendix A. Localization and Estimation of Loads Based on Castigliano's Theorem

Based on Castigliano's second theorem [40,41], for a linearly elastic structure under loads $P_1, P_2, P_3, \ldots, P_m$ at points $1, 2, 3, \ldots, n$, the deflection W_j of point j can be calculated as follows:

$$W_j = \sum_m a_{n \times m} P_m \tag{A1}$$

where $a_{n \times m}$ is the deflection at point n due to a unit load applied at point m, normalized by the magnitude of the load. Based on this theorem, given loads at m points, one can calculate deflections at n points, as given in matrix form in Equation (A2). In order to construct A matrix, the concentrated loads mentioned in Section 2.1 were placed at eight different positions of the wing and the corresponding deflections D were recorded. Afterward, these measured deflections were normalized by the magnitude of the loads (L) acting on them to calculate the A matrix as given in Equation (A3). Moreover, this was performed for all load magnitudes i.e, [2.45, 4.9, 7.35, 9.81, 12.26, 14.71] N. Afterward, the average of all these magnitudes was calculated to be used as the final A matrix due to the nonlinear nature of the wing material as shown in Figure A1 and the inevitable sensor noise.

$$W = AP \tag{A2}$$

$$A = \frac{D^T}{L} \tag{A3}$$

where $W \in \Re^n$, $A \in \Re^{nm}$, $P \in \Re^m$, $D \in \Re^{mn}$, and $L \in \Re$.

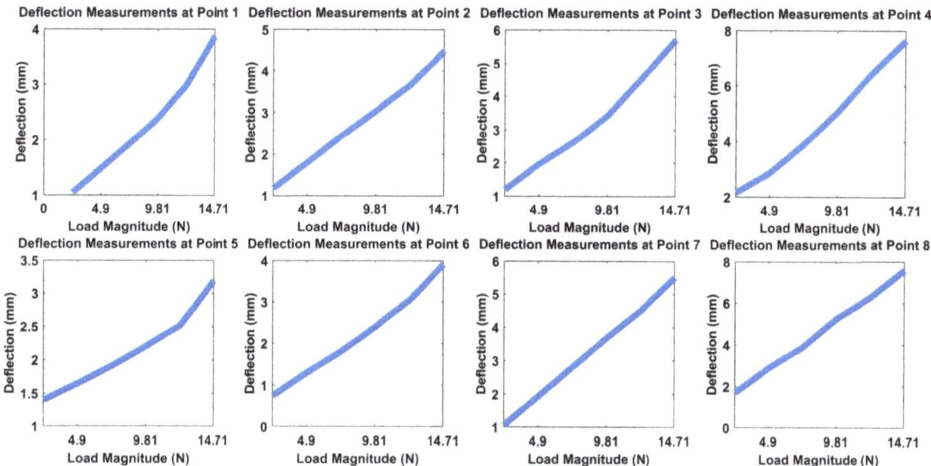

Figure A1. Deflection measurements for loads acting at point 2 of the wing.

However, the objective of this work is the inverse of the above mentioned problem i.e., given deflections at n points, we want to calculate loads at m points. To do that, one can calculate the pseudo inverse of A to obtain K. The calculated K is shown in Table A1. This is valid since Castigliano's first theorem is basically the inverse of its second theorem. Therefore, loads can be estimated by using the following equation:

$$P = KW \tag{A4}$$

$$\begin{bmatrix} P_1 \\ \vdots \\ P_8 \end{bmatrix} = \begin{bmatrix} 4244.729867 \cdots -1615.667686 \\ \vdots \ddots \vdots \\ -2459.317023 \ldots 1233.037717 \end{bmatrix} \begin{bmatrix} W_1 \\ \vdots \\ W_8 \end{bmatrix} \tag{A5}$$

where $K = A^{-1}$ is the unit load at point m due to the deflection measured at point n, normalized by the magnitude of the deflection.

Table A1. Calculated K matrix.

Deflection Measurement Points	Load Position							
	1	2	3	4	5	6	7	8
1	4244.72987	−698.685	553.1295	1970.979	−2260.34	−4685.71	−167.085	−1615.67
2	−6541.85525	1100.622	−640.224	−3025.09	3043.874	6964.995	49.58206	2560.928
3	−2694.30173	545.9633	−648.338	−1104.24	804.3221	2842.263	388.2263	979.052
4	6206.63609	−1303.48	833.578	2568.318	−2257.68	−6721.99	−652.854	−2450.1
5	3874.76045	−314.587	580.265	1808.919	−1436.91	−3625.9	344.6381	−1384.48
6	13,631.5189	−2141.29	2321.193	6042.803	−5624.8	−13,741	−375.178	−4919.03
7	−12,850.7951	3280.897	−1219.55	−5191.01	4515.127	14,360.46	1428.491	4495.81
8	−2459.31702	−152.446	−956.594	−1151.77	1369.812	2172.997	−43.5148	1233.038

After calculation of K matrix, one can now estimate loads at different points of the wing by providing the deflections measured at various points of the wing. Below are the results for the cases when the load is acting at point 2 of the wing with varying magnitudes of [2.45, 4.9, 7.35, 9.81, 12.26, 14.71] N, and the loads at all of the 8 points were calculated using Equations (A4) and (A5), for which the results are plotted in Figure A2.

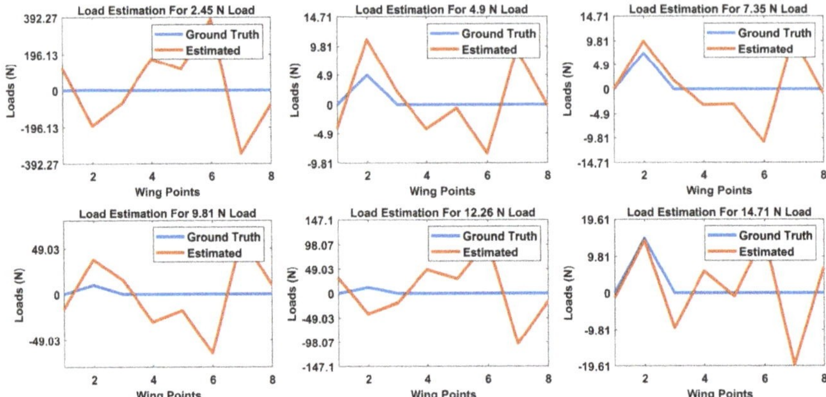

Figure A2. Load estimations for concentrated loads acting at point 2 of the wing.

Appendix A.1. Load Localization Based on Castigliano's Theorem

From the above results, it is seen that the load estimation fails as it is since it is estimating that loads with different magnitudes at different points of the wing are responsible for the measured deflections. In order to overcome this shortcoming, this work proposes the loads to be first localized, and then their magnitudes can be calculated. In order to localize the loads, the measured deflections at any time are compared with the constructed deflection D matrix and then the Pearson correlation[42] is used for finding the maximum correlation between them as follows:

- Given a new measurement as the one in Table A2.

Table A2. New deflection measurements (mm) at 8 points of the UAV wing.

Load Position	Deflection Measurements (mm) at 8 Points							
	1	2	3	4	5	6	7	8
2	3.65	4.58	5.84	7.71	2.28	4.01	5.32	7.40

- Compare the new measurements with the constructed deflection D matrix given in Table A3.

Table A3. Constructed deflection (D) matrix.

Load Position	Deflection Measurements (mm) at 8 Points							
	1	2	3	4	5	6	7	8
1	2.12	2.21	2.23	3.26	1.06	1.67	2.21	3.01
2	2.54	3.09	3.67	5.19	1.30	2.50	3.70	5.19
3	3.42	4.47	5.88	8.24	1.98	3.75	5.88	7.74
4	4.19	5.75	7.87	10.98	2.84	5.13	8.12	10.50
5	2.38	2.26	2.66	3.51	1.74	1.96	2.95	3.56
6	3.13	3.03	3.91	4.79	2.32	3.07	4.49	5.14
7	3.43	4.15	5.82	7.68	2.95	4.29	6.72	8.08
8	4.61	5.52	7.99	10.76	3.79	5.68	8.76	11.01

- Find the maximum Pearson correlation coefficient between the current measurement given in Table A2 and the constructed deflection matrix D given in Table A3, which in turn will correspond to the actual load position as shown in Table A4.

Table A4. Calculated Pearson correlation coefficients.

	Load Position							
	1	2	3	4	5	6	7	8
Pearson Correlation Coefficient	0.9498	0.9920	0.9914	0.9859	0.9324	0.9354	0.9461	0.9611

From these results, it is seen that the load can be successfully localized using this method. Besides its obvious usage for load localization, this information is vital for the correct estimation of load magnitudes based on the proposed methodology for load estimation based on Castigliano's theorem as formulated in the following subsection.

Appendix A.2. Correction of Load Estimation Based on Castigliano's Theorem via Localization and Optimization

Now that the loads are localized, one can correct the estimated loads by using the localization information. This can be done by first eliminating all the loads other than the localized one. Second, since the theory estimates that the measured deflections are due to a combination of loads acting at different points, one can conclude that the actual load applied at a single point is the cumulative sum of all the loads acting on the wing. Therefore, once the load is localized, its magnitude will be equal to the sum of all the estimated loads. The effectiveness of this can be seen in the results obtained in Figure A3 for loads acting again at point 2 of the wing.

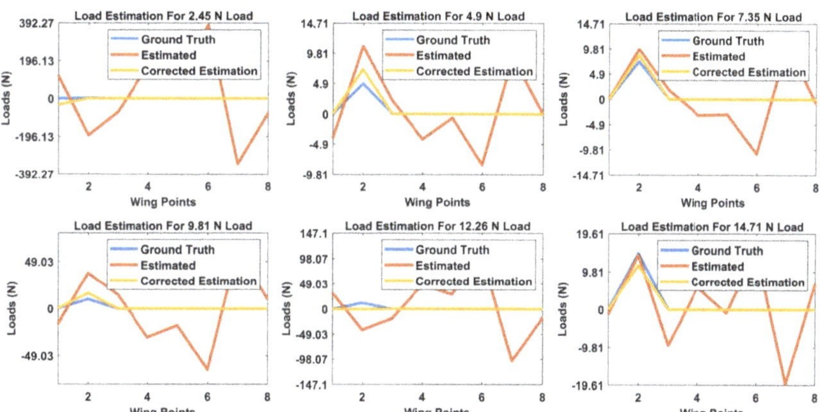

Figure A3. Corrected load estimations for concentrated loads acting at point 2 of the wing.

However, as seen from the obtained results in Figure A3, the estimated loads' accuracy is low in general. This is due to the calculated K matrix. Since the K matrix was calculated through pseudo inverse, it may not provide the optimal relationship between the measured deflections and the applied loads. Sensor noise also affects this, since the A matrix was constructed using measurements acquired at different loading cases. Therefore, this work proposes to optimize the calculated K matrix using a backpropagation algorithm such as Adam [20] optimizer. The initial solution to the optimizer was provided by the calculated pseudo inverse, and the optimization was defined as follows:

$$\hat{F} = \hat{K}V \tag{A6}$$

$$CF = ||(F - \hat{F})||_2 \tag{A7}$$

where $\hat{K} \in \Re^{mn}$ is the matrix to be optimized, $V \in \Re^{nN}$ is the deflection matrix with N samples measured by the depth sensor, $\hat{F} \in \Re^{mN}$ is the estimated load matrix for N samples, $F \in \Re^{mN}$ is the ground truth load matrix obtained from the applied loads' magnitudes and CF is the cost function defined to be the Frobenius norm of the matrix.

The optimized $\hat{K}$ is shown in Table A5. Using this optimized matrix, the loads can now be estimated quite accurately, as shown in Figure A4. The same procedure can be applied to distributed loads by considering them as new load locations and constructing a 21 × 21 deflection matrix. A sample result obtained for distributed loads is shown in Figure A5, in which the deflection matrix was extended to include the new 13 distributed points.

Table A5. Optimized $\hat{K}$ matrix.

Deflection Measurement Points	Load Position							
	1	2	3	4	5	6	7	8
1	24.7421	−11.9939	−17.0887	30.0612	−93.1855	−65.4709	−12.1050	−17.1733
2	38.5250	−9.2518	9.0438	−6.6004	27.5545	−3.4873	−13.5474	31.7971
3	14.5861	79.1131	60.0194	23.6086	−31.8746	−5.7411	24.9256	4.4628
4	−10.9021	36.3568	−3.8216	−22.3221	24.4179	−6.5668	−4.5749	−25.8135
5	15.0917	42.9939	47.0306	35.2446	67.4271	104.4526	96.2090	18.4261
6	31.7370	−31.4353	23.4873	−21.0367	18.1937	59.4679	69.4669	21.4363
7	11.7676	135.0571	89.8542	37.6748	−9.2130	29.2120	37.8665	2.0163
8	−15.6636	−23.5698	−59.6106	−5.5923	−25.2610	70.9806	27.5923	56.5209

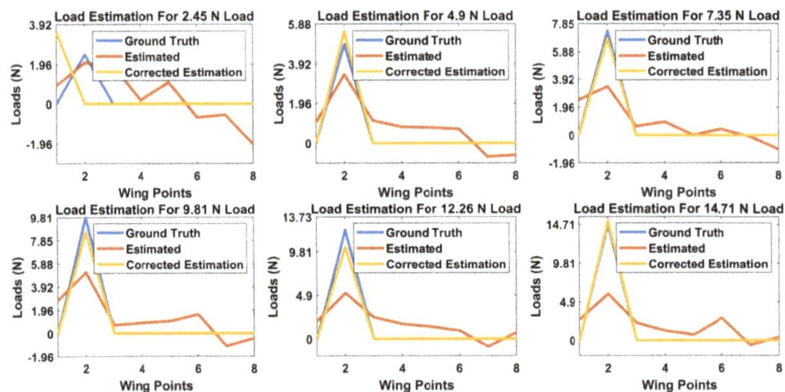

Figure A4. Optimized and corrected load estimations for concentrated loads acting at point 2 of the wing.

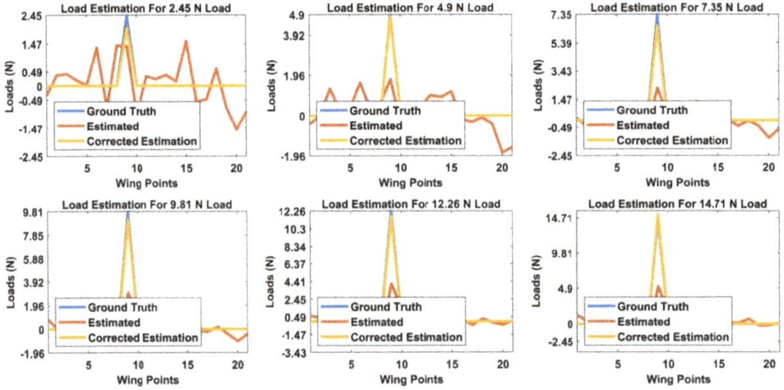

Figure A5. Optimized and corrected load estimations for distributed loads acting at point 9 of the wing.

References

1. Hong, C.Y.; Zhang, Y.F.; Zhang, M.X.; Leung, L.M.G.; Liu, L.Q. Application of FBG sensors for geotechnical health monitoring, a review of sensor design, implementation methods and packaging techniques. *Sens. Actuators A. Phys.* **2016**, *244*, 184–197. [CrossRef]
2. Ma, Z.; Chen, X. Fiber Bragg gratings sensors for aircraft wing shape measurement: Recent applications and technical analysis. *Sensors* **2019**, *19*, 55. [CrossRef] [PubMed]
3. Burner, A.W.; Liu, T. Videogrammetric model deformation measurement technique. *J. Aircr.* **2001**, *38*, 745–754. [CrossRef]
4. Burner, A.W.; Lokos, W.A.; Barrows, D.A. In-Flight Aeroelastic Measurement Technique Development. In Proceedings of the Optical Diagnostics for Fluids, Solids, and Combustion II, San Diego, CA, USA, 10 November 2003.
5. Helios–NASA. Available online: https://www.nasa.gov/centers/dryden/news/ResearchUpdate/Helios/ (accessed on 11 June 2020).
6. Marqués, P.; Da Ronch, A. *Advanced UAV Aerodynamics, Flight Stability and Control*; John Wiley & Sons, Inc.: Hoboken, NJ, USA, 2017.
7. Lizotte, A.; William L. Deflection based aircraft structural loads estimation with comparison to flight. In Proceedings of the 46th AIAA/ASME/ASCE/AHS/ASC Structures, Structural Dynamics and Materials Conference, Austin, TX, USA, 18–21 April 2005.
8. Richards, L.; Parker, A.R.; Ko, W.L.; Piazza, A. Fiber Optic Wing Shape Sensing on NASA's Ikhana UAV. In Proceedings of the NAVAIR Meeting, Edwards, CA, USA, 7 February 2008.
9. Alvarenga, J.; Derkevorkian, A.; Pena, F.; Boussalis, H.; Masri, S. Fiber optic strain sensor based structural health monitoring of an uninhabitated air vehicle. In Proceedings of the International Astronautical Congress, Naples, Italy, 1–5 October 2012.
10. Ciminello, M.; Flauto, D.; Mennella, F. In proceedings of the Sensors and Smart Structures Technologies for Civil, San Diego, CA, USA, 2013. *Int. Soc. Opt. Photonics* **2013**, *8692*, 869221.
11. Nicolas, M.J.; Sullivan, R.W.; Richards, W.L. Large scale applications using FBG sensors: Determination of in-flight loads and shape of a composite aircraft wing. *Aerospace* **2016**, *3*, 18. [CrossRef]
12. Kim, S.W.; Kang, W.R.; Jeong, M.S., Lee, I.; Kwon, I.B. Deflection estimation of a wind turbine blade using FBG sensors embedded in the blade bonding line. *Smart Mater. Struct.* **2013**, *22*, 125004. [CrossRef]
13. Brotherton, T.; Jahns, G.; Jacobs, J.; Wroblewski, D. Prognosis of faults in gas turbine engines. In Proceedings of the 2000 IEEE Aerospace Conference, Big Sky, MT, USA, 25 March 2000.
14. Hornik, K.; Stinchcombe, M.; White, H. Multilayer feedforward networks are universal approximators. *Neural Netw.* **1989**, *2*, 359–366. [CrossRef]
15. Mucha, W.; Kuś, W.; Viana, J.C.; Nunes, J.P. Operational Load Monitoring of a Composite Panel Using Artificial Neural Networks. *Sensors* **2020**, *20*, 2534. [CrossRef] [PubMed]
16. Augustauskas, R.; Lipnickas, A. Improved Pixel-Level Pavement-Defect Segmentation Using a Deep Autoencoder. *Sensors* **2020**, *20*, 2557. [CrossRef] [PubMed]
17. Li, S.; Zuo, X.; Li, Z.; Wang, H. Applying Deep Learning to Continuous Bridge Deflection Detected by Fiber Optic Gyroscope for Damage Detection. *Sensors* **2020**, *20*, 911. [CrossRef] [PubMed]
18. Hinton, G.E.; Salakhutdinov, R.R. Reducing the dimensionality of data with neural networks. *Science* **2006**, *313*, 504–507. [CrossRef] [PubMed]
19. Ng, A. Sparse Autoencoder. CS294A Lecture Notes. Available online: http://ailab.chonbuk.ac.kr/seminar_board/pds1_files/sparseAutoencoder.pdf (accessed on 11 June 2020).
20. Kingma, D.P.; Ba, J. Adam: A method for stochastic optimization. In Proceedings of the 3rd International Conference for Learning Representations, San Diego, CA, USA, 7–9 May 2015.
21. Srivastava, N.; Hinton, G.; Krizhevsky, A.; Sutskever, I.; Salakhutdinov, R. Dropout: A simple way to prevent neural networks from overfitting. *J. Mach. Learn. Res.* **2014**, *15*, 1929–1958.
22. Simard, P.Y.; LeCun, Y.A.; Denker, J.S.; Victorri, B. Transformation invariance in pattern recognition—tangent distance and tangent propagation. In *Neural Networks: Tricks of the Trade*; Springer: Berlin/Heidelberg, Germany, 1998.
23. Mathworks. Available online: https://www.mathworks.com/help/ident/ref/goodnessoffit.html (accessed on 11 June 2020).

24. System Identification Toolbox, Mathworks. Available online: https://www.mathworks.com/products/sysid (accessed on 11 June 2020).
25. Cetinsoy, E.; Dikyar, S.; Hançer, C.; Oner, K.T.; Sirimoglu, E.; Unel, M.; Aksit, M.F. Design and construction of a novel quad tilt-wing UAV. *Mechatronics* **2012**, *22*, 723–745. [CrossRef]
26. Microsoft Kinect V1. Available online: https://docs.microsoft.com/en-us/archive/msdn-magazine/2012/november/kinect-3d-sight-with-kinect (accessed on 11 June 2020).
27. SwissRanger. Available online: https://www.itcs.com.pk/product/mesa-imaging-swissranger-sr4500/ (accessed on 11 June 2020).
28. Gokturk, S.B.; Yalcin, H.; Bamji, C. A time-of-flight depth sensor-system description, issues and solutions. In Proceedings of the 2004 Conference on Computer Vision and Pattern Recognition Workshop, Washington, DC, USA, 27 June 2004.
29. Shotton, J.; Fitzgibbon, A.; Cook, M.; Sharp, T.; Finocchio, M.; Moore, R.; Blake, A. Real-time human pose recognition in parts from single depth images. In Proceedings of the CVPR 2011, Providence, RI, USA, 20–25 June 2011.
30. Khoshelham, K.; Elberink, S.O. Accuracy and resolution of kinect depth data for indoor mapping applications. *Sensors* **2012**, *12*, 1437–1454. [CrossRef] [PubMed]
31. MultiSense S21B—Carnegie Robotics LLC. Available online: https://carnegierobotics.com/multisense-s21b (accessed on 11 June 2020).
32. Arcure Omega. Available online: https://arcure.net/omega-stereo-camera-for-indoor-outdoor-applications (accessed on 11 June 2020).
33. MYNT EYE. Available online: https://www.mynteye.com/products/mynt-eye-stereo-camera (accessed on 11 June 2020).
34. Garrido-Jurado, S.; Muñoz-Salinas, R.; Madrid-Cuevas, F.J.; Marín-Jiménez, M.J. Automatic generation and detection of highly reliable fiducial markers under occlusion. *Pattern Recognition* **2014**, *47*, 2280–2292. [CrossRef]
35. Kato, H.; Billinghurst, M. Marker tracking and hmd calibration for a video-based augmented reality conferencing system. In Proceedings of the 2nd IEEE and ACM International Workshop on Augmented Reality (IWAR'99), San Francisco, CA, USA, 20–21 October 1999.
36. Wagner, D.; Schmalstieg, D. Artoolkitplus for pose tracking on mobile devices. In Proceedings of the 12th Computer Vision Winter Workshop CVWW07, St. Lambrecht, Austria, 6–8 February 2007.
37. Fiala, M. Designing highly reliable fiducial markers. *IEEE Trans. Pattern Anal. Mach. Intell.* **2009**, *32*, 1317–1324. [CrossRef] [PubMed]
38. Leica DISTO X310. Available online: https://www.leicadisto.co.uk/shop/leica-disto-x310/ (accessed on 11 June 2020).
39. TensorFlow. Available online: https://www.tensorflow.org/ (accessed on 11 June 2020).
40. Bauchau O.A.; Craig, J.I. Energy methods. Structural Analysis. In *Solid Mechanics and Its Applications;* Springer: Dordrecht, The Netherlands, 2009.
41. Gharibi, A.; Ovesy, H.R.; Khaki, R. Development of wing deflection assessment methods through experimental ground tests and finite element analysis. *Thin Walled Struct.* **2016**, *108*, 215–224. [CrossRef]
42. Pearson Correlation Coef. Available online: https://www.mathworks.com/help/matlab/ref/corrcoef.html (accessed on 11 June 2020).

Article

Mechanical Properties of Optical Fiber Strain Sensing Cables under γ-Ray Irradiation and Large Strain Influence

Arianna Piccolo [1,2,*], Sylvie Delepine-Lesoille [1], Etienne Friedrich [3], Shasime Aziri [3], Yann Lecieux [2] and Dominique Leduc [2]

1 French National Radioactive Waste Management Agency (Andra), 92298 Chatenay-Malabry, France; sydelepine@yahoo.fr
2 Laboratoire GeM UMR 6183, Université de Nantes, 44000 Nantes, France; yann.lecieux@univ-nantes.fr (Y.L.); dominique.leduc@univ-nantes.fr (D.L.)
3 Solifos AG, Fiber Optic Systems Klosterzelgstrasse 41, CH-5210 Windisch, Switzerland; etienne.friedrich@solifos.com (E.F.); shasime.aziri@solifos.com (S.A.)
* Correspondence: arianna.piccolo@andra.fr

Received: 18 December 2019; Accepted: 21 January 2020 ; Published: 27 January 2020

Abstract: Optical fiber strain sensing cables are widely used in structural health monitoring; however, the impact of a harsh environment on them is not assessed despite the huge importance of the stable performances of the monitoring systems. This paper analyzes (i) the impact of the different constituent layers on the behavior of a strain sensing cable whose constitutive materials are metal and polyamide, (ii) the radiation influence on the optical fiber strain sensing cable response (500 kGy of γ-rays), and (iii) the behavior of the cable under high axial strain (up to 1%, 10,000 $\mu\varepsilon$). Radiation impact on strain sensitivity is negligible for practical application, i.e., the coefficient changes by 4% at the max. The influence of the composition of the cable is also assessed: the sensitivity differences remain under 15%, a standard variation range when different cable compositions and structures are considered. The elasto-plastic behavior is at the end evaluated, highlighting the residual strain (about 1600 $\mu\varepsilon$ after imposing 10,000 $\mu\varepsilon$) of the cable (especially for metallic parts).

Keywords: distributed optical fiber strain sensing cable; Brillouin scattering; Rayleigh scattering; strain sensing cable characterization; elasto-plastic behavior; strain sensitivity coefficients

1. Introduction

Structural health monitoring (SHM) is an important topic in society currently, as buildings' maintenance must be the most cost and time effective as possible. Distributed optical fiber sensing, to monitor strain and other parameters, is found to be a useful tool thanks to optical fiber's desirable features (versatility, dimensions, measurement range, insensitivity to electromagnetic fields, etc.) [1,2]. Distributed optical fiber sensors are sensitive to strain and temperature thanks to the backscattering mechanisms occurring inside the fiber, namely Brillouin and Rayleigh scatterings. These mechanisms allow us to use optical fibers as distributed sensors, reaching many sensing points in the distance with only one sensor. The light that propagates inside the core of the fiber is partially backscattered to its origin, carrying back information on the thermal and mechanical state of the fiber. Once the backscattered light frequency is recollected, it is possible to obtain the frequency shift $\Delta\nu$ at every point along the fiber between the injected and backscattered light, which depends on the fiber conditions as:

$$\Delta\nu = C_T \Delta T + C_\varepsilon \Delta\varepsilon, \tag{1}$$

where C_T and C_ε are the thermal and strain sensitivity coefficients of the considered sensor. Optical fibers' sensitivity coefficients are in the order of $C_T^B = 1$ MHz/°C for temperature and $C_\varepsilon^B = 0.05$ MHz/µε for strain sensitivity for Brillouin scattering [3], while for Rayleigh scattering, the coefficients are respectively in the order of $C_T^R = -1.5$ GHz/°C and $C_\varepsilon^R = -0.15$ GHz/µε [4].

In SHM applications, optical fibers are often inserted into cables to improve the sensitivity performances, enhance their mechanical robustness, and protect the fiber from the harsh environment in which it is employed. The structure of the cable may however have an influence in both the short and the long term sensing characteristics of the fiber. A harsh environment, in fact, can affect not only measurement results, but also the mechanical behavior of the sensor itself and its durability. In the majority of the applications, however, the maintenance over the monitoring period is possible only on the interrogation instrument, while the sensor is definitively embedded in its environment and cannot be accessed to be repaired nor replaced. It is therefore important to select not only the best interrogation method, to reduce at maximum the measurement errors, but the sensing cable as well, able to resist the application's load level and harsh environment over the needed monitoring period. In nuclear structures' monitoring, such as nuclear power plants' operation and dismantlement phases, physics reactors (CERN, etc.), or the space industry, sensing systems face radiation while the monitoring period must exceed 50 years. For the French project of the deep geological disposal facility for high level and intermediate level long lived radioactive waste (known as Cigéo), the application includes the presence of radiation, high temperatures (up to 90 °C), and chemicals (H_2). Investigations on the impact of these harsh conditions on optical fibers in their primary coating have been and are still being carried out, in order to select the best optical fiber composition and interrogation method [5,6]. Nevertheless, as distributed optical fiber sensors (DOFS) are often put into a cable when employed on-field, their composition and the structure of the cable must be carefully selected to be sufficiently resistant and, at the same time, keep as much as possible the elasticity of the sensor. The external sheath should also be chosen in order to maintain or even enhance its sensitivity to the measured variable. For this reason, tests should be carried out in order to analyze the physical and sensitivity characteristics of pre-existent or brand new optical fibers and optical fiber cables. Much work has already been devoted to optical fibers (in primary coating): for example, in [7], the mechanical properties and strain transferring mechanism of optical fiber sensors were analyzed, on the different layers of a fiber Bragg grating (FBG) , while in [8], the role of the coating was studied for strain transfer. In [9], the concept of thermal stability in optical fibers was clarified, and in [10], the coating thermal stability and mechanical strength at elevated temperatures of optical fibers were evaluated on different samples. The physical properties of the coating were also evaluated in [11], where the elasto-plastic bond mechanics of the fiber coating were evaluated, while in [12], a fatigue test was carried out assessing the performance stability of DOFS over two million load cycles.

Regarding cables, however, the bibliography is not so wide. As strain sensing cables are deployed to measure the strain of the structure, many papers are focused on the analysis of the strain transfer function, i.e., to know how much of the structure's strain is transferred to the fiber: $\varepsilon_{FO}(s) = \varepsilon_{struct}(s) \otimes MTF(s)$. The mechanical transfer function $MTF(s)$, which translates the strain of the structure $\varepsilon_{struct}(s)$ in the sensor strain $\varepsilon_{FO}(s)$, represents the behavior of the sensor without the need to specify its physical and mechanical characteristics. The strain transfer function of different kinds of cables is already assessed (for example, [13,14]); however, the physical and sensitivity characteristics of strain sensing cables, as the elasto-plastic behavior and the impact of the protection layers on the optical fiber measurements, are not well considered in the literature. One exception is [15], where the strain sensitivity of different strain sensing cables was analyzed, with attention to the initial residual hysteresis.

The French national agency for radioactive waste management (Andra), for its monitoring needs, has selected some strain sensing cables, of which one has been already used for convergence monitoring tests [16]. This cable was then analyzed in order to understand its physical behavior under traction, as well as its mechanical characteristics and sensitivity. Another important aspect is the durability of the sensors in a harsh environment. The impact of radiation, alone or in combination with other environmental conditions (as high temperature), is well assessed for optical fibers [5,6,17,18]; however, it is not known for commercial strain sensing cables. For this reason, part of the tested strain sensing cable samples has also been irradiated, in order to check the radiation impact on the cable's different layers and mechanical behavior.

In this paper, the mechanical properties and in particular elasticity characteristics of the same optical fiber strain sensing cable are analyzed under diverse conditions. The sensor was tested in its whole, with part of the protection removed and without other protection than the primary coating (i.e., bare fiber). Moreover, the cable's characteristics are also evaluated under irradiation in order to determine the radiation impact on the behavior of the sensor. Lastly, a distinguishing aspect of this study is testing the response of an optical cable subjected to large strain. As any other sensor, an optical fiber cable is designed to operate in the elastic domain, i.e., for strain levels often less than 0.2% (2000 $\mu\varepsilon$). However, in the long run or due to unforeseen loads, this threshold may be exceeded. It is important then to understand if the acquired measures are reliable; to provide an answer, the optical fiber cables are tested up to 1% in strain (10,000 $\mu\varepsilon$), which is beyond what a sensor is expected to undergo on-site. After reporting the specimen's characteristics (comprised of a summary of the sensors' sensitivities) and the test procedure, the elasto-plastic behavior of the selected optical fiber strain sensing cable is shown.

2. Cable Mechanical Characterization: Motivation

In Cigéo, two kinds of long lived radioactive wastes, intermediate level (IL-LL) and high level (HL), will be hosted inside cylindrical repository cells with concrete and steel liners, respectively. These cells are located 500 m deep, inside a 120 m thick Callovo-Oxfordian claystone layer. For the French law, the repository must be reversible for the first one hundred years at least; therefore, a monitoring program will be implemented from the construction phase and throughout its operating life, to keep track of repository safety related parameters. What is more, it will also contribute to ensure the safety of the waste and the surroundings. The particular application however reduces the choice of the possible sensing systems, mostly due to the environmental conditions: the sensor must endure radiation up to 1 MGy (total dose received at the external surface of the metallic liner of "HL0" repository cells after a century of monitoring), maximum temperature around 90 °C, hydrogen presence, cells' convergence (i.e., reduction of the cell's vertical diameter) of 10 mm, and an orthoradial strain level (compression and traction) of about $\pm 0.3\%$ around the cell's circumference ($\sim$2700 $\mu\varepsilon$). Moreover, it has to be embedded in or fixed on the liner in order to avoid (i) most of the harsh environment's impact and (ii) limiting space dedicated to the waste and handling spaces.

Following (1), temperature impacts optical fiber by simply shifting the frequency. Radiation effects on light propagation (frequency shift and attenuation) are attenuated if the fiber is fluorine doped [18], while it is known that a carbon coating helps the fiber to be hermetic to hydrogen absorption, which would otherwise lead to additional losses [19]. In order to limit the risk of the breaking of the sensor, which could not be replaced, the fiber should be protected in a cable to better endure the compression of the surrounding rock and last over time. For this reason, cables with a metallic structure are preferred as it is more durable than plastic, while a rough surface might help the strain transfer of the cable in concrete thanks to a higher level of bonding between the two materials. In the framework of the Cigéo project, to reduce cost and save time, it is preferred to focus on commercial products, when possible. If the market does not offer performances that fulfill the requirements, it is necessary to start specific developments. For this reason, Andra investigated many suppliers and analyzed different cables from over seven companies, and a good tradeoff between tensile strength and minimum curvature radius was found to be given by the BRUsens V9 type from Solifos AG. The V9 type is a 3.2 mm mini armored fiber optic strain sensing cable with an $\sim$0.9 mm central metal tube (FIMT, fiber in metal tube), a structured polyamide (PA) outer sheath, and one optical single mode fiber (SMF) inside. The design of the V9 type cable is depicted in Figure 1.

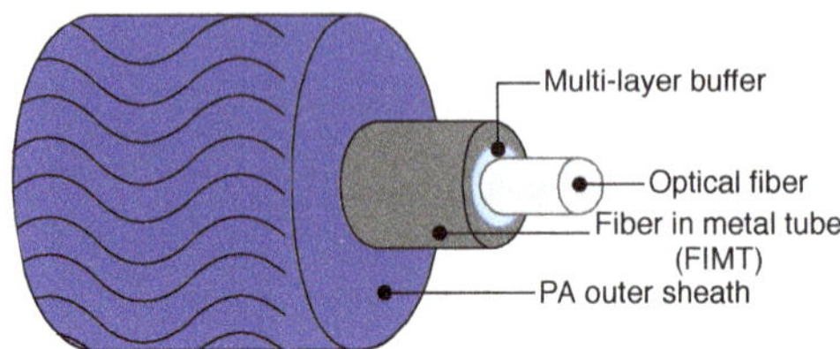

Figure 1. Schematic of the V9 type strain sensing cable, composed by a polyamide sheath of 3.2 mm in diameter, a steel tube (FIMT) of ∼0.9 mm in diameter, a multi-layer buffer, which helps the strain transfer, and a single mode fiber (SMF) (250 µm).

This cable was already tested by Andra in a surface test [16], revealing the suitability for convergence and strain monitoring. The same cable, which normally has a standard single mode fiber with higher curvature resistance inside, was used to develop a custom cable for Andra's application. The resulting cable was the outcome of the insertion of a custom SMF, carbon coated for hydrogen hermeticity, fluorine doped for radiation hardening, in a commercial strain sensing cable (in this case, the V9 type). This custom SMF had a 0.3 wt% F doped core and a 2.3 wt% F doped cladding, a numerical aperture of 0.14, and a core diameter of 7.4 µm, with an attenuation at 1550 nm of 0.40 dB/km and an effective refractive index of about 1.439. Before using this cable on site, however, it is necessary to know its characteristics in sensitivity and durability. Once the fiber is protected in a cable, its characteristics may change as the composition (materials, dimensions, etc.) of the sensor changes. For these reasons, the tests presented in this paper were meant to assess (i) the sensitivity of the newly developed sensor, (ii) the influence of the different protective layers on the behavior of the sensor, and (iii) the impact of a harsh environment (radiation, in this case) on its performances. Radiation impacts both optical fibers' attenuation and frequency shift. The radiation induced attenuation (RIA) and SNR issues were already thoroughly evaluated in [5] on optical fibers in their primary coating. As attenuation poorly depends on the presence of the cable and the total absorbed dose here considered was lower, those results could be seen as a worst case scenario and were therefore supposed to remain valid in this case. For this reason, the radiation impact on optical fiber sensing cables was focused only on the induced frequency shift.

3. Materials

In order to characterize the sensor, the tests were conducted not only on custom V9 type cable samples (Figure 2a), but also on their constitutive parts: the FIMT (with the custom radiation hard fiber inside; Figure 2b) and the naked fiber itself (only primary coating). Besides, the same analysis was carried out on standard commercial samples of the same types (V9, FIMT and bare fiber), which had an SMF G657 with acrylate coating. It is in fact interesting to assess whether the different fibers inside the cable influence in different ways the performances of the sensors. Furthermore, as the goal was also to assess the impact of radiation, part of the V9 and FIMT samples were previously irradiated up to 500 kGy, which is half of the absorbed total dose during the first 100 years of monitoring. The dose was the result of an irradiation campaign where bare optical fibers were irradiated up to 1 MGy [5]. The different distances of the cables from the irradiation source and their support in metal allowed them absorb less dose, i.e., up to 500 kGy. The dose rate was about 1.5 kGy/h. For the sake of comprehension, the different samples under tests are synthesized in Table 1 and will be so addressed from now on.

(a) V9

(b) FIMT

Figure 2. Tested cables: V9 (**a**) and FIMT (**b**) types.

Table 1. Tested samples.

	V9	FIMT	Fiber
Standard SMF G657 acrylate coating	*a*	*d*	*g*
Custom (not irradiated) SMF F doped carbon acrylate coating	*b*	*e*	*h*
Custom (irradiated) SMF F doped carbon acrylate coating	*c*	*f*	-

4. Methods

The samples were tested under traction cycles: the frequency shifts of the samples were acquired, using the Neubrescope NBX-7020 from Neubrex Co., Ltd., which is able to exploit Brillouin and Rayleigh scatterings with the pulse pre-pump Brillouin optical time domain analysis (PPP-BOTDA) and the tunable wavelength coherent optical time domain reflectometry (TW-COTDR), respectively, with the highest spatial resolution of 2 cm.

Traction tests were performed by fixing the samples on a 10 m manual traction bench. The samples were elongated using a winch, checking the new length with a ruler and a laser distance meter (millimetric precision). Measurements were acquired every 500 με (nominal value, 5 mm of elongation), while each 1000 με step, the sample was taken back to the initial position (no elongation) in order to check whether there was (or not) residual strain (i.e., to analyze the plastic strain of the sample). Another measurement was then acquired. This was performed up to 10,000 με, while for the FIMT type samples, the measurements back to zero were performed up to 7000 με. Once the maximum strain range was reached (1% of strain by the datasheet, i.e., 10,000 με), measurements were acquired every 1000 με (10 mm in elongation), without taking the sample back to its original position (no elongation), up to 30,000 με or up to the breaking point. The generalized traction cycle is depicted in Figure 3.

This kind of measurement cycle was conceived of to reach various goals: (i) practically measure the impact of elongation on the materials of the samples (i.e., the elasto-plastic behavior) and (ii) obtain the most representative sensitivity coefficient of the sample after pre-straining it, avoiding hysteresis (as explained in [15]). Moreover, the maximum strain expected in the Cigéo application is ±2700 με, which would be progressively reached over a period of 100 years in a monotonous fashion, in the reference scenario where radioactive waste packages are inserted right after the construction of the cells. However, as the implementation scenario is not already settled, it could be possible that the growth in strain and convergence is not fully monotonic. For example, the strain is non-monotonic if waste packages are inserted long after the construction, especially for concrete liners (IL-LL waste repository cells). The insertion of the packages would cause an immediate change in the strain of the structure, which would be already influenced by creep and shrinkage. Along with this, the concrete would also be affected by seasonal thermal cycles, taking to an even more non-monotonous behavior. Nevertheless, after the insertion of the waste packages, there would be a rise in temperature, causing a dilatation of the structure (especially the steel liner of HL waste repository cells). For all these reasons and in order to anticipate extraordinary situations, as well as to extend the study beyond the nominal conditions, this research explored the effect on the sensors of both "non-monotonous" strain and of very large strains, up to 10,000 με.

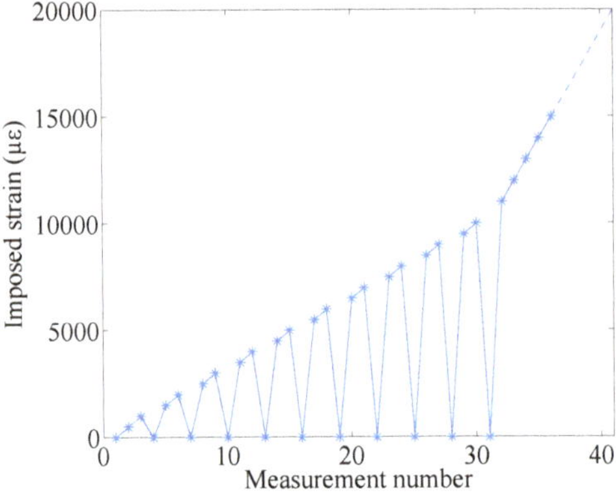

Figure 3. Traction cycle.

During elongation, the cabled samples, especially of the V9 type, slipped from the anchoring due to difficulties in the fixation of samples with a diameter greater than a millimeter. This led to an error (between the desired strain value and the one obtained in reality after the slippage) that remained however under 7%–8%, and it grew in a distributed and homogeneous way from 0 με to the maximum elongation reached by the sample. In this way, the correct analysis of the samples was guaranteed.

Measurements were taken with a resolution of 20 cm and a sampling of 10 cm, obtaining about 100 measurement points over the 10 m bench (excluding connection cables). It has also to be specified that, as all tests were performed in the same period of time and in the same place, the temperature was supposed to be stable (differences in the order of ±2°C). Therefore, the measured frequency shift was attributed solely to the imposed traction. The results presented in the following are the outcome of the analysis of one sample of each specimen summarized in Table 1. In this case, only measurements up to 10,000 με were considered.

5. Strain Sensitivity Coefficients

The strain sensitivities of the tested samples were obtained by averaging the frequency shifts $\Delta\nu$, obtained by interrogating the samples with Brillouin and Rayleigh scatterings, over the central 9 m of the samples (to avoid measurements on the fixations) and dividing them by the imposed strain range. The curves $\Delta\nu$ over strain so obtained for Brillouin and Rayleigh scatterings are plotted in Figure 4. The strain sensitivities of both types of bare fibers were the ones that were closer to the standard values, $C_\varepsilon^B = 0.050$ MHz/με for Brillouin and $C_\varepsilon^R = -0.15$ GHz/με for Rayleigh, while for the cables, the coefficients stayed respectively around $C_\varepsilon^B = 0.045$ MHz/με and $C_\varepsilon^R = -0.13$ GHz/με.

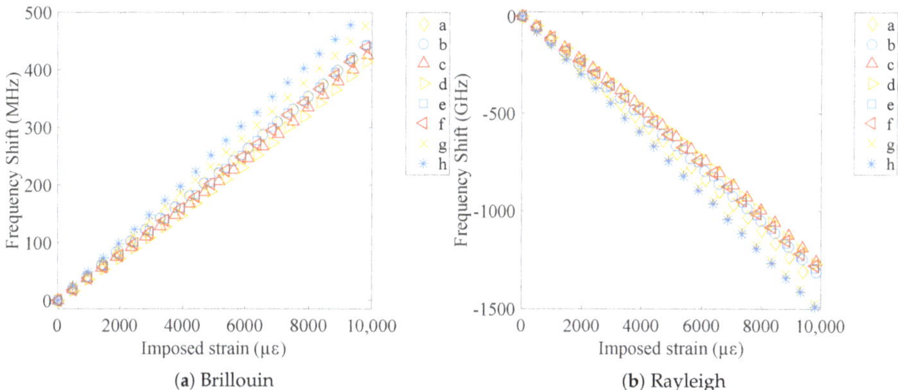

(a) Brillouin (b) Rayleigh

Figure 4. Frequency shift over strain curves for all the tested samples, for Brillouin and Rayleigh scatterings.

Observing the results, especially the cabled samples (V9 and FIMT types), it is noticeable how the strain sensitivities do not differ very much from one sample to another. The strain sensitivity differences between the final sensor (V9 type) and the original bare fiber went from 9% (standard type) to 12% (custom radiation hard), revealing the possibility to insert the desired fiber into the cable, keeping the information on its strain sensitivity as much as possible. The sensitivity differences remains thus under 15%, a standard variation range when different cable compositions and structures are considered. When it comes to analyzing the impact of radiation on the sensor, the sensitivity difference between irradiated and not irradiated samples was as low as 1% (FIMT type) and 4% (V9 type), which means that the sensitivity remained stable. This is a very promising result: if we suppose radiation influence as linear, the 4% in error over 500 kGy would mean an error in strain of 8% in 100 years (1 MGy), i.e., only about 220 με over 2700 με. Moreover, for the F doped fiber, the radiation influence is not linear; it tends to saturate in a parabolic way [5], which means that the error in reality would be even less. In every case, sensitivity coefficient values for irradiated samples were lower than for those not irradiated. Most of the radiation impact was exerted on the physical elasticity of the cable: the PA outer sheath became more rigid and less elastic due to radiation [20], leading to more cracks during handling and elongation and, therefore, to break sooner than non-irradiated samples (Figure 5).

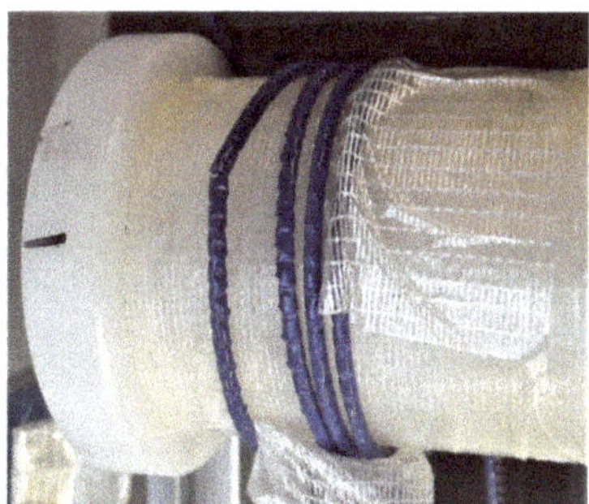

Figure 5. Impact of the radiation on the cable: radiation reduces the elasticity of the plastic, causing cracks when curved.

With the exception of the fibers, the standard samples, and the not irradiated V9 type, some cables broke during the test. The custom irradiated V9 type broke, reaching the nominal value of 12,000 με, while custom FIMT type cables broke at 21,000 με and 29,000 με, respectively, for the pristine and the irradiated one. In practice, the only cable that suffered from radiation influence was the V9 type due to the impact on the polyamide, while the FIMT type cables broke mainly due to their structure. In all cases, if breaks occurred during traction, it was always after the 10,000 με; therefore, the datasheet guaranteed strain range was maintained. These results are very promising for the use of such a cable in an application where radiation is present.

6. Elasto-Plastic Behavior

As the cable is partially composed of steel, which is the most rigid component of the cable and tends to show a plastic behavior after a certain strain (typically, 0.2% [21], 2000 με), it was interesting to look for the possible plastic behavior of the tested cabled samples (FIMT and V9 types). Let us take as an example the standard FIMT (sample *d* in Table 1), in order to directly observe the behavior of the steel protecting the fiber. In Figure 6a,b, the frequency shift over the strain curve is plotted, respectively for Brillouin and Rayleigh scatterings.

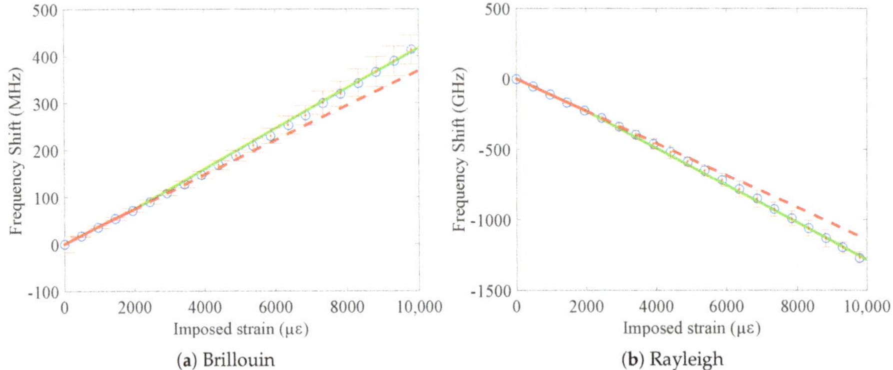

(**a**) Brillouin (**b**) Rayleigh

Figure 6. Detail of the mechanical behavior of the FIMT standard type sample (Sample d). The curve presents two zones with different slopes (highlighted in red and green): this represents the plasticity of the steel of which the FIMT is made.

The uncertainty of the measurements is reported as error bars, while the slope of the curve is calculated separating the strain behavior into two zones: before and after 2000 με. It is visible how the linear regressions of these two zones are different: before 2000 με (red line), the strain coefficient (i.e., the slope) is smaller than afterwards (green line), showing a possible plastic behavior of the sample due to traction. This is valid also for the other samples, the strain coefficient values of which are reported in Table 2.

Table 2. Strain sensitivity coefficients of the different tested samples.

B: MHz/με R: GHz/με	Custom				Standard	
	Not irradiated		Irradiated			
	<0.2%	>0.2%	<0.2%	>0.2%	<0.2%	>0.2%
V9	B: 0.045 R: −0.13	B: 0.046 R: −0.14	B: 0.040 R: −0.12	B: 0.045 R: −0.13	B: 0.042 R: −0.14	B: 0.045 R: −0.14
FIMT	B: 0.041 R: −0.12	B: 0.047 R: −0.14	B: 0.040 R: −0.12	B: 0.047 R: −0.14	B: 0.037 R: −0.11	B: 0.045 R: −0.14
Fiber	B: 0.050 R: −0.15	B: 0.052 R: −0.15	B: - R: –	B: - R: –	B: 0.046 R: −0.15	B: 0.049 R: −0.15

To make it more immediate to evaluate, the strain sensitivities are plotted in Figure 7.

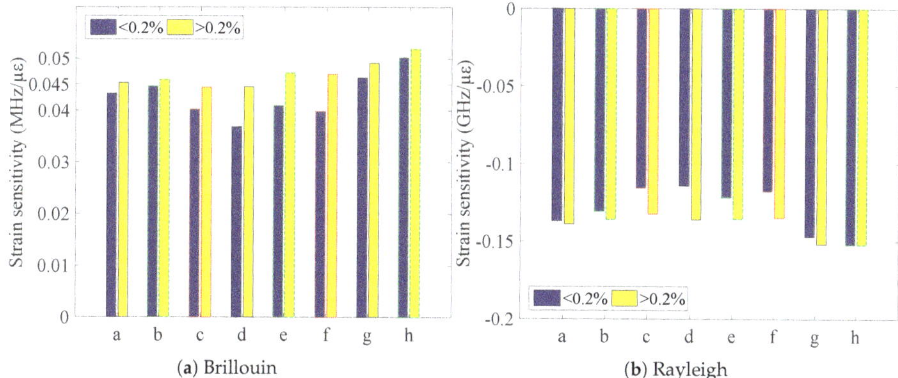

(**a**) Brillouin (**b**) Rayleigh

Figure 7. Strain sensitivity coefficients for different linear fittings: "<0.2%" and ">0.2%" represent the fit done considering only the values before or after 2000 με. The different contours of the bars define the type of sample: the black and straight line are the standard samples, the green and dashed the custom not irradiated, and the red and dotted the custom irradiated.

The biggest difference between the sensitivities of the two identified zones (and therefore, the biggest plastic effect) appears to be exerted on the FIMT, as the V9 type is composed also of the external PA layer, which limited the plastic behavior of the steel.

This behavior should be then confirmed looking at the measurements performed when the samples were at their original position, i.e., when they were not elongated. The frequency shifts of the samples, obtained by interrogating the samples via Brillouin and Rayleigh scatterings at their original position, are plotted in Figure 8a and Figure 8b respectively.

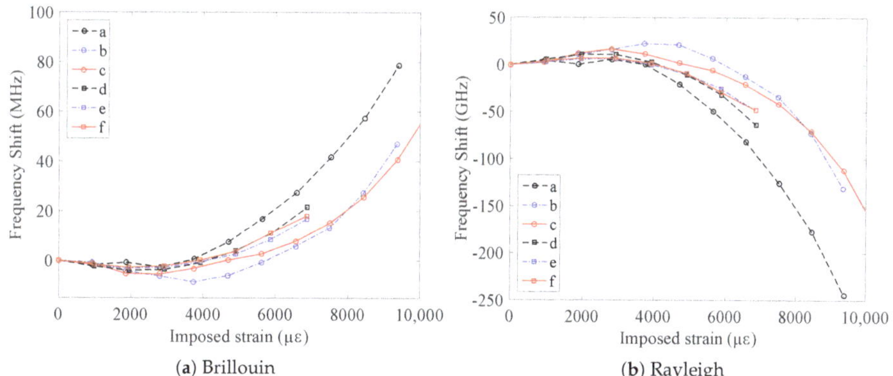

(**a**) Brillouin (**b**) Rayleigh

Figure 8. Residual frequency shift of the samples after elongation (represented in the x axis) of the cabled samples (V9 and FIMT) for the three tested types of condition (standard fiber, custom radiation hard fiber, and custom radiation hard fiber irradiated).

It is remarkable how the cabled samples (V9 and FIMT type) show a permanent frequency shift, i.e., residual strain, after being elongated. This does not happen for the fiber, which undergoes only a slight relaxation (Figure 9). This is related to the multilayered nature of the cable: parts of it underwent permanent strain, and there may have also been slippage at the interface between the layers. The steel layer of the two cable types samples has in fact an elasto-plastic behavior and a higher Young's modulus compared to the polyamide layer, whose behavior is visco-elastic. For this reason, steel led the mechanical behavior of the whole cable, limiting the relaxation effect of the PA, especially when it reached its plastic zone (imposed strain >2000 με). Moreover, looking at raw measurements (Figure 10), we do not observe any relaxation caused by the imposed strain; otherwise, a significant and negative slope while increasing strain should have been observed. Furthermore, as each strain level was maintained for about five minutes, a consistent noise would be visible.

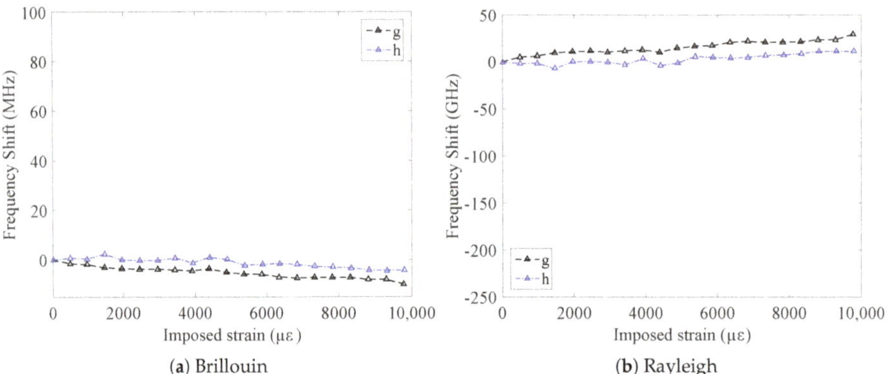

(a) Brillouin (b) Rayleigh

Figure 9. Residual frequency shift of the samples after elongation (represented in the x axis) of the fiber samples for the two tested types of conditions (standard fiber, custom radiation hard fiber).

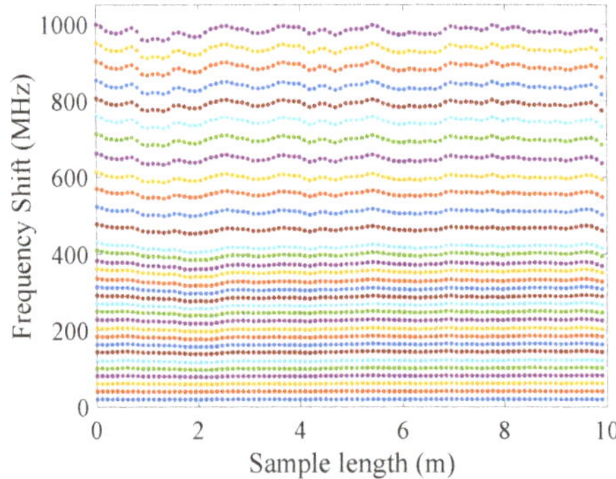

Figure 10. Frequency shift traces during traction (higher frequency shift for higher traction), Sample a, showing no relaxation of the sample.

This underlines the importance of characterizing the whole sensing cable and not only the fiber in primary coating for the sensitivity.

This is similar whatever the fiber (custom radiation hard or standard), as the main actors in this behavior are the protective layers of the fiber (polyamide, steel tube) and whether the samples were irradiated or not. From about 2000 με, the samples were increasingly and permanently deformed, reaching about 60 MHz for Brillouin and −170 GHz for Rayleigh scatterings, which correspond to about 1400 με, a non-negligible value (if the strain range reaches 10,000 με, plasticity is an important phenomenon to consider in the design phase). For the Cigéo reference scenario (the mechanical monitoring of radioactive waste repository cells), as the foreseen maximum strain would be around ±3000 με, the error due to this permanent strain is practically none (around 100 με in compression).

Using the strain sensitivity coefficients previously calculated, it was possible to check whether the sensors, interrogated with two different scatterings, measured the same strain values. Using the values in Table 2, it is possible to transform the frequency shift into strain following (1). The results are plotted in Figure 11.

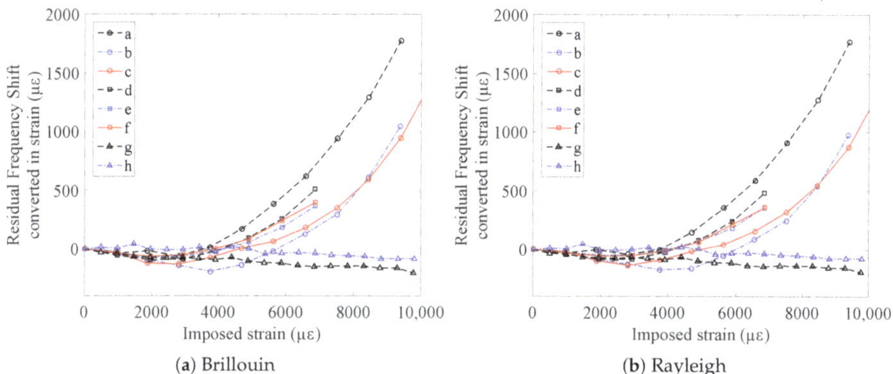

Figure 11. Residual strain of the samples after elongation (represented in the x axis) of all tested samples.

The two strain profiles, deriving from Brillouin and Rayleigh scatterings, are çery close to each other. The results are in general in agreement with the conclusions drawn from Figure 7: FIMT type samples (d, e, and f) are the most plasticized, with a higher residual strain with respect to samples b and c (the custom V9, not irradiated, and irradiated). However, sample a is the one that shows the highest permanent strain between all, even if it is of the V9 type. This behavior, which has yet to be explained, could be due to the different adhesion between the fiber and the internal surface of the FIMT. In any case, as for the sensitivity, there is practically no difference between not irradiated and irradiated samples (b and c, e and f), considering as negligible the impact of the absorbed dose. This is very important for an environment where radiation is present, as this means that the behavior of the sensor is practically not impacted by it, then being suitable to work in such applications.

Even if Cigéo repository cells are not concerned with the plasticity of the cable during the monitoring phase, it is very important to keep in mind that its conditions may change with an unexpected rise in the strain. Nevertheless, this is a general useful reminder for all kinds of applications where the strain is over 4000 με.

The difference between the residual strains obtained with Rayleigh and Brillouin scattering is lower than the uncertainty on the Brillouin measurements (of the order of 20 με) up to 6000 με of imposed strain and remains smaller than 10% in relative value for higher imposed strains. This is very positive, as it shows that the results are the same despite the use of two interrogation methods based on two different scatterings, underlining the interoperability of the two. Moreover, since the measurement principles were different, it proves that the residual strain is related only to the modification of the cable's structure (i.e., not on the backscattering properties).

At this point, (i) strain results acquired from the tested cables, i.e., the strain sensitivities obtained fitting the results in the imposed strain zones <2000 με and >2000 με, (ii) the independence of the cable's behavior from radiations influence, as well as (iii) the results regarding the residual strain showed that the considered cable and its components could be operable for long term measurement. In fact, if the history of the stresses suffered by the cable were known, it would be possible to discriminate the residual strains of the cable from the elastic strains underwent at a given moment. In addition, the work done on the calibration of the strain sensitivity coefficients allowed to choose the most appropriate frequency-strain conversion coefficient for the actual state of the optical cable at the time of measurement. Thus, it would still be possible to perform strain measurements on structures, even at large strain levels, with however a slightly degraded accuracy related to the loss of linearity of the sensor. This opens up the perspective of using the instrumentation deployed, for example in Cigéo, well beyond the conditions for which the system was originally designed.

7. Conclusions

This paper assessed the mechanical characteristics of an optical fiber strain sensing cable, composed of layers of steel and polyamide external sheath. The considered samples, i.e., the cable at its whole, the steel tube alone, and the optical fiber in primary coating, were tested in order to analyze their strain sensitivity and the elasto-plastic behavior. These two topics were examined under two aspects: (i) the influence of the different layers of the cable and (ii) the impact of radiation on the mechanical behavior on the samples. In fact, the protective layers that composed the cable influenced both the strain sensitivity and the elasto-plastic behavior. The strain sensitivity changed at most 12% going from the one of the bare fiber (external primary coating only) to the complete cable (fiber, steel tube, and PA external sheath), while the cable itself was more plasticized than the fiber due to the presence of the steel tube, which was more ductile than glass. Radiation impact was not significant, at least for the tested total dose of 500 kGy: at most, the change in strain sensitivity coefficient was 4% (between irradiated and not irradiated samples), which is a very promising result for monitoring nuclear structure. The same was observed for the elasto-plastic behavior, which was practically unchanged in accelerated aging conditions whether the sample absorbed a radiation dose or not. The presence of other materials than the glass allowed the sensor to show a plastic behavior after 0.2% of strain (2000 $\mu\varepsilon$), which is the elasto-plastic limit for steel. After reaching 10,000 $\mu\varepsilon$ of imposed strain, the cabled samples underwent from 1300 to 2000 $\mu\varepsilon$ of residual strain even if the sample was relaxed and not elongated, which is not negligible for some applications. For our case, as the maximum strain to be reached is around $\pm$3000 $\mu\varepsilon$, this paper confirmed the feasibility to use this kind of optical fiber strain sensing cable. In the future, a similar analysis is to be done for the thermal sensitivity of strain sensing cables, while a deeper investigation on the origins of the residual strain of the samples and on other mechanical behaviors (as under curvature) might be planned.

Author Contributions: Conceptualization, A.P., S.D.-L. and E.F.; Data curation, A.P.; Formal analysis, A.P.; Funding acquisition, S.D.-L.; Investigation, A.P.; Methodology, A.P., S.A. and Y.L.; Project administration, S.D.-L., E.F. and S.A.; Resources, E.F. and S.A.; Supervision, S.D.-L., E.F., Y.L. and D.L.; Validation, S.D.-L., E.F., Y.L. and D.L.; Writing – original draft, A.P.; Writing – review & editing, A.P., S.D.-L., E.F., Y.L. and D.L. All authors have read and agreed to the published version of the manuscript.

Funding: Arianna Piccolo's Ph.D. program is in the framework of ITN-FINESSE, an innovative training network funded by the European Union's Horizon 2020 research and innovation program under Marie Skłodowska-Curie, GA No. 722509.
Andra coordinates the Modern2020 project that received funding from the Euratom research and training program 2014–2018, GA No. 662177.

Conflicts of Interest: The authors declare no conflict of interest.

References

1. Leung, C.K.Y.; Wan, K.T.; Inaudi, D.; Bao, X.; Habel, W.; Zhou, Z.; Ou, J.; Ghandehari, M.; Wu, H.C.; Imai, M. Review: Optical fiber sensors for civil engineering applications. *Mater. Struct.* **2015**, *48*, 871–906. [CrossRef]
2. Barrias, A.; Casas, J.; Villalba, S. A Review of Distributed Optical Fiber Sensors for Civil Engineering Applications. *Sensors* **2016**, *16*, 748. [CrossRef] [PubMed]
3. Nikles, M.; Thevenaz, L.; Robert, P.A. Brillouin gain spectrum characterization in single-mode optical fibers. *J. Lightwave Technol.* **1997**, *15*, 1842–1851. [CrossRef]
4. Kishida, K.; Yamauchi, Y.; Guzik, A. Study of optical fibers strain-temperature sensitivities using hybrid Brillouin-Rayleigh system. *Photonic Sens.* **2014**, *4*, 1–11. [CrossRef]
5. Piccolo, A.; Delepine-Lesoille, S.; Landolt, M.; Girard, S.; Ouerdane, Y.; Sabatier, C. Coupled temperature and γ-radiation effect on silica-based optical fiber strain sensors based on Rayleigh and Brillouin scatterings. *Opt. Express* **2019**, *27*, 21608–21621. [CrossRef] [PubMed]

6. Planes, I.; Girard, S.; Boukenter, A.; Marin, E.; Delepine-Lesoille, S.; Marcandella, C.; Ouerdane, Y. Steady γ-Ray Effects on the Performance of PPP-BOTDA and TW-COTDR Fiber Sensing. *Sensors* **2017**, *17*, 396. [CrossRef] [PubMed]
7. Li, D.; Ren, L.; Li, H. Mechanical Property and Strain Transferring Mechanism in Optical Fiber Sensors. In *Fiber Optic Sensors*; Yasin, M., Harun, S.W., Arof, H., Eds.; IntechOpen: Rijeka, Croatia, 2012; Chapter 18. [CrossRef]
8. Her, S.C.; Huang, C.Y. Effect of Coating on the Strain Transfer of Optical Fiber Sensors. *Sensors* **2011**, *11*, 6926–6941. [CrossRef] [PubMed]
9. Stolov, A.A.; Simoff, D.A.; Li, J. Thermal Stability of Specialty Optical Fibers. *J. Lightwave Technol.* **2008**, *26*, 3443–3451. [CrossRef]
10. Li, J.; Sun, X.; Huang, L.; Stolov, A. Optical fibers for distributed sensing in harsh environments. In Proceedings of the SPIE Commercial and Scientific Sensing and Imaging, Orlando, FL, USA, 15–19 April 2018.
11. Li, Q.; Li, G.; Wang, G. Elasto-plastic bond mechanics of embedded fiber optic sensors in concrete under uniaxial tension with strain localization. *Smart Mater. Struct.* **2003**, *12*, 851–858. [CrossRef]
12. Barrias, A.; Casas, J.R.; Villalba, S. Fatigue performance of distributed optical fiber sensors in reinforced concrete elements. *Constr. Build. Mater.* **2019**, *218*, 214–223. [CrossRef]
13. Billon, A.; Henault, J.M.; Quiertant, M.; Taillade, F.; Khadour, A.; Martin, R.P.; Benzarti, K. Quantitative Strain Measurement with Distributed Fiber Optic Systems: Qualification of a Sensing Cable Bonded to the Surface of a Concrete Structure. Available online: https://hal.inria.fr/hal-01022049/ (accessed on 24 January 2020).
14. Henault, J.M. Methodological Approach for Performance and Durability Assessment of Distributed Fiber Optic Sensors: Application to a Specific Fiber Optic Cable Embedded in Concrete. Ph.D. Thesis, Université Paris-Est, Champs-sur-Marne, France, 2013.
15. Monsberger, C.; Woschitz, H.; Lienhart, W.; Račanský, V.; Hayden, M. Performance assessment of geotechnical structural elements using distributed fiber optic sensing. In Proceedings of the SPIE Smart Structures and Materials + Nondestructive Evaluation and Health Monitoring, Portland, OR, USA, 25–29 March 2017.
16. Piccolo, A.; Lecieux, Y.; Delepine-Lesoille, S.; Leduc, D. Non-invasive tunnel convergence measurement based on distributed optical fiber strain sensing. *Smart Mater. Struct.* **2019**, *28*, 045008. [CrossRef]
17. Berghmans, F.; Fernandez, A.F.; Brichard, B.; Vos, F.; Decreton, M.C.; Gusarov, A.I.; Deparis, O.; Megret, P.; Blondel, M.; Caron, S.; et al. Radiation hardness of fiber optic sensors for monitoring and remote handling applications in nuclear environments. In Proceedings of the Process Monitoring with Optical Fibers and Harsh Environment Sensors, Boston, MA, USA, 1–6 November 1998.
18. Girard, S.; Kuhnhenn, J.; Gusarov, A.; Brichard, B.; Uffelen, M.V.; Ouerdane, Y.; Boukenter, A.; Marcandella, C. Radiation Effects on Silica-Based Optical Fibers: Recent Advances and Future Challenges. *IEEE Trans. Nucl. Sci.* **2013**, *60*, 2015–2036. [CrossRef]
19. Lemaire, P.J.; Kranz, K.S.; Walker, K.L.; Huff, R.G.; Dimarcello, F.V. Hydrogen permeation in optical fibres with hermetic carbon coatings. *Electron. Lett.* **1988**, *24*, 1323–1324. [CrossRef]
20. Porubská, M. Radiation Effects in Polyamides. In *Radiation Effects in Materials*; Monteiro, W.A., Ed.; IntechOpen: Rijeka, Croatia, 2016; Chapter 10, doi:10.5772/62464. [CrossRef]
21. Smith, W.F.; Hashemi, J. *Foundation of Material Science and Engineering*; McGraw-Hill: New York, NY, USA, 2010.

Article

Determination of the Real Cracking Moment of Two Reinforced Concrete Beams through the Use of Embedded Fiber Optic Sensors

Julián García Díaz [1,*], Nieves Navarro Cano [1] and Edelmiro Rúa Álvarez [2]

1 School of Building Engineering, Universidad Politécnica de Madrid, 28040 Madrid, Spain; nieves.navarro@upm.es

2 School of Civil Engineering, Universidad Politécnica de Madrid, 28040 Madrid, Spain; edelmiro.rua@upm.es

* Correspondence: julian.gdiaz@alumnos.upm.es

Received: 18 December 2019; Accepted: 8 February 2020; Published: 10 February 2020

Abstract: This article investigates the possibility of applying weldable optic fiber sensors to the corrugated rebar in reinforced concrete structures to detect cracks and measure the deformation of the steel. Arrays have initially been designed comprised of two weldable optic fiber sensors, and one temperature sensor to compensate its effect in measuring deformations. A series of tests were performed on the structures to evaluate functioning of the sensors, and the results obtained from the deformation measures shown by the sensors have been stored using specific software. Two reinforced concrete beams simply resting on the support have been designed to perform the tests, and they have been monitored in the zones with maximum flexion moment. Different loading steps have been applied to the beams at the center of the span, using a loading cylinder, and the measurement of the load applied has been determined using a loading cell. The analysis of the deformation measurements of the corrugated rebar obtained by the optic fiber sensors has allowed us to determine the moment at which the concrete has cracked due to the effect of the loads applied and the deformation it has suffered by the effect of the different loading steps applied to the beams. This means that this method of measuring deformations in the corrugated rebar by weldable optic fiber sensors provides very precise results. Future lines of research will concentrate on determining an expression that indicates the real cracking moment of the concrete.

Keywords: Fiber Bragg grating; fiber optic sensors embedded in concrete; strain measurement; monitoring; cracking; weldable fiber optic sensors

---

1. Introduction

It is fundamental to know the steel deformation in order to correctly calculate concrete structures. Although it deforms slightly before the concrete has not yet cracked, as shall be seen below, it acquires its greatest deformation when the concrete cracks, as it is not able to absorb traction. This article examines the appearance of the first cracks in reinforced concrete, embedding optic fiber sensors welded to the corrugated rebar steel, and it evaluates their deformation, comparing them with the values obtained from traditional material resistance calculation.

Although there is a lot of literature regarding crack detection in structural elements of reinforced concrete, the novelty of this article lies in the use of optic fiber sensors based on Bragg gratings (FBGs) welded to the corrugated steel rebars, which allows, on the one hand, determination of the precise moment when the crack appears, as the deformation of the steel shows a highly significant leap at the moment when it takes place, due to the concrete ceasing to collaborate in traction and the steel beginning to do so and, on the other hand, the deformation the steel suffers during the successive loading steps.

Diverse procedures have been used since methods have existed to detect cracks in reinforced concrete elements: infrared thermography [1–3], acoustic emission [4,5], fiber Bragg grating (FBG) [6–12], and digital image correlation [13,14]. One must emphasize the study of propagation and determination of the width of cracks in concrete, applying novel microwave sensors for crack detection in four reinforced concrete beam specimens [15]. By use of diffusing ultrasonic sensors, it has been possible to locate micro-cracks within the beam [16]. The stress wave technique with embedded smart aggregates in three samples of FRP reinforced concrete beams have provided satisfactory results in crack detection in the samples [17]. The use of plastic optical fibers to detect hairline cracks and ultimate failure crack in civil engineering structures has also obtained good results, even detecting the moment at which the structural element begins to crack and its evolution until the ultimate failure [18]. Carbon nanotube sensors embedded into concrete beams have also been used and were able to detect the initiation of cracks at an early stage of loads [19]. A novel sensing skin for monitoring cracks in concrete structures is capable of detecting, localizing and quantifying cracks in post-tensioned concrete specimens [20]. On the other hand, there are methods to determine cracks by images, such as the deep fully convolutional network (FCN). Images extracted from a video of a cyclic loading test on a concrete specimen is a reasonably method for crack detection [21], or use of a fully convolutional neural network [22]. A crack monitoring technique based on oblique fiber optic sensing network can accurately measure concrete cracks with a precision of 0.05 mm [23].

Fiber optic sensors based on Bragg gratings (FBGs) have been chosen for this study as they are more durable than conventional electric gages. They also provide long term signal stability and system stability, even under very unfavorable conditions, such as the vibration caused by roads. The cable length has no impact on the precision of the measurement. Multiplexing use allows different sensors to be placed on the same fiber optic cable, a much lighter cable than the conventional one of electric extensometric gages. The optic sensors are immune to electromagnetic and radiofrequency (EMI/RFI) interference, and resist hostile environments in the presence of water, salt, extreme temperatures, pressure (up to 400 bar), potentially explosive atmospheres and high voltage zones. Definitively, fiber optic sensors based on Bragg grating offer greater economic advantages, performance and precision than traditional electric gages [24].

FBG have been used both to detect cracks on the surface as well as by embedding them in the concrete. They have also been used to measure temperature and strain [25], biaxial-bending structural deformations [26], stress on the post-tensioned rod, detect moisture ingress in concrete based building structures [27].

In this study, we investigate the use of optic fiber sensors welded to the corrugated rebar to detect the moment at which the first crack takes place and the deformation the steel suffers from that moment onward during the whole process of loading the structural element. Section 2 analyses the materials used to perform the tests, the shape and dimensions of the beams studied. A tour is made through the characteristics of the equipment used and the software used. In Section 3, the test results are displayed and evaluated to determine the cracks and deformation of the steel in the two reinforced concrete beams, and conclusions are discussed in Section 4.

2. Materials and Methods

There are different devices on the market to measure deformations. The sensors chosen, within the group of Bragg grating (FBG) sensors, are weldable sensors, as these are to be integrated with the structural steel of the reinforced concrete structural elements to be tested.

The fiber optic sensors are highly temperature sensitive, so in order to compensate these effects, a temperature sensor will be included within the array. An array is a chain of sensors linked by a fiber optic cable; custom made for the structural element that is to be tested. The array to be installed on our beams shall contain two weldable deformation sensors and a temperature sensor. Each one of the deformation sensors shall be welded to the two bars that reinforce the beam under flexion and are to be stressed. As the temperature sensor only measures this, it will not be welded to the steel bars and will be located near to the deformation sensors.

Two beams of reinforced concrete are to be tested, with sections of 200 × 300 mm and 300 × 500 mm, and a length of 3000 mm. These sections have been chosen so they are sensitive to the loads to be applied.

2.1. Fiber Optic Sensors

The fiber optic deformation sensors used are weldable (Figure 1). These are welded to the corrugated steel bars that constitute the reinforcement of the beam. Their position is that of the maximum effort that it will receive according to the different loading steps applied. Measurement of the deformation is reliably obtained in micrometers per meter.

Figure 1. Weldable deformation sensor.

The sensor chain was custom manufactured for each beam. Each chain has a temperature sensor (Figure 2) to compensate the temperature effect on the deformation sensors. The sensor chains have two terminals to which the interrogator may be connected. That redundancy effect is important as if a part of the chain were to be damaged, the terminal could always be measured in the undamaged area of the array.

Figure 2. Temperature sensor.

2.2. Reinforced Concrete Beams

The first structural element we are going to test is a reinforced concrete beam with a section of 200 × 300 mm. Its reinforcement is of four rebars with a diameter of 12 mm, with stirrups with a diameter of 8 mm every 200 mm. The steel quality is B500S. The beam length between resting points is 3000 mm. The reinforcement scheme and placement of the fiber optic sensors is illustrated in Figures 3 and 4.

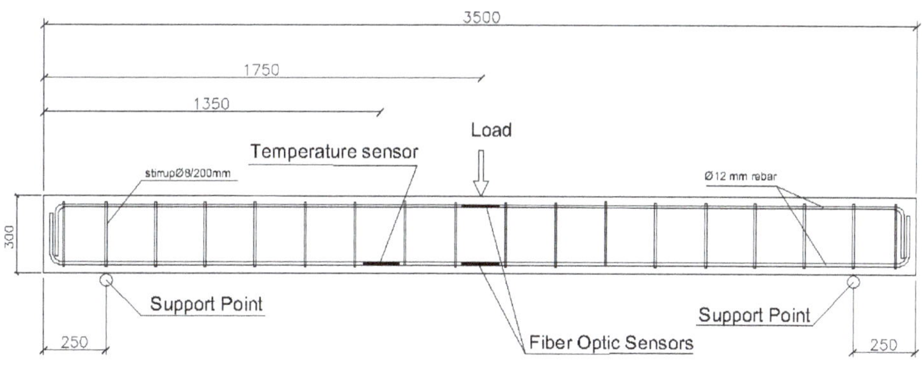

Figure 3. Reinforcement scheme of the 200 × 300 mm beam and application point of the loading steps.

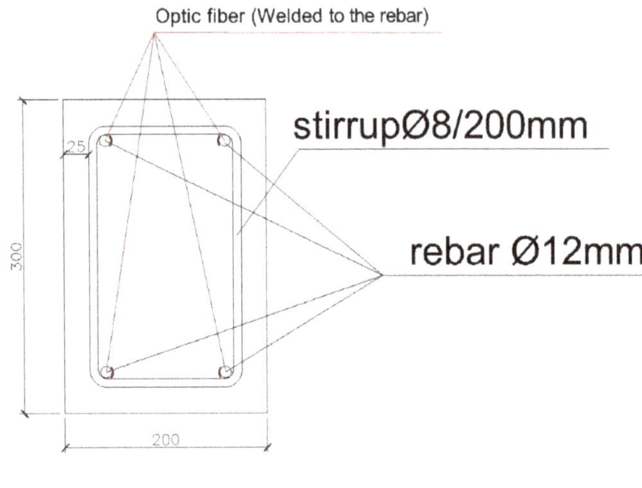

Figure 4. Beam section 200 × 300 mm.

The choice of this beam with this section and reinforcement is due to the loads that are to be applied to it, so it is deformed significantly, and one may measure the deformations the steel suffers, as well as observe the cracks in the structural element, something that will be decisive in the section study.

The sensors are placed in the center of the beam, which is where the different loading steps are to be applied and where, as it is a beam that is simply resting on the support, there will be the maximum deformations in the structural element.

The temperature sensor has been placed attached to one of the corrugated steel bars, near the center zone, so it may compensate the effects the temperature has on the deformation sensors.

The second structural element we are going to test is a reinforced concrete beam with a section of 300 × 500 mm. It is reinforced by two rebars with a diameter of 25 mm on the lower face and 2 rebars with a diameter of 12 mm on the upper face. The frames are rebar with a diameter of 8 mm, every 20 mm. As with the 200 × 300 mm beam, the length between support points is 3000 mm. Figures 5 and 6 show schemes of the structural elements.

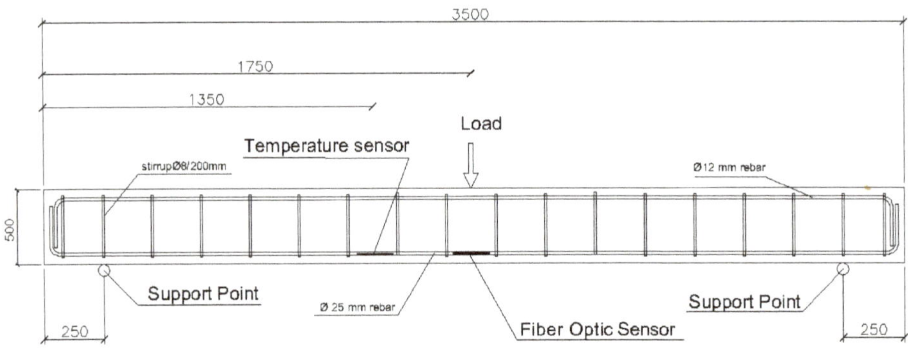

Figure 5. Reinforcement scheme of the 300 × 500 mm beam and application point of the loading steps.

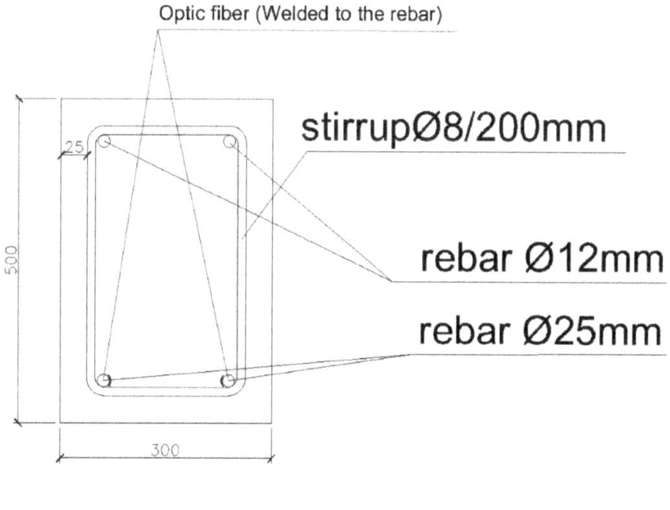

Figure 6. Beam section 300 × 500 mm.

In this case, the sensors have only been placed on the lower bars, with a diameter of 25 mm, as the maximum deformation will take place on these and in the center of the span. As with the 200 × 300 mm beam, the temperature sensor has been placed near to the deformation sensors.

2.3. Design of the Experiment

In order to perform the experiment, the beams will be placed under different loading steps. The successive loads will be applied by a hydraulic press that will press on an RTN type loading cell of 10 Tn, with a ring torsion for the 200 × 300 mm beam. The device specifications are as follows: Nominal load: 10 t; Precision class: 0.05; Body measured: stainless steel; Protection class: IP68 to EN 60529; Cable type: shielded round cable, four wires in the case of the 200 × 300 mm beam, and for the 300 × 500 mm beam under RTN 100 Tn maximum ring torsion loading, Nominal load: 100 t; Precision class: 0.05; Body measured: stainless steel; Protection class: IP68 to EN 60529; Cable type: round shielded cable, four wires

Both loading cells are HBM brand (Hottinger Brüel & Kjaer Ibérica, S.L.U. San Sebastián de los Reyes, Madrid, Spain) with European Union Declaration of Conformity No. 238/2017-07 connected to an HBM QuantumX MX1615 data acquisition system with 16 channels, compatible with the following transducer technologies in all the channels:

- Extensometric gages on a circuit of 1⁄4, 1⁄2 with full bridge, variable bridge power (DC or bearing frequency of 1200 Hz), internal terminal resistance on 1/4 bridge (120 or 350 ohm); - Voltage (± 10 V);
- Pt100, resistance
- Potentiometer
- Sampling speed: max. 20 kS/s
- Automatic transducer identification: TEDS
- European Union Declaration of Conformity No. 263/2017-07.

In the process of successively loading the beam, the sag acquired by the structural elements will be measured by a linear potentiometer displacement transducer of 20 mm, 0.1% precision, compatible with MX1615B amplifier. The displacement transducer will also be connected to the QuantumX MX1615 data acquisition system.

The QuantumX MX1615 data acquisition system is connected to a computer in which software is installed to collect all the information, both from the loading cells as well as the displacement transducer. The software is Catman Easy, by the commercial brand HBM.

Figure 7 shows the loading cell device and displacement transducer installed to perform the test, as well as the loading bridge with the beam located in the position to commence testing.

(**a**) 200 × 300 beam

(**b**) Loading cell and displacement transducer

Figure 7. Loading cell and displacement transducer on the 200 × 300 mm beam.

Data acquisition from the elements to measure applied force and deformation have been connected to the aforementioned QuantumX data acquisition system. Both the deformation as well as temperature sensors are connected to another device called interrogator. An optic interrogator is an optoelectrical instrument able to read fiber sensors with Bragg grating (FBG) in static and dynamic monitoring applications.

The same interrogator may obtain readings from an ample network of sensors of various types (deformation, temperature, displacement, acceleration, slope, etc.) connected through different fiber lines. All the data may be acquired simultaneously and with different sampling frequencies.

During the data acquisition, the interrogator measures the bandwidth associated with the light reflected by the optic sensors and converts it to technical units.

The interrogator model we are going to use is the FS22 (HBM), a device designed to interrogate Bragg grating based sensors. Its technology is continuous laser scan. This includes a reference to scannable bandwidth that provides continuous calibration and guarantees the long-term precision of the system. These interrogators are executed in an operating system in real time to acquire high quality data from a large number of sensors provided by the combination of broadband tuning range and simultaneous and parallel acquisition.

The interrogator is connected to the computer, which uses specific Catman Easy software, providing us the data on the deformations suffered by the beams in real time. The data acquisition system installed is shown in Figure 8.

Figure 8. Data acquisition system installed. QuantumX, Interrogator and computer with Catman Easy software.

3. Experimental Results and Discussion

The deformations we are going to measure are those of the steel, as that is the material the optic fiber is measuring. The cracking moment is a fundamental datum, as we know the moment at which the concrete ceases to absorb traction, in order for the steel to begin to work.

We must take into account, quoting Calavera [28] that "*Between the crack lips, the steel takes on the full traction strain on its own, but between the cracks, there is the anchorage of the reinforcement in the concrete and part of the traction force on the steel is transferred to it. If the traction on the concrete equals its resistance to traction, a new crack is formed*".

This means that there is, between cracks, part of the concrete that absorbs deformations. At the exact point where there is a crack, the concrete does not collaborate and the whole deformation is absorbed by the steel. That fact is fundamental to understand how the structural element works (Figure 9).

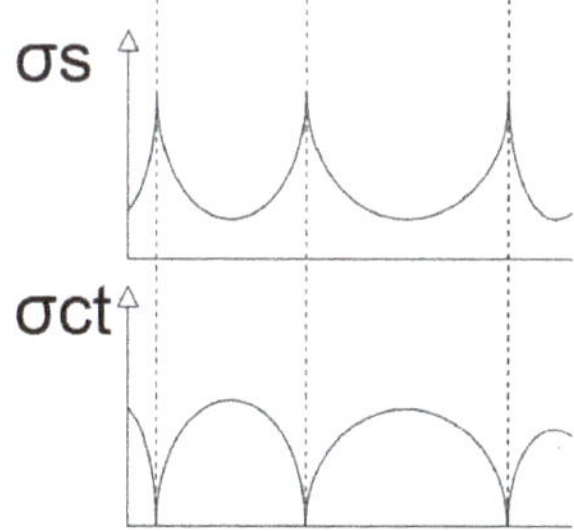

Figure 9. Variation in the tensions in concrete (σ_{ct}) and steel (σ_s) between cracks. In CALAVERA, J. (2008), Proyecto y cálculo de estructuras de hormigón. Tomo II, p. 372. (Designing and calculating concrete Structures, Volume II, p. 372). Give as [28].

In the process of loading the structural elements, a visual inspection of the cracks that appear is carried out, to subsequently compare them with the theoretical calculations (Figure 10).

(**a**) Cracks detail

(**b**) General view of the cracked beam

Figure 10. Visual inspection of cracks in 200 × 300 mm and 300 × 500 mm beams.

Laboratory test were carried out on the two beams studied, obtaining the following figures: For the 200 × 300 mm beam, the laboratory data is: Resistance to flexion-traction of concrete (f_{ct}): 5.4 MPa; Resistance to compression of the concrete (f_{ck}): 54.6 MPa; Concrete elasticity module (E): 33.400 MPa; and for the 300 × 500 mm beam: Resistance to flexion-traction of the concrete (f_{ct}): 8.9 MPa; Resistance to compression of the concrete (f_{ck}): 58.8 MPa; Elasticity module of the concrete (E): 33.053 MPa.

We used the values obtained in the laboratory for resistance of flexion-traction of the concrete (f_{ct}) to calculate the cracking moment. The cracking moments are calculated by applying the Equation (1), obtaining the following results:

$$M_{fis} = f_{ct} * (b * h^2)/6 \quad (1)$$

For the 200 × 300 mm beam this gave M_{fis} = 16.2 mkN, and for the 300 × 500 mm beam: M_{fis} = 111.25 mkN. Once these values were known, the beams underwent different loading steps, applied in the center of the span, linking the loads applied to the deformations caused in the corrugated steel according to the tables included in the relevant sections. Visual inspections were performed to control the moment when the first cracks appear in these.

If we analyze the cracks in the section, and when they take place, we observe that these have taken place much before the cracking moment obtained by calculation. For the 200 × 300 mm beam, it is observed that the cracks nearest to the center of the span take place with a load of 10.4 kN, and a moment of 7.80 mkN, which is much further from the theoretical value obtained. The value of 10.4 kN has been obtained by visual inspection (Figure 11), but as we shall see, the real load for the beam to begin to crack is 8.5 kN.

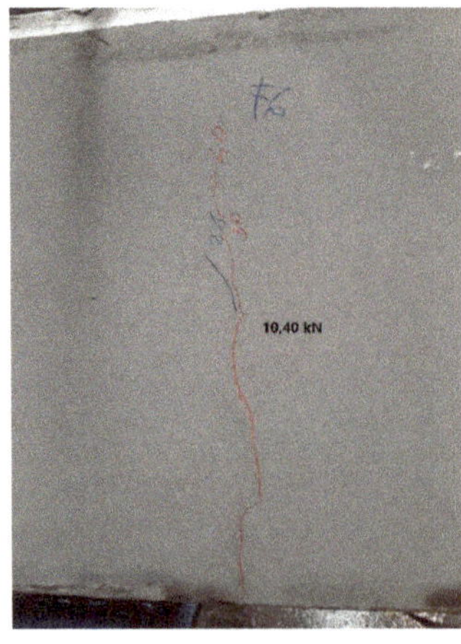

Figure 11. Visual inspection of cracks on the 200 × 300 mm beam.

In the case of the beam of 300 × 500 mm, although the cracking moment obtained by calculation is 111.25 mkN, the real cracking moment is 40.5 mkN, that corresponds to a load of 58 kN, as we may see below (Figure 12).

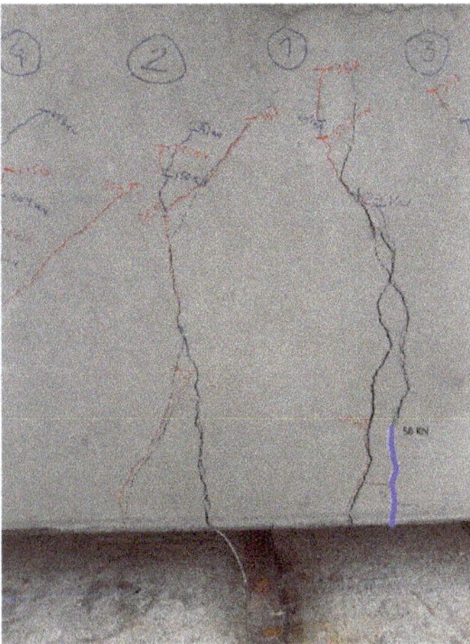

Figure 12. Visual inspection of cracks on the 300 × 500 mm beam.

3.1. 200 × 300 mm Beam

The section studied is shown in Figure 13. The corrugated steel bars correspond to those located in the lower part of the beam, which will be subject to traction.

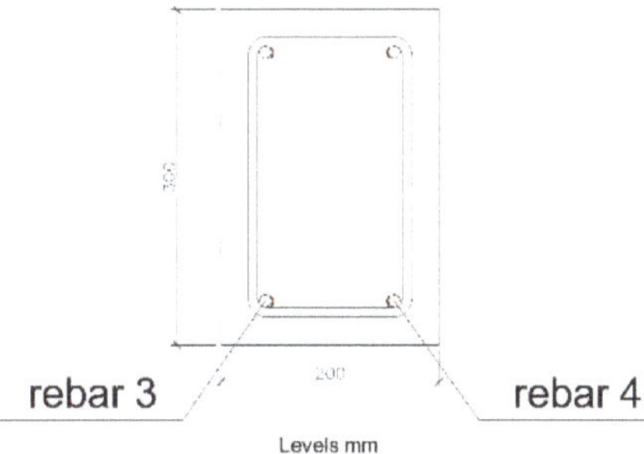

Figure 13. Position of the rebars studied in the 200 × 300 mm beam.

3.1.1. Determining the Moment When the Crack Takes Place

The loads began to be applied in the center of the beam span, within an interval ranging from 0.56 kN to 42.24 kN. Using Catman software, we obtained the deformation that takes place in corrugated steel bars according to the loads applied in Table 1.

Table 1. Relation between loads applied and deformation of the rebars 3 and 4.

Load (kN)	Moment (mkN)	Rebar Deformation 3 (µm/m)	Rebar Deformation 4 (µm/m)
0.56	0.42	−0.18	−0.17
2.49	1.87	2.71	3.02
3.85	2.89	2.54	3.28
5.08	3.81	7.79	7.61
6.88	5.16	13.4	12.70
8.07	6.05	20.05	21.61
9.05	6.79	39.05	38.03
13.59	10.19	106.46	134.48
14.15	10.61	119.68	149.43
15.34	11.51	147.52	179.24
17.69	13.27	156.19	189.27
19.09	14.32	177.99	213.38
20.01	15.01	198.74	234.04
22.71	17.03	277.62	301.36
25.42	19.06	453.07	447.94
26.64	19.98	526.35	513.97
29.11	21.83	740.68	725.88
30.45	22.84	800.39	802.02
33.80	25.35	988.09	1027.15
35.57	26.67	1060.50	1130.08
37.75	28.31	1121.17	1205.10

As shown in Figure 14, the steel deformation is about five or six µm/m for increases in load between one and two kN, while it is 20 µm/m when the load is from 8.07 to 9.05 kN.

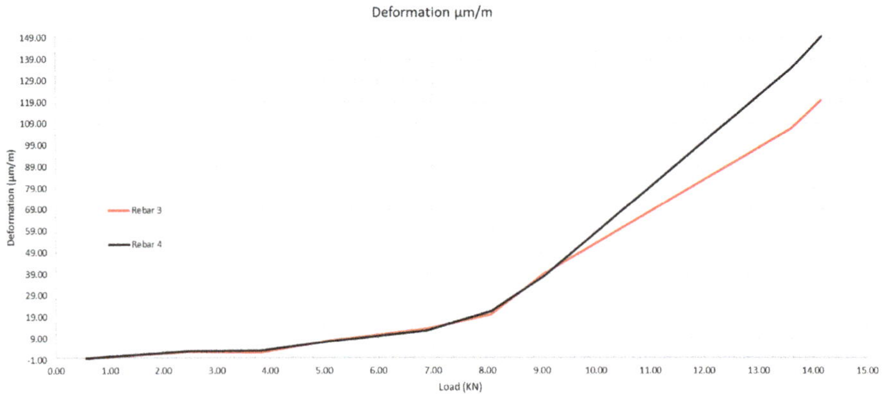

Figure 14. Deformation of rebars according to the loads applied.

These results imply the need to reconsider the moment of the concrete cracking and the steel deformation, as the moment of real cracking is less than that obtained by calculation as well as by visual inspection. It is evident that micro-cracks that are invisible to the human eye are formed, but that the optic fiber is able to detect. One may thus determine that a load of 8.5 kN is what makes the concrete crack.

3.1.2. Steel Deformation with Laboratory Data on the Concrete Compared with Steel Deformation Obtained by Optic Fiber Sensors

A comparison shall be made between the theoretical deformation obtained by calculation and the real deformation process indicated by the sensors. To that end, a concrete with the characteristics obtained in the laboratory has been created, with a resistance to traction of 5.4 MPa, which provides a cracking moment of 16.2 mkN. Although all the moments acting have been checked in FAGUS, Table 2 shows the deformation of the steel for the moment of 31.68 mkN.

Table 2. Deformation of the rebars 3 and 4 for a moment of 31.68 mkN by FAGUS.

Formula/Result	Name	Max.	Min	Value
ε Rebar 3	Rebar 3	0.3414	0.3414	‰
ε Rebar 3 Value in crack	Rebar 3	0.4878	0.4878	‰
ε Rebar 4	Rebar 4	0.3414	0.3414	‰
ε Rebar 4 Value in crack	Rebar 4	0.4878	0.4878	‰

The rest of the deformation values are shown in Table 3 and Figure 15.

Table 3. Theoretical deformation of the rebars 3 and 4 for M_{fis} = 16.2 mkN and f_{ct} = 5.4 MPa with FAGUS.

Load (kN)	Moment (mkN)	Theoretical Deformation (μm/m)	Deformation Rebar 3 (μm/m)	Deformation Rebar 4 (μm/m)
0.56	0.42	2.60	−0.18	−0.17
2.49	1.87	12.50	2.71	3.02
3.85	2.89	19.10	2.54	3.28
5.08	3.81	25.00	7.79	7.61
6.88	5.16	34.20	13.4	12.70
8.07	6.05	39.40	20.05	21.61
9.05	6.79	44.70	39.05	38.03
13.59	10.19	66.90	106.46	134.48
14.15	10.61	69.50	119.68	149.43
15.34	11.51	75.40	147.52	179.24
17.69	13.27	87.20	156.19	189.27
19.09	14.32	93.70	177.99	213.38
20.01	15.01	98.30	198.74	234.04
22.71	17.03	158.80	277.62	301.36
25.42	19.06	179.80	453.07	447.94
26.64	19.98	190.40	526.35	513.97
29.11	21.83	215.10	740.68	725.88
30.45	22.84	231.10	800.39	802.02
33.80	25.35	282.10	988.09	1027.15
35.57	26.67	313.60	1060.50	1130.08
37.75	28.31	359.10	1121.17	1205.10

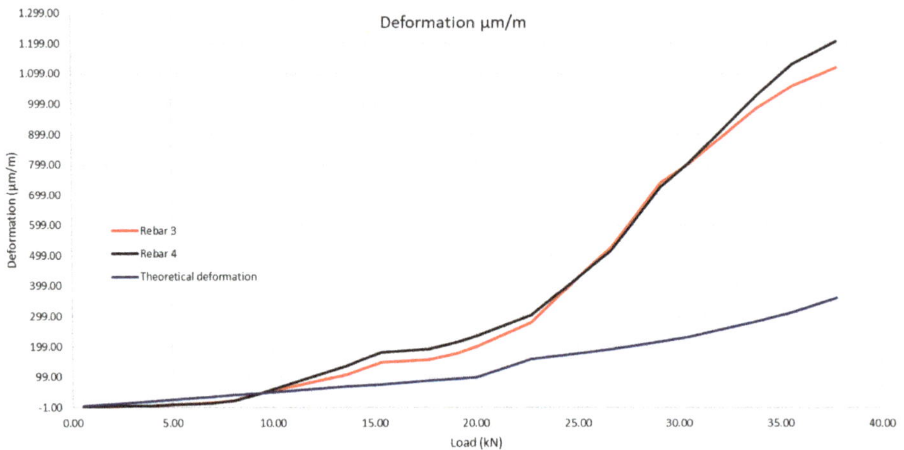

Figure 15. Theoretical deformation of the rebars compared with real deformation obtained with the optic fiber sensors.

It is evident that the deformation of the steel is in fact greater than what is stated in the theoretical calculations. This is caused by cracking of the beam that, as aforementioned, happened before it was expected.

3.1.3. Deformation of the Steel with the Real Cracking Moment of the Concrete Compared with the Deformation Obtained Using Optic Fiber Sensors

We shall now see how the steel in the beam is deformed at theoretical level with the real data for traction resistance of the concrete and the real cracking moment. We have already noted that the beam cracks under a load of 8.5 kN, which corresponds to a cracking moment of 6.38 mkN. With that moment, and applying Equation (1) given above.

We obtain the real resistance to traction of the concrete, that shall be 2.13 MPa. With that data, we input the value in the characteristics of our concrete in the FAGUS program. The program provides a value of the steel deformation in the crack, and another in the uncracked concrete. The section studied is between two cracks, so that figure must be averaged. Table 4 provides the values with and without cracking, for a moment of 28.31 mkN. The rest of the values have been obtained the same way.

Table 4. Deformation of rebars 3 and 4 for a moment of 28.31 mkN by FAGUS.

Formula/Result	Name	Max.	Min	Value
ε Rebar 3	Rebar 3	1.3895	1.3895	‰
ε Rebar 3 Value in crack	Rebar 3	1.9850	1.9850	‰
ε Rebar 4	Rebar 4	1.3895	1.3895	‰
ε Rebar 4 Value in crack	Rebar 4	1.9850	1.9850	‰

In Figure 16 one observes, on the one hand, the position of the optic fiber, the distance between cracks, which is 300 mm, and the position related to the crack on the left side of the optic fiber, which is 180 mm. That means that the concrete between fissures contributes to traction of the beam, and to the steel becoming deformed, but not if it is in the crack.

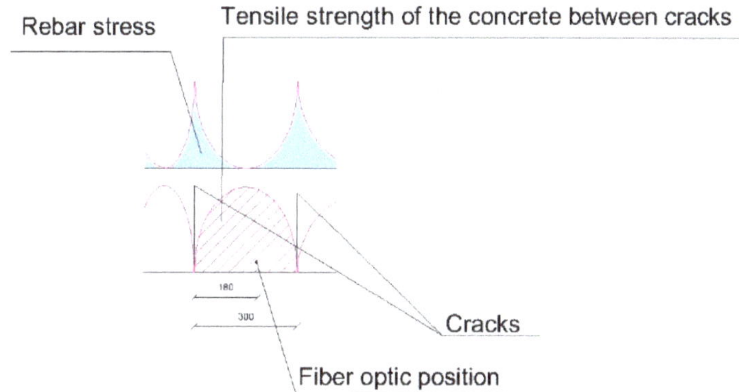

Figure 16. Determining deformation of the steel between cracks on 200 × 300 mm beam.

Table 5 includes the steel deformation without cracking and the steel deformation with cracking. The deformation adopted shall be an interpolation between both figures.

Table 5. Deformation of the rebars 3 and 4 for M_{fis} = 8.5 mkN and f_{ct} = 2.13 MPa with FAGUS.

Load (kN)	Moment (mkN)	Deformation without Crack (μm/m)	Cracking Deformation (μm/m)	Interpolation Value (μm/m)	Deformation Rebar 3 (μm/m)	Deformation Rebar 4 (μm/m)
0.56	0.42	2.60		2.60	−0.18	−0.17
2.49	1.87	12.50		12.50	2.71	3.02
3.85	2.89	19.10		19.10	2.54	3.28
5.08	3.81	25.00		25.00	7.79	7.61
6.88	5.16	34.20		34.20	13.4	12.70
8.07	6.05	39.40		39.40	20.05	21.61
9.05	6.79	44.70	63.70	46.60	39.05	38.03
13.59	10.19	79.50	113.60	82.91	106.46	134.48
14.15	10.61	85.90	122.80	89.59	119.68	149.43
15.34	11.51	102.80	146.90	107.21	147.52	179.24
17.69	13.27	147.80	211.10	154.13	156.19	189.27
19.09	14.32	179.70	256.80	187.41	177.99	213.38
20.01	15.01	205.20	293.10	213.99	198.74	234.04
22.71	17.03	291.70	416.70	304.20	277.62	301.36
25.42	19.06	404.30	577.50	421.62	453.07	447.94
26.64	19.98	459.50	656.40	479.19	526.35	513.97
29.11	21.83	583.80	834.00	608.82	740.68	725.88
30.45	22.84	662.10	945.80	690.47	800.39	802.02
33.80	25.35	910.10	1300.10	949.10	988.09	1027.15
35.57	26.67	1075.10	1535.80	1121.17	1060.50	1130.08
37.75	28.31	1389.50	1985.00	1449.05	1121.17	1205.10

Considering these results, Figure 17 shows that the theoretical deformation of the steel is in keeping with that obtained by the optic fiber sensors. The contribution by the concrete between cracks plays an important role in determining the steel deformation.

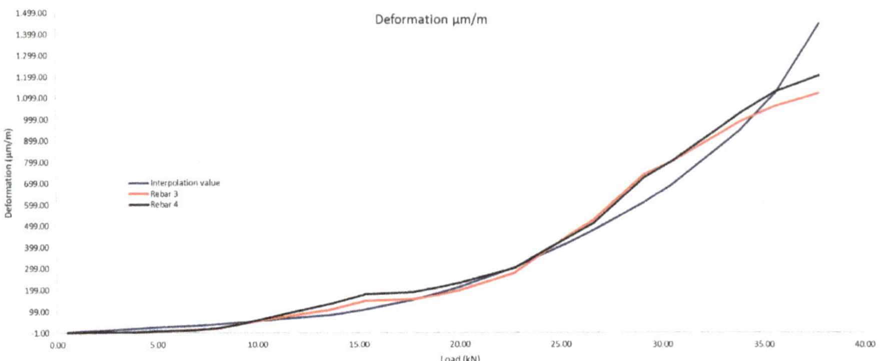

Figure 17. Theoretical deformation of the rebars with its real resistance to traction of 2.13 MPa compared with the real deformation obtained with the optic fiber sensors.

We observe that the theoretical deformation value interpolated grows over a soft curve. This is due to the sensor being approximately in the center between cracks, which makes the concrete between cracks contribute to less deformation of the steel than if the sensor were to be very near to a crack or in the actual crack. That is precisely what happens in the following beam studied, where we observe that the interpolation curve suffers a major leap at the moment of the cracks taking place.

In order to be able to determine the precise moment when the crack takes place and how this grows by application of the successive loads, we shall transform Figure 17 into a graphic, Figure 18 that shows, in an equivalent manner to Figure 17, the deformation slope curves according to the loads. The greater the slope, the greater the deformation.

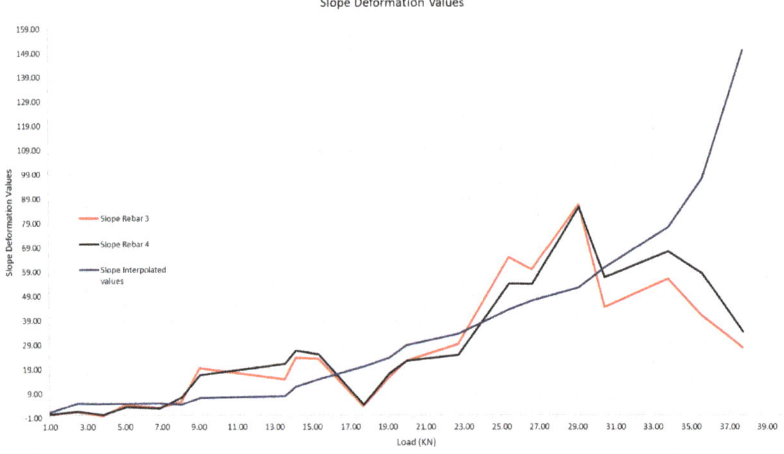

Figure 18. Slope curves of the theoretical deformation of the corrugated steel (interpolated values) and of rebars 3 and 4.

In Figure 18 we observe how the first significant leap in loading takes place, which is in the interval [8.07;9.05], that is, that the first crack begins to form at a load of 8.07 kN, and it cracks until reaching 9.05 kN. The slope values of these curves are shown in Table 6.

Table 6. Slope values of the theoretical deformation curves and of rebars 3 and 4.

Load (kN)	Interpolation Value (μm/m)	Deformation Rebar 3 (μm/m)	Deformation Rebar 4 (μm/m)	Slope Interpolation Value	Slope Rebar 3	Slope Rebar 4
0.56	2.60	−0.18	−0.17	0.00	0.00	0.00
2.49	12.50	2.71	3.02	5.13	1.50	1.66
3.85	19.10	2.54	3.28	4.85	-0.13	0.19
5.08	25.00	7.79	7.61	4.80	4.27	3.52
6.88	34.20	13.4	12.70	5.11	3.12	2.83
8.07	39.40	20.05	21.61	4.37	5.59	7.49
9.05	46.60	39.05	38.03	7.35	19.39	16.76
13.59	82.91	106.46	134.48	8.00	14.85	21.24
14.15	89.59	119.68	149.43	11.95	23.64	26.74
15.34	107.21	147.52	179.24	14.77	23.34	25.00
17.69	154.13	156.19	189.27	19.99	3.69	4.27
19.09	187.41	177.99	213.38	23.77	15.57	17.22
20.01	213.99	198.74	234.04	28.85	22.52	22.42
22.71	304.20	277.62	301.36	33.43	29.23	24.95
25.42	421.62	453.07	447.94	43.40	64.85	54.18
26.64	479.19	526.35	513.97	47.01	59.84	53.92
29.11	608.82	740.68	725.88	52.48	86.77	85.79
30.45	690.47	800.39	802.02	60.75	44.43	56.65
33.80	949.10	988.09	1027.15	77.29	56.10	67.28
35.57	1121.17	1060.50	1130.08	97.43	41.00	58.28
37.75	1449.05	1121.17	1205.10	150.14	27.78	34.35

It is evident that concrete is a material regarding which we cannot do more than approach its behavior by experience, although with embedded sensors we are able to know the moment at which the material cracks, and with that result, know its behavior much better. Figure 19 shows the interval in which the crack arises in greater detail.

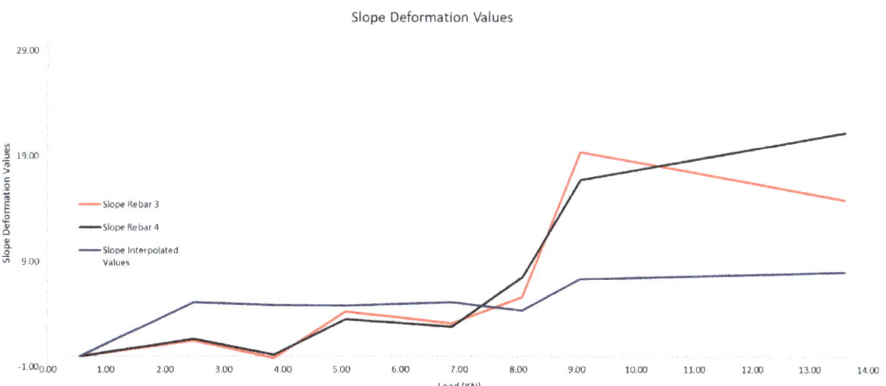

Figure 19. Detail of the first crack forming in the 200 × 300 mm beam.

One may see that, while the beam has not cracked, the steel is gradually deformed, on the contrary to what traditional structure calculation theory says as, in this, the traction is absorbed by the concrete, a fact that is proven not to be the case in these graphs. Once the cracks start, when we go from 8.07 to 9.05 kN applied load, the deformation of the steel is much more significant, as the concrete contributes to a lesser extent to absorb the traction. The greater or lesser contribution by the concrete depends on the position of the sensor between the cracks.

3.2. 300 × 500 mm Beam

The section of the beam that will be analyzed below is shown in Figure 20. The corrugated steel bars studied correspond to the lower part of the beam and are those that will be placed under traction.

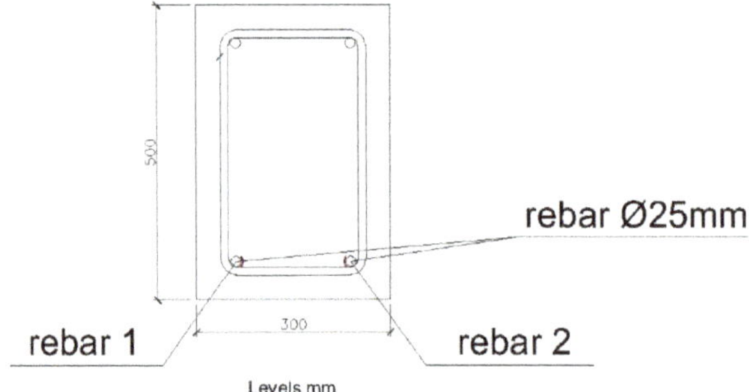

Figure 20. Position of bars studied in 300 × 500 mm beam.

3.2.1. Determination of the Moment the Crack Takes Place

Loading begins in the center of the beam span, at an interval ranging from 5.20 kN to 275 kN. After applying these loads, and using Catman software, we obtain the deformation that takes place in the corrugated steel bars according to the loads applied (Table 7).

Table 7. Relation between loads applied and deformation of the rebars 1 and 2.

Load (kN)	Moment (mkN)	Deformation Rebar 2 (μm/m)	Deformation Rebar 1 (μm/m)
5.20	3.90	12.45	9.84
10.41	7.81	14.14	9.94
15.24	11.43	17.10	12.40
20.26	15.20	24.48	17.21
25.11	18.83	28.39	20.20
30.02	22.52	33.77	24.26
35.01	26.26	37.04	26.19
40.32	30.24	45.91	33.03
43.42	32.57	45.27	32.71
54.00	40.50	70.28	50.66
60.54	45.41	252.00	138.00
70.00	52.50	269.20	150.90
80.37	60.28	454.20	306.10
90.00	67.50	612.70	490.10
101.20	75.90	749.70	630.70
110.30	82.73	814.30	700.40
120.80	90.60	900.30	786.80
131.30	98.48	1042.00	925.70
142.10	106.58	1167.00	1055.00
150.00	112.50	1166.00	1056.00
170.50	127.88	1338.00	1233.00
182.10	136.58	1514.00	1401.00
201.20	150.90	1688.00	1559.00
210.00	157.50	1927.00	1783.00
252.20	189.15	2212.00	2066.00
275.20	206.40	2397.00	2255.00

Once more, the beam has cracked before reaching the calculated cracking moment. The cracking process of the beam was observed, as with the 200 × 300 mm beam, obtaining perceptible cracking on visual inspection with a load of 58 kN. Figure 21 shows the visual control of cracking of the beam throughout the whole process.

Figure 21. Visual control of cracking in the 300 × 500 mm beam.

In Figure 22, we may observe the moment at which the beam cracks, as the steel deformation grows in an obvious manner. The loading at which the cracking takes place, according to the deformation measure obtained by the optic fiber is 54 kN, compared with the 58 observed in the visual inspection. Thus, it would be those 54 kN that would be taken as the loading value to calculate the cracking moment.

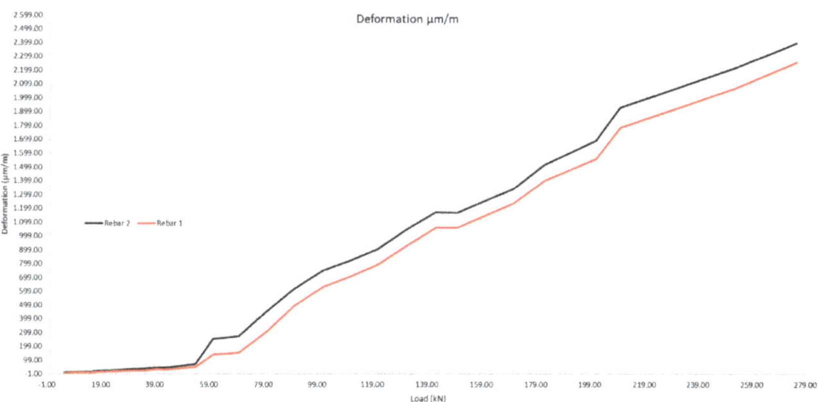

Figure 22. Deformation of the rebars 1 and 2 according to the loads applied.

3.2.2. Steel Deformation Using the Concrete Laboratory Data Compared with Steel Deformation Obtained From the Optic Fiber Sensors

The steel deformation shall first be analyzed using the data obtained in the laboratory, compared with that provided by the optic fiber. To do so, a concrete has been created with the characteristics described above, with fct = 8.9 MPa, and M_{fis} = 111.25 mkN.

The steel deformation is obtained using the program FAGUS. We shall show the result provided by FAGUS for a moment of 127.88 mkN, the rest of the results having been obtained in the same way (Table 8).

Table 8. Deformation of the rebars 1 and 2 for a moment of 127.88 mkN by FAGUS.

Formula/Result	Name	Max.	Min	Value
ε Rebar 1	Rebar 1	0.2058	0.2058	‰
ε Rebar 1 Value in crack	Rebar 1	0.2940	0.2940	‰
ε Rebar 2	Rebar 2	0.2058	0.2058	‰
ε Rebar 2 Value in crack	Rebar 2	0.2940	0.2940	‰

The same method shall be applied as for the 200 × 300 mm beam, considering the steel deformation for resistance to traction of the concrete obtained by the laboratory tests, and subsequently that obtained with the optic fiber (Table 9 and Figure 23).

Table 9. Theoretical deformation of the rebars 1 and 2 for M_{fis} = 111.25 mkN and f_{ct} = 8.90 MPa with FAGUS.

Load (kN)	Moment (mkN)	Theoretical Deformation (μm/m)	Deformation Rebar 2 (μm/m)	Deformation Rebar 1 (μm/m)
5.20	3.90	6.30	12.45	9.84
10.41	7.81	12.70	14.14	9.94
15.24	11.43	18.50	17.10	12.40
20.26	15.20	24.70	24.48	17.21
25.11	18.83	30.50	28.39	20.20
30.02	22.52	36.50	33.77	24.26
35.01	26.26	42.70	37.04	26.19
40.32	30.24	49.00	45.91	33.03
43.42	32.57	52.90	45.27	32.71
54.00	40.50	65.70	70.28	50.66
60.54	45.41	73.60	252.00	138.00
70.00	52.50	85.10	269.20	150.90
80.37	60.28	97.70	454.20	306.10
90.00	67.50	109.30	612.70	490.10
101.20	75.90	122.90	749.70	630.70
110.30	82.73	133.90	814.30	700.40
120.80	90.60	146.60	900.30	786.80
131.30	98.48	159.40	1042.00	925.70
142.10	106.58	172.30	1167.00	1055.00
150.00	112.50	259.30	1166.00	1056.00
170.50	127.88	294.00	1338.00	1233.00
182.10	136.58	315.90	1514.00	1401.00
201.20	150.90	358.30	1688.00	1559.00
210.00	157.50	380.90	1927.00	1783.00
252.20	189.15	523.10	2212.00	2066.00
275.20	206.40	630.20	2397.00	2255.00

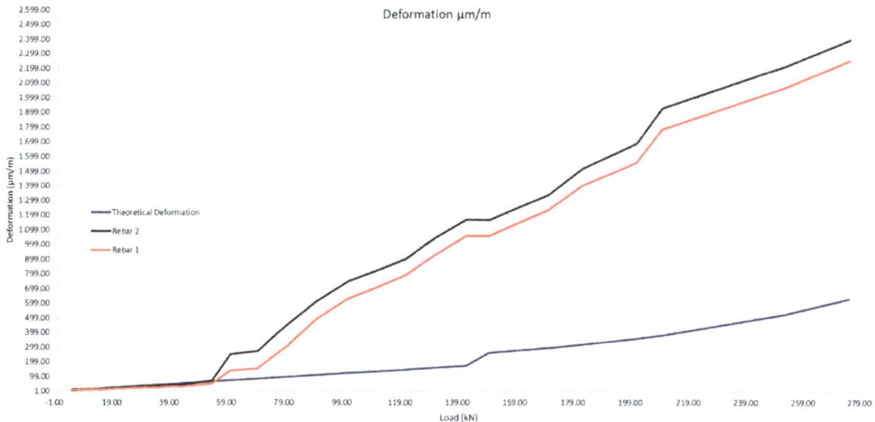

Figure 23. Theoretical deformation of the rebars compared with the real deformation obtained using the optic fiber sensors.

3.2.3. Deformation of the Steel at the Real Cracking Moment of the Concrete Compared with Deformation of the Steel Obtained Using Optic Fiber Sensors

For a load of 54 kN, the moment corresponds to 40.5 mkN, and applying Equation (1) we obtain a resistance to traction of the concrete of 3.24 MPa, compared with the 8.9 MPa obtained in the laboratory. In this case, the crack has opened at a point very near to the optic fiber, so the concrete can barely contribute to avoid deformation of the steel (Figure 24).

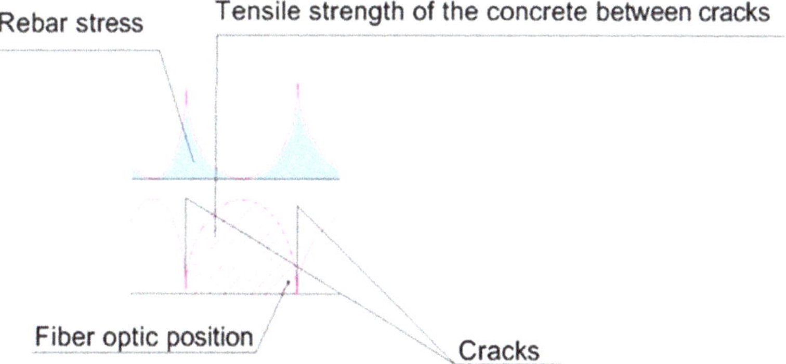

Figure 24. Determining deformation of the steel between cracks on 300 × 500 mm beam.

We see what happens when the resistance to traction of the concrete with which we perform the calculations using the FAGUS program corresponds to the 3.24 MPa. Table 10 shows the deformation of the corrugated steel for a moment of 127.88 mkN.

Table 10. Deformation of the rebars 1 and 2 for a moment of 127.88 mkN by FAGUS.

Formula/Result	Name	Max.	Min	Value
ε Rebar 1	Rebar 1	1.0294	1.0294	‰
ε Rebar 1 Value in crack	Rebar 1	1.4706	1.4706	‰
ε Rebar 2	Rebar 2	1.0294	1.0294	‰
ε Rebar 2 Value in crack	Rebar 2	1.4706	1.4706	‰

Table 11 includes the deformation of the steel without cracking, and the deformation of the steel with cracking. The deformation adopted shall be an interpolation of both values.

Table 11. Deformation of the rebars 1 and 2 for M_{fis} = 40.5 mkN and f_{ct} = 3.24 MPa with FAGUS.

Load (kN)	Moment (mkN)	Deformation without Crack (µm/m)	Cracking Deformation (µm/m)	Interpolation Value (µm/m)	Deformation Rebar 2 (µm/m)	Deformation Rebar 1 (µm/m)
5.20	3.90	6.30		6.30	12.45	9.84
10.41	7.81	12.70		12.70	14.14	9.94
15.24	11.43	18.50		18.50	17.10	12.40
20.26	15.20	24.70		24.70	24.48	17.21
25.11	18.83	30.50		30.50	28.39	20.20
30.02	22.52	36.50		36.50	33.77	24.26
35.01	26.26	42.70		42.70	37.04	26.19
40.32	30.24	49.00		49.00	45.91	33.03
43.42	32.57	52.90		52.90	45.27	32.71
54.00	40.50	64.00	91.50	88.75	70.28	50.66
60.54	45.41	71.80	102.50	99.43	252.00	138.00
70.00	52.50	389.70	556.70	540.00	269.20	150.90
80.37	60.28	460.70	658.10	638.36	454.20	306.10
90.00	67.50	523.90	748.40	725.95	612.70	490.10
101.20	75.90	596.10	851.50	825.96	749.70	630.70
110.30	82.73	653.70	933.90	905.88	814.30	700.40
120.80	90.60	720.10	1028.70	997.84	900.30	786.80
131.30	98.48	786.10	1123.00	1089.31	1042.00	925.70
142.10	106.58	853.40	1219.10	1182.53	1167.00	1055.00
150.00	112.50	902.20	1288.90	1250.23	1166.00	1056.00
170.50	127.88	1029.40	1470.60	1426.48	1338.00	1233.00
182.10	136.58	1101.00	1572.90	1525.71	1514.00	1401.00
201.20	150.90	1218.60	1740.80	1688.58	1688.00	1559.00
210.00	157.50	1272.80	1818.20	1763.66	1927.00	1783.00
252.20	189.15	1532.80	2189.70	2124.01	2212.00	2066.00
275.20	206.40	1870.40	2672.00	2591.84	2397.00	2255.00

Figure 25 is the graphic representation of figures in Table 11.

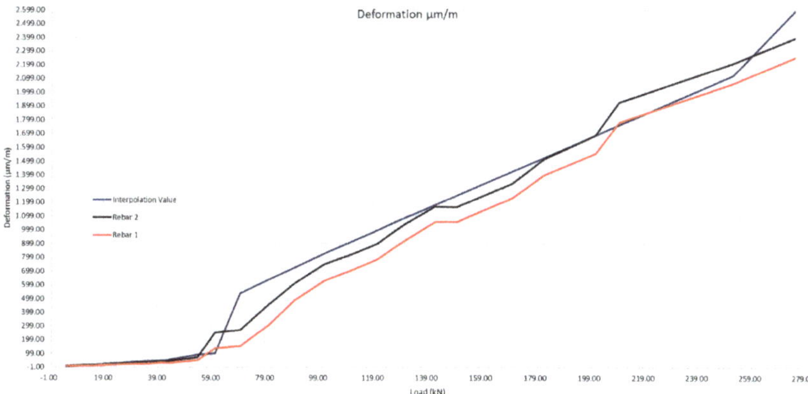

Figure 25. Deformation of the rebars with its Resistance to real traction of 2.13 MPa compared with real deformation obtained using optic fiber sensors.

We observe that by inputting the behavior real values of the concrete, the beam deformation curve corresponds to the data provided by the optic fiber sensors. In this case, as the optic fiber is very near to a crack, there is a sharp leap at the moment it takes place. As we shall see in the slope figures, these are much steeper than in the previous case, in which the optic fiber was approximately in the center between the cracks, which caused these slopes to be less steep. The slope values of these curves are recorded in Table 12.

Table 12. Slope values of theoretical deformation curves and of rebars 1 and 2.

Load (kN)	Interpolation Value (µm/m)	Deformation rebar 2 (µm/m)	Deformation rebar 1 (µm/m)	Slope Interpolation Value	Slope Rebar 2	Slope Rebar 1
5.20	6.30	12.45	9.84	1.23	2.36	1.86
10.41	12.70	14.14	9.94	1.23	0.32	0.02
15.24	18.50	17.10	12.40	1.20	0.61	0.51
20.26	24.70	24.48	17.21	1.24	1.47	0.96
25.11	30.50	28.39	20.20	1.20	0.81	0.62
30.02	36.50	33.77	24.26	1.22	1.10	0.83
35.01	42.70	37.04	26.19	1.24	0.66	0.39
40.32	49.00	45.91	33.03	1.19	1.67	1.29
43.42	52.90	45.27	32.71	1.26	-0.21	-0.10
54.00	88.75	70.28	50.66	3.39	2.36	1.70
60.54	99.43	252.00	138.00	1.63	27.79	13.35
70.00	540.00	269.20	150.90	46.57	1.82	1.36
80.37	638.36	454.20	306.10	9.49	17.84	14.97
90.00	725.95	612.70	490.10	9.10	16.46	19.11
101.20	825.96	749.70	630.70	8.93	12.23	12.55
110.30	905.88	814.30	700.40	8.78	7.10	7.66
120.80	997.84	900.30	786.80	8.76	8.19	8.23
131.30	1089.31	1042.00	925.70	8.71	13.50	13.23
142.10	1182.53	1167.00	1055.00	8.63	11.57	11.97
150.00	1250.23	1166.00	1056.00	8.57	-0.13	0.13
170.50	1426.48	1338.00	1233.00	8.60	8.39	8.63
182.10	1525.71	1514.00	1401.00	8.55	15.17	14.48
201.20	1688.58	1688.00	1559.00	8.53	9.11	8.27
210.00	1763.66	1927.00	1783.00	8.53	27.16	25.45
252.20	2124.01	2212.00	2066.00	8.54	6.75	6.71
275.20	2591.84	2397.00	2255.00	20.34	8.04	8.22

In Figure 26, we observe how there is a significant first leap at a load that is in the interval [54.00;60.54], that is, that the first crack begins to form with a load of 54 kN, and it cracks until reaching 60.54 kN.

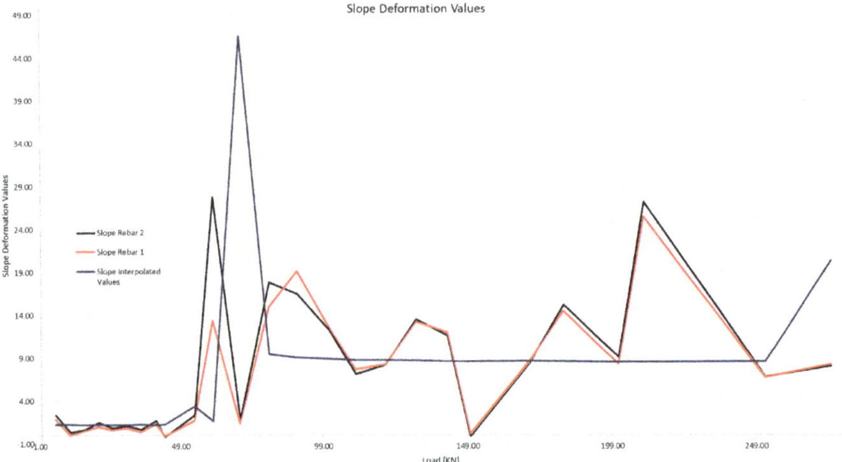

Figure 26. Slope theoretical deformation curves of the rebars (interpolated values) and of rebars 1 and 2.

Figure 27 shows the interval within which the crack takes place in greater detail

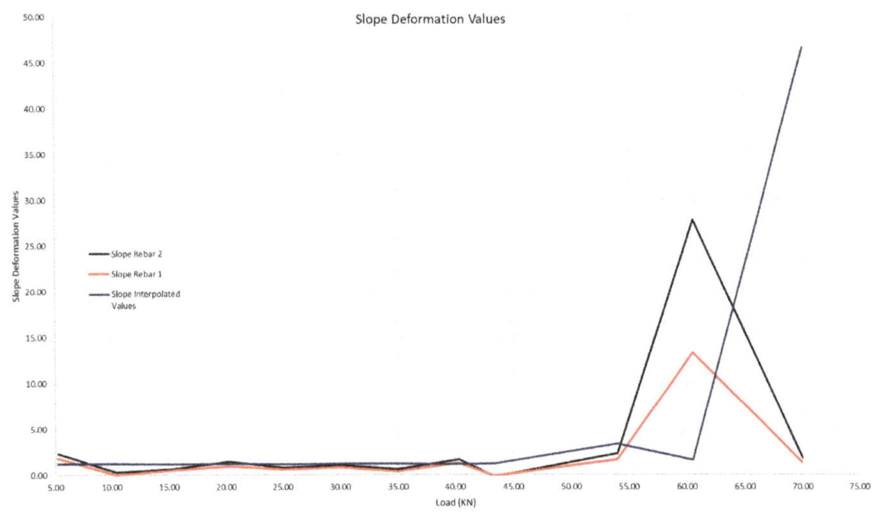

Figure 27. Detail of the first crack forming in the 300 × 500 mm beam.

The moment at which the crack in the beam takes place is evident, as the deformation of the steel grows evidently. The load at which the crack takes place is 54 kN, compared with the 58 kN observed during the visual inspection.

As happened with the 200 × 300 mm beam, the theoretical deformation of the beam reinforcement is far from appearing like its real behavior. In this case, the crack has opened up at a point very near to the optic fiber, so the concrete shall not collaborate when its traction tension is reached.

In this case, we must observe what happens at the loading point of 54 kN, where the slope of the curve is much steeper, something that did not happen in the graph of beam 200 × 300 mm. This is due to the crack having opened very near to the location of the optic fiber, so that, in analysis of the section, the concrete barely collaborates in traction of the beam, and practically all the traction is borne by the steel.

It is noted that the optic fiber shows us the precise moment at which the beam cracks, that the cracks open much earlier than what the laboratory tests say, and that taking the data provided by the optic fiber, we may determine the behavior of the structural element in a much more precise way.

It is evident that when calculating a structural element, we do not know what will happen to it, when the piece will really crack. Considering an expression that draws that value closer to reality shall be a matter to be studied in future lines of investigation.

4. Conclusions

Two beams with a rectangular section in which fiber optic sensors were embedded have been tested to analyze the real deformation of the steel when they are submitted to different loading stages.

Appearance of the first cracks has been observed in both cases. These appear much earlier than the calculation predicts. The appearance of the first cracks is a fundamental matter to understand the real behavior of the structures. Fiber optic sensors were used to observe how a sudden change in deformation of the steel takes place. Moreover, with the advantage of the measurements being in real time, a fact that provides greater value to evaluation of the structural health of the elements tested. It is evident that this sudden change leads to it being deformed to greater extent as a consequence of the concrete cracking.

Thus, considering the results obtained, we may know the precise moment at which the beam cracks through embedded fiber optic sensors. On studying the deformations, it has been noted that even when test pieces extracted at the moment of concrete pouring were tested, and they were tested on the day when the tests were to be carried out, these values do not match the behavior of the concrete under traction. In both cases, the beam cracked much before the laboratory tests indicated.

Thus, placement of sensors welded on corrugated steel bars within reinforced concrete structural elements, at their maximum effort points, is a precise, reliable method to determine the moment at which the first cracks take place, as it has been possible to prove according to the results obtained.

After ascertaining the real cracking moment of the concrete, we precisely obtained its resistance to traction and, thus, the real deformation of the corrugated steel during application of the loads.

As stated in the introduction, the existing studies on concrete cracking use diverse methods to detect cracks. As is known, concrete is a material that resists compression well, but that is not the case with traction efforts. The reinforcement of the structural elements is placed, among other reasons, to bear the traction the concrete is not able to bear. The method proposed herein provides, as a novelty, detection of cracks that is observed thanks to the optic fiber sensors welded to the corrugated steel bars, at the precise moment when the steel begins to deform significantly. This causes a leap in its deformation, which is detected for the relevant load applied. Moreover, once the steel begins to deform, it is possible to know the deformation it will suffer during the whole period of application of the different loading steps to which the structural element is submitted. It has been proven that the deformation of the steel measured with optic fiber sensors corresponds to the theoretical values of the traditional materials resistance calculation, as long as the real cracking moment of the concrete is taken as the starting point.

We may conclude, considering these graphs obtained from the experiment carried out, on the one hand that laboratory tests to determine flexion-traction resistance of concrete provide very conservative results, that have nothing to do with what really happens in the structural element. And on the other, that deformation of the steel, obtained with these tests, are quite far from its real behavior. This method of evaluating the structural health of a simple element of reinforced concrete may be transferred to more complex structural elements of buildings in construction to know the behavior of the structure when the formwork removal takes place and their actual weight begins to bear down on the structures, and subsequently the application of deadweight and overburdens in use, when the building is put into operation.

A monitored building may provide us information on what overburdens it is able to bear, a highly important factor when one wishes to change the use of a building and the overburdens it is to be subject to are higher than those initially designed. In this case, and according to the data obtained, one might even be able to avoid possible structural reinforcement, as we would know what the building may really bear, with the financial savings that would involve.

Author Contributions: Conceptualization and methodology, J.G.D. and E.R.Á.; validation and formal analysis, N.N.C. and E.R.Á.; investigation, J.G.D.; writing—original draft preparation, J.G.D.; supervision, N.N.C. and E.R.Á. All authors have read and agree to the published version of the manuscript.

Funding: This research was funded by CDTI (Centro para el Desarrollo Tecnológico Industrial), assigned to the Ministry of Science, Innovation and Universities, under the R&D project "SENSOSMART" IDI-20171055.

Acknowledgments: The authors wish to thanks Ortiz Construcciones y Proyectos, S.A., INDAGSA and HBM for their support in the process of constructing the concrete beams and setting up the testing facility.

Conflicts of Interest: The authors declare no conflict of interest.

References

1. Szymanik, B.; Frankowski, P.K.; Chady, T.; Chelliah, C.R.A.J. Detection and Inspection of Steel Bars in Reinforced Concrete Structures Using Active Infrared Thermography with Microwave Excitation and Eddy Current Sensors. *Sensors* **2016**, *16*, 234. [CrossRef] [PubMed]
2. Abu Dabous, S.; Yaghi, S.; Alkass, S.; Moselhi, O. Concrete bridge deck condition assessment using IR Thermography and Ground Penetrating Radar technologies. *Autom. Constr.* **2017**, *81*, 340–354. [CrossRef]
3. Tran, Q.H.; Huh, J.; Mac, V.H.; Kang, C.; Han, D. Effects of rebars on the detectability of subsurface defects in concrete bridges using square pulse thermography. *NDT E Int.* **2018**, *100*, 92–100. [CrossRef]
4. Ohno, K.; Ohtsu, M. Crack classification in concrete based on acoustic emission. *Constr. Build. Mater.* **2010**, *24*, 2339–2346. [CrossRef]
5. Goszczyńska, B.; Świt, G.; Trąmpczyński, W.; Krampikowska, A.; Tworzewska, J.; Tworzewski, P. Experimental validation of concrete crack identification and location with acoustic emission method. *Arch. Civ. Mech. Eng.* **2012**, *12*, 23–28. [CrossRef]
6. Miller, S.; Chakraborty, J.; Van Der Vegt, J.; Brinkerink, D.; Erkens, S.; Liu, X.; ANUPAM, K.; Sluer, B.; Mohajeri, M. Smart sensors in asphalt monitoring key process parameters during and post construction. *Spool* **2017**, *4*, 45–48.
7. Rodrigues, C.; Félix, C.; Lage, A.; Figueiras, J. Development of a long-term monitoring system based on FBG sensors applied to concrete bridges. *Eng. Struct.* **2010**, *32*, 1993–2002. [CrossRef]
8. Mao, J.; Xu, F.; Gao, Q.; Liu, S.; Jin, W.; Xu, Y. A Monitoring Method Based on FBG for Concrete Corrosion Cracking. *Sensors* **2016**, *16*, 1093. [CrossRef]
9. Luo, D.; Yue, Y.; Li, P.; Ma, J.; Zhang, L.L.; Ibrahim, Z.; Ismail, Z. Concrete beam crack detection using tapered polymer optical fiber sensors. *Meas.* **2016**, *88*, 96–103. [CrossRef]
10. Wang, L.; Xin, X.; Song, J.; Wang, H.; Sai, Y. Finite element analysis-based study of fiber Bragg grating sensor for cracks detection in reinforced concrete. *Opt. Eng.* **2018**, *57*, 1.
11. Wang, L.; Song, J.; Sai, Y.; Wang, H. Crack Width Analysis of Reinforced Concrete Using FBG Sensor. *IEEE Photon- J.* **2019**, *11*, 1–8. [CrossRef]
12. Yao, Y.; Li, S.; Li, Z. Structural Cracks Detection Based on Distributed Weak FBG. In Proceedings of the 26th International Conference on Optical Fiber Sensors, Lausanne, Switzerland, 24–28 September 2018.
13. Smith, G.M.; Higgins, O.; Sampath, S. In-situ observation of strain and cracking in coated laminates by digital image correlation. *Surf. Coat. Technol.* **2017**, *328*, 211–218. [CrossRef]

14. Gajewski, T.; Garbowski, T. Calibration of concrete parameters based on digital image correlation and inverse analysis. *Arch. Civ. Mech. Eng.* **2014**, *14*, 170–180. [CrossRef]
15. Gkantou, M.; Muradov, M.; Kamaris, G.S.; Hashim, K.; Atherton, W.; Kot, P. Novel Electromagnetic Sensors Embedded in Reinforced Concrete Beams for Crack Detection. *Sensors* **2019**, *19*, 5175. [CrossRef]
16. Chakraborty, J.; Katunin, A.; Klikowicz, P.; Salamak, M. Early Crack Detection of Reinforced Concrete Structure Using Embedded Sensors. *Sensors* **2019**, *19*, 3879. [CrossRef] [PubMed]
17. Jiang, T.; Hong, Y.; Zheng, J.; Wang, L.; Gu, H. Crack Detection of FRP-Reinforced Concrete Beam Using Embedded Piezoceramic Smart Aggregates. *Sensors* **2019**, *19*, 1979. [CrossRef] [PubMed]
18. Kuang, K.S.C.; Akmaluddin; Cantwell, W.; Thomas, C. Crack detection and vertical deflection monitoring in concrete beams using plastic optical fibre sensors. *Meas. Sci. Technol.* **2003**, *14*, 205–216. [CrossRef]
19. Saafi, M. Wireless and embedded carbon nanotube networks for damage detection in concrete structures. *Nanotechnol.* **2009**, *20*, 395502. [CrossRef]
20. Yan, J.; Downey, A.; Cancelli, A.; Laflamme, S.; Chen, A.; Li, J.; Ubertini, F. Concrete Crack Detection and Monitoring Using a Capacitive Dense Sensor Array. *Sensors* **2019**, *19*, 1843. [CrossRef]
21. Islam, M.M.M.; Kim, J.-M. Vision-Based Autonomous Crack Detection of Concrete Structures Using a Fully Convolutional Encoder-Decoder Network. *Sensors* **2019**, *19*, 4251. [CrossRef]
22. Dung, C.V.; Anh, L.D. Autonomous concrete crack detection using deep fully convolutional neural network. *Autom. Constr.* **2019**, *99*, 52–58. [CrossRef]
23. Wu, C.; Sun, K.; Xu, Y.; Zhang, S.; Huang, X.; Zeng, S. Concrete crack detection method based on optical fiber sensing network and microbending principle. *Saf. Sci.* **2019**, *117*, 299–304. [CrossRef]
24. Barbosa, C. Optical Fiber Sensors vs. Conventional Electrical Strain Gauges for Infrastructure Monitoring. Available online: https://www.hbm.com/fileadmin/mediapool/local/usa/white_paper_optical_infrastructure_monitoring.pdf (accessed on 12 December 2019).
25. Fernando, C.; Bernier, A.; Banerjee, S.; Kahandawa, G.; Eppaarchchi, J. An Investigation of the Use of Embedded FBG Sensors to Measure Temperature and Strain Inside a Concrete Beam During the Curing Period and Strain Measurements under Operational Loading. *Procedia Eng.* **2017**, *188*, 393–399. [CrossRef]
26. Chen, Z.; Zheng, D.; Shen, J.; Qiu, J.; Liu, Y. Research on distributed optical-fiber monitoring of biaxial-bending structural deformations. *Measurement* **2019**, *140*, 462–471. [CrossRef]
27. Ghimire, M.; Wang, C.; Dixon, K.; Serrato, M. In situ monitoring of prestressed concrete using embedded fiber loop ringdown strain sensor. *Measurement* **2018**, *124*, 224–232. [CrossRef]
28. Calavera, J. *Proyecto y Cálculo de Estructuras de Hormigón. Tomo II*; Intemac: Madrid, Spain, 2008.

Article

Rotating Stall Induced Non-Synchronous Blade Vibration Analysis for an Unshrouded Industrial Centrifugal Compressor

Xinwei Zhao [1], Qiang Zhou [1], Shuhua Yang [2] and Hongkun Li [1,*]

1 School of Mechanical Engineering, Dalian University of Technology, No. 2 Linggong Road, Dalian 116024, China; xinweizhao@mail.dlut.edu.cn (X.Z.); zhouqdut@mail.dlut.edu.cn (Q.Z.)

2 Shenyang Blower Works Group, No.16 Development Road, Shenyang 110869, China; yangshuhua@shengu.com.cn

* Correspondence: lihk@dlut.edu.cn

Received: 23 September 2019; Accepted: 13 November 2019; Published: 16 November 2019

Abstract: Rotating stall limits the operating range and stability of the centrifugal compressor and has a significant impact on the lifetime of the impeller blade. This paper investigates the relationship between stall pressure wave and its induced non-synchronous blade vibration, which will be meaningful for stall resonance avoidance at the early design phase. A rotating disc under a time-space varying load condition is first modeled to understand the physics behind stall-induced vibration. Then, experimental work is conducted to verify the model and reveal the mechanism of stall cells evolution process within flow passage and how blade vibrates when suffering such aerodynamic load. The casing mounted pressure sensors are used to capture the low-frequency pressure wave. Strain gauges and tip timing sensors are utilized to monitor the blade vibration. Based on circumferentially distributed pressure sensors and stall parameters identification method, a five stall cells mode is found in this compressor test rig and successfully correlates with the blade non-synchronous vibration. Furthermore, with the help of tip timing measurement, all blades vibration is also evaluated under different operating mass flow rate. Analysis results verify that the proposed model can show the blade forced vibration under stall flow condition. The overall approach presented in this paper is also important for stall vibration and resonance free design with effective experimental verification.

Keywords: rotating stall; non-synchronous blade vibration; blade tip timing; centrifugal compressor

---

1. Introduction

Centrifugal compressors have the advantages of high single-stage pressure ratio, wide working range, and compact structure. They are highly valued and broadly applied, such as turbochargers in automotive engines, the processing of natural gas, aerospace, gas turbine engines and so on [1]. With the improvement of industrial requirements and aerodynamic design, the structural integrity of rotating impeller meets great challenges. During the operation of the compressor, the blade vibrates due to mechanical parts and unsteady aerodynamic loads. The unsteady aerodynamic load is inherently and can cause large blade vibration and even high cycle fatigue (HCF) failure. The most common flow-induced vibration is due to rotor-stator interaction. The variable inlet guide vanes (VIGVs) and diffuser vanes (DVs) are two main typical excitation sources in centrifugal compressor stage. However, the impeller resonance caused by these exciting sources can mostly be avoided by utilizing Campbell diagram during the design phase [2]. Currently, the most challenging problems occur at off-design conditions. Blade failures can also occur and are caused by rotating stall effects as Haupt et al. experimentally found [3]. The rotating stall cells are also the possible blade excitation source. It is a localized phenomenon and the compressor can still give acceptable aerodynamic performance.

However, it will result in circumferential non-uniform and periodic pressure pulsation (stall mode) in blade rows and form a transient rotating stall dominated exciting force applying on the entire impeller. In some specific cases, the impeller modes may be excited and resonance occurs, which is undesirable for blade structure. The unstable operating behavior of rotating stall and induced blade forced response mechanism are not yet fully understood [4]. The impact of the rotating cells on blade vibration amplitude and stress level is also unknown. It is still a key aspect in the academic and industrial fields of turbomachinery.

A great number of numerical and experimental studies have been conducted to reveal the mechanism of rotating stall [4]. However, compared to the axial compressor [5–8], researches about rotating stall happened in centrifugal compressors start later [9,10]. Abdelhamid et al. [11,12] investigated the effects of vaneless diffusers with different geometries on rotating stall. Frequency components were analyzed in detail and they found that the stall frequency was dependent on the diffuser radius ratio and the complex coupling between the impeller and the diffuser. In addition to Abdelhamid's work, many experimental tests investigating the influence of different geometrical configurations were also conducted by Ferrara [13]. These results are plentiful and give a more detailed benchmark for a large number of geometry configurations. Fujisawa et al. [14,15] combined experimental and numerical methods to study the unsteady flow phenomenon of rotating stall in diffuser passage. Numerical results showed a vortex generated at the diffuser leading-edge is the main cause of the stall. Stall occurrence in centrifugal compressor with vaned and vaneless diffusers has different features. In addition, both the impeller and the diffuser, even the return channel [16], will have rotating stall and origin differs. All of these raise the uncertainty and randomness of the rotating stall happened in the centrifugal compressor.

The literature mentioned above contributes to the causes and evolution of the rotating stall from an aerodynamic perspective. Rotating stall induced non-synchronous vibration (NSV) and resonance should be paid more attention. Currently, only a small part of the research work focuses on the stall induced vibration mechanisms and quantitative assessment of real applied industrial compressors. J. Chen [17] used dynamic pressure transducers and strain gauges to record the unsteady pressure fields and blade vibration. Two different stall modes were reported in this work. Although the flow mechanism of the rotating stall was explained, the blade vibration was still not analyzed in detail. Seidel et al. [18,19] observed large amplitudes of vibration occurred in centrifugal compressor impeller. Identified seven stall cells which induce dangerous blade vibration was also reported in recent work by Jenny [20] using the IGV sweep method. Higher harmonics of rotating stall disturbance in Ref. [21] was also found to be the cause of blade vibration. Vahdati et al. [22] conducted numerical simulation to investigate the axial compressor stall process observed in the experiments. The predicted stall cells number was quite close to the identified results. The numerical method they used provided a promising approach to study the stall and blade vibration. Others are concerned about rotor vibration caused by rotating stall, such as Reference [23], which gave a discussion centered on forced vibration of rotors depending on the vibration characteristics. Moreover, Ferrari et al. [24,25] put great efforts to experimentally quantify the force coming from rotating stall. Identified amplitude and frequency of the external force acting on the impeller is further included in the rotodynamic model of the test rig to get more accurate results.

Currently, the stall imposed aerodynamic forcing function is not theoretically built up. The general resonance condition of impeller blade under rotating stall excitation is also not modeled and deduced, which will be highly meaningful to understand the physics. In addition, the impact of the stall cells on all blade vibration is not quantitatively accessed, which is the basis for aeroelastic design of impeller blade. Related results are still far from providing a complete explanation of this phenomenon. Mechanisms behind this complex fluid-structure coupling phenomenon should also be interpreted more in detail. Based on the above aims, the remainder of the paper is organized as follows to achieve this goal. In Section 2, the rotating stall forcing function traveling around the impeller is built up. Then, the asynchronous traveling excitation of the impeller structure is modeled. The resonance condition is deduced using steady part of the blade forced response function. Based on the theoretical model, Section 3 introduces the test rig and experimental method in order to acquire the substantial data which is needed to verify the model and understand the rotating stall phenomenon. In Section 4, experimental results are analyzed in detail. The identified rotating stall pattern is further parameterized and correlated with the blade non-synchronous vibration. Finally, all blade vibration is monitored and quantified with a large amount of tip timing data for structure assessment purpose. These vibration data will also be useful for aeroelastic design of impeller blade under stall condition. At last, the conclusions are given in Section 5.

2. Rotating Stall Identification and Resonance Condition for Impeller

2.1. Parameter Characterization of Rotating Stall

According to its real behavior recorded by the pressure sensor, the rotating stall can be assumed to have a nearly uniform distribution after the stall cells stably formed in the flow passage. Figure 1 gives a brief picture of typical rotating stall happened in a centrifugal compressor. Considering rotating stall will also occur in the vaned diffuser, these stall cells are only drawn for illustration and used to describe the stall region presented circumferentially. The rotating stall state can be uniquely defined by the number of stall cells NB_C, their rotation speed f_{cell} and propagate direction relative to the impeller. These parameters can be identified by proper interpretation of the pressure fields within the compressor during rotating stall.

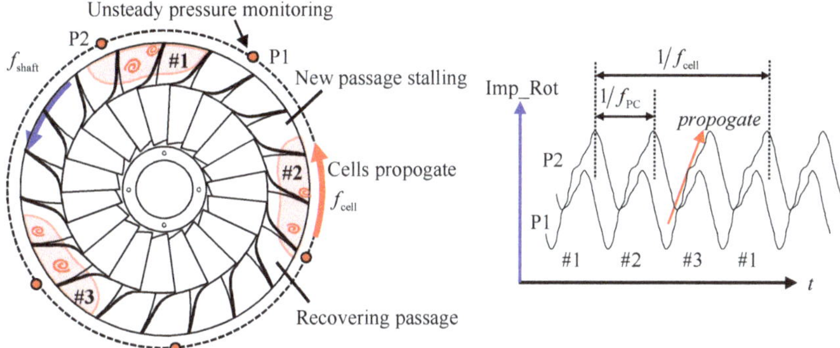

Figure 1. Dynamic characteristics of rotating stall in a centrifugal compressor impeller.

Since these stall cells are self-similar and cannot be directly identified by the FFT spectrum. Several pressure transducers should be used to monitor circumferential pressure fluctuating and combined time-frequency analysis method is needed. Assume that there are 2 monitoring points in the circumferential direction, namely P1 and P2. For the stall cell #1, it will pass through P1 and P2 successively, causing the time-delay characteristics of the signals collected in the two sensors. The time delay and cell propagation are also shown in the right picture of Figure 1. The frequency f_{PC} detected by the unsteady pressure sensor mounted on the casing corresponds to the product of the number of stall cells NB_C and the rotating frequency f_{cell} of the individual stall cell in stationary coordinate

$$f_{PC} = NB_C \cdot f_{cell} \tag{1}$$

2.2. Identification of Stall Induced Impeller Vibration and Resonance

Since the rotating stall is related to flow separation, the pressure loading is distorted circumferentially and can be described as a rotating wave arising and decaying intermittently. The induced non-uniform force is approximated as a sinusoidal signal here and other forms can refer to Ref. [26]. Corresponding aerodynamic force (thick black line in Figure 2) is then built up in rotating coordinate system (RCS). In RCS, the impeller is static while the aerodynamic force is traveling around it. The relative moving speed between impeller and stall cells corresponds to the frequency difference between shaft frequency f_{shaft} and individual cell rotating frequency f_{cell}. If the stall cells rotate in the same direction as the impeller, the exciting frequency will be $f_{\mathrm{shaft}} - f_{\mathrm{cell}}$, otherwise, the value is $f_{\mathrm{shaft}} + f_{\mathrm{cell}}$. It can be found through simple analysis. If the stall cells rotate in the same direction, the impeller must rotate more in order to form a periodic load (Stall cells also move a certain distance in the same direction during the impeller rotation), thus making this period longer than shaft rotating period $1/f_{\mathrm{shaft}}$. Relationship between pressure pulsation frequency of rotating stall and blade vibration can be calculated by the following expression

$$f_e = n_e \cdot \mathrm{NB_C}(f_{\mathrm{shaft}} \pm f_{\mathrm{cell}}) = n_e \cdot (\mathrm{NB_C} \cdot f_{\mathrm{shaft}} \pm f_{\mathrm{PC}}),\ n_e \in \mathbb{Z}_+ \tag{2}$$

where n_e is the harmonic index, f_{shaft} is the rotating speed of impeller, and f_e is the frequency of the exciting force.

Equation (2) gives the relationship between rotating stall parameters and non-integer engine order excitation. However, it is still independent of the impeller vibration mode and cannot explain when the impeller will resonate and which mode is excited. Thus, forced response modal should be further built up so that the resonance conditions for impeller under rotating stall condition can be derived. The pressure pulsation produced by the rotating stall is a periodic time-varying function. In stationary coordinate system (SCS), the aerodynamic force $F^s_{stall}(\phi)$ along the circumference of the rotating impeller can be described as

$$\begin{cases} F^s_{stall}(\phi, t) = F_v(\phi)\cos(2\pi \mathrm{NB}_c \cdot f_{\mathrm{cell}} t),\ \phi \in [0, 2\pi] \\ F_v(\phi) = \sum\limits_k (a_k \cos[k \cdot \mathrm{NB}_c \cdot \phi] + b_k \sin[k \cdot \mathrm{NB}_c \cdot \phi]) \end{cases} \tag{3}$$

where $F_v(\phi)$ corresponds to the distribution of force, a_k and b_k are the corresponding Fourier coefficients. NB_c and f_{cell} are rotating stall parameters mentioned above. Then, the distributed force can be further transformed into rotating coordinate

$$\begin{cases} F^R_{stall}(\theta, t) = F_v(\phi)\cos\left[n_e \cdot \mathrm{NB}_c \cdot \Omega_{stall}(t + t_\phi)\right]\delta[\theta - \Omega_{stall} t - \phi] \\ t_\phi = \phi / \Omega_{stall},\ \Omega_{stall} = 2\pi(f_{\mathrm{shaft}} \pm f_{\mathrm{cell}}) \end{cases} \tag{4}$$

where $\delta[\cdot]$ is the Dirac delta function, θ is the angle measured in impeller rotating coordinate, t_ϕ is a time delay due to the phase shift between $F^R_{stall}(\theta, t)\big|_{t=0}$ and $F_v(\phi)$. Relationship between $F_v(\phi)$ and $F^R_{stall}(\theta, t)$ at two different times is drawn in Figure 2a,b. The thick black line in Figure 2 corresponds to the aerodynamic force of the rotating stall. The circular disc corresponds to the impeller. The dynamical behavior of the distributed force described by Equation (4) has been well presented. It shows the relative rotation of the stall cells on the impeller. Furthermore, with such circumferential moving, the phase between the impeller and aerodynamic load changes accordingly.

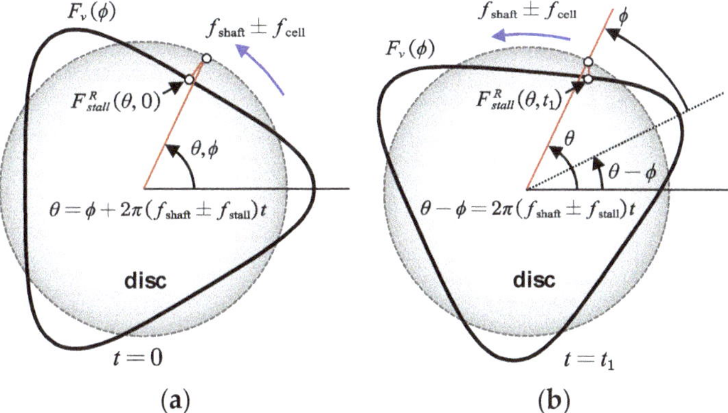

Figure 2. Rotating stall induced impeller forced vibration model based on simplified disc vibration behavior at the time $t = 0$ (**a**) and $t = t_1$ (**b**) in RCS.

The impeller eigenmodes with ND-th nodal diameters of blade vibration can be defined as sine mode with the phase θ in RCS

$$\Phi(\theta) = \sin \text{ND}\theta \ (\text{ND} = 0, 1, 2 \ldots, N - 1/2 \text{ or } N/2) \tag{5}$$

which is the modeling and analysis treatment method for vibration of mechanical structures with cyclic symmetry property. And, the mistuning of the real impeller is ignored here.

So far, the traveling excitation force corresponding to typical rotating stall condition has been derived. And, the impeller mode is also characterized into different waves. Qualitative analysis of the forced response of a rotating disc can be derived step by step according to Ref. [27]. The generalized force $Q(t)$ for the ND nodal diameter mode can be first acquired by

$$\begin{aligned} Q(t) &= \int_0^{2\pi} F^R_{stall}(\theta, t)\Phi(\theta)\, d\theta \\ &= \int_0^{2\pi} F_v(\phi) \cos\left[n_e \cdot \text{NB}_c \cdot \Omega_{stall}(t + t_\phi)\right] \delta[\theta - \Omega_{stall} t - \phi] \sin \text{ND}\theta\, d\theta \\ &= F_v(\phi) \cos\left[n_e \cdot \text{NB}_c \cdot \Omega_{stall}(t + t_\phi)\right] \sin \text{ND}(\Omega_{stall} t + \phi) \end{aligned} \tag{6}$$

Based on Equation (6), convolution integral is then used to acquire normal response on the disc for this lightly-damped system, that is

$$x_\phi(\phi, \theta, t) = \frac{1}{m_{ND}\omega_{ND}} \int_0^t Q(\tau) \sin \omega_{ND}(t - \tau) d\tau \tag{7}$$

where m_{ND} and ω_{ND} are the corresponding modal mass and damped natural frequency. Substituting the generalized force in Equation (6) into the above formula, we can get the disc normal response

$$\begin{aligned} x_\phi(\phi, \theta, t) &= \frac{1}{m_{ND}\omega_{ND}} \int_0^t F_v(\phi) \cos\left[n_e \cdot \text{NB}_c \cdot \Omega_{stall}(\tau + t_\phi)\right] \sin \text{ND}(\Omega_{stall}\tau + \phi)\, \sin \omega_{ND}(t - \tau) d\tau \\ &= \frac{F_v(\phi)}{4 m_{ND}\omega_{ND}} \left(\begin{array}{c} -\frac{\sin(-\phi \text{NB}_C n_e - \omega_{ND} t + \text{ND}\phi)}{\Omega_{stall}(\text{ND} - \text{NB}_C n_e) + \omega_{ND}} - \frac{\sin(-\phi \text{NB}_C n_e + \omega_{ND} t + \text{ND}\phi)}{\Omega_{stall}(\text{NB}_C n_e - \text{ND}) + \omega_{ND}} \\ -\frac{\sin(\phi \text{NB}_C n_e - \omega_{ND} t + \text{ND}\phi)}{\Omega_{stall}(\text{NB}_C n_e + \text{ND}) + \omega_{ND}} + \frac{\sin(\text{NB}_C n_e(\Omega_{stall} t + \phi) + \text{ND}(\Omega_{stall} t + \phi))}{\Omega_{stall}(\text{NB}_C n_e + \text{ND}) + \omega_{ND}} \\ + \frac{2\omega_{ND}}{\omega_{ND}^2 - \Omega_{stall}^2(\text{ND} - \text{NB}_C n_e)^2} \sin[\text{ND}(\Omega_{stall} t + \phi) - \text{NB}_C n_e(\Omega_{stall} t + \phi)] \\ + \frac{\sin(\text{NB}_C n_e(\Omega_{stall} t + \phi) + \text{ND}(\Omega_{stall} t + \phi))}{\omega_{ND} - \Omega_{stall}(\text{NB}_C n_e + \text{ND})} - \frac{\sin(\phi \text{NB}_C n_e + \omega_{ND} t + \text{ND}\phi)}{\omega_{ND} - \Omega_{stall}(\text{NB}_C n_e + \text{ND})} \end{array} \right) \end{aligned} \tag{8}$$

In the preceding equation, terms with $\omega_{ND}t$ correspond to the transient response of the disc and are assumed to decay over time since damping is considered. Only terms with $\Omega_{stall}t$ are of interest. Finally, the steady-state response of the disc under rotating stall disturbance has the following form

$$x_\phi(\phi,\theta,t) = \frac{F_v(\phi)}{2m_{ND}}\left(\frac{\sin[(\mathrm{ND}-n_e\mathrm{NB}_c)(\phi+\Omega_{stall}t)]}{\omega_{ND}^2-(\mathrm{ND}-n_e\mathrm{NB}_c)^2\Omega_{stall}^2}+\frac{\sin[(\mathrm{ND}+n_e\mathrm{NB}_c)(\phi+\Omega_{stall}t)]}{\omega_{ND}^2-(\mathrm{ND}+n_e\mathrm{NB}_c)^2\Omega_{stall}^2}\right) \tag{9}$$

The response over the entire pressure distribution should be further integrated by

$$x(\theta,t) = \int_0^{2\pi} x_\phi(\phi,\theta,t)\,d\phi \tag{10}$$

Combined with $F_v(\phi)$ and the integral properties of trigonometric functions, two cases can be identified in order to have $x(\theta,t)\neq 0$. First, for the case of $k\mathrm{NB}_c = |\mathrm{ND}-n_e\mathrm{NB}_c|$, it follows from Equation (10) that

$$x^{C1}(\theta,t) = \frac{\pi}{2m_{ND}A_1}(a_k\sin[(\mathrm{ND}-n_e\mathrm{NB}_c)\Omega_{stall}t]+b_k\cos[(\mathrm{ND}-n_e\mathrm{NB}_c)\Omega_{stall}t]) \tag{11}$$

Second, for the case of $k\mathrm{NB}_c = \mathrm{ND}+n_e\mathrm{NB}_c$,

$$x^{C2}(\theta,t) = \frac{\pi}{2m_{ND}A_2}(a_k\sin[(\mathrm{ND}+n_e\mathrm{NB}_c)\Omega_{stall}t]+b_k\cos[(\mathrm{ND}+n_e\mathrm{NB}_c)\Omega_{stall}t]) \tag{12}$$

A_1 and A_2 in Equations (11) and (12) have the following form

$$\begin{cases} A_1 = \omega_{ND}^2-(\mathrm{ND}-n_e\mathrm{NB}_c)^2\Omega_{stall}^2 \\ A_2 = \omega_{ND}^2-(\mathrm{ND}+n_e\mathrm{NB}_c)^2\Omega_{stall}^2 \end{cases} \tag{13}$$

The discussion of these two cases is due to the different directions of rotation of the stall cells. Due to the different propagation nature, Equations (11)–(13) is treated separately. These equations can be combined to describe the steady response of the impeller structure under different stall conditions. It is obvious that the resonance will occur if A_1 and A_2 in Equations (11) and (12) become zero. Thus, two different frequency relationship between impeller mode and dynamic stall mode can be built up based on the cell propagation direction

$$\omega_{ND} = \begin{cases} |n_e\mathrm{NB}_c-\mathrm{ND}|\Omega_{stall},\ k\mathrm{NB}_c = |n_e\mathrm{NB}_c-\mathrm{ND}| \\ (n_e\mathrm{NB}_c+\mathrm{ND})\Omega_{stall},\ k\mathrm{NB}_c = n_e\mathrm{NB}_c+\mathrm{ND} \end{cases} \tag{14}$$

$k\mathrm{NB}_c = |n_e\mathrm{NB}_c-\mathrm{ND}|$ and $k\mathrm{NB}_c = n_e\mathrm{NB}_c+\mathrm{ND}$, which should not be dropped, are two prerequisites of A_1 and A_2, respectively. Further, Equation (14) can be simplified as

$$\begin{cases} \mathrm{ND} = m\mathrm{NB}_c,\ m\in\mathbb{Z}_+ \\ f_{ND} = m\mathrm{NB}_c(f_{\mathrm{shaft}}\pm f_{\mathrm{cell}}) \end{cases} \tag{15}$$

where f_{ND} is the natural frequency of impeller in Hz. Equation (15) points out that for $\mathrm{ND} = \mathrm{NB}_c$ nodal diameter mode, the impeller resonance will occur if the natural frequency coincides with $\mathrm{NB}_c(f_{\mathrm{shaft}}\pm f_{\mathrm{cell}})$. This is the most common situation happened during a lot of authors' experimental research [18–21]. The number of stall cells is very limited, which means excitation frequency will not be very high for industrial centrifugal compressors. In these cases, the first-order vibration mode will be quite close to the excitation frequency and needs to be considered. In addition to the basic disturbance coming from rotating stall, the higher harmonic component will also induce the impeller resonance with higher nodal diameter modes $m\mathrm{NB}_c$. However, the actual number of nodal diameters needs to be further calculated based on the specific number of blades. Although this situation is not common and very rare, it has also been found to exist by Mischo [21]. Since the circumferential non-uniform load caused by rotating stall with different origins (diffuser stall or impeller stall) is consistent, the derived equation is a general relationship and will be useful for impeller resonance check considering the rotating stall effects.

3. Test Facilities and Measurement Procedure

The experiments with full-size single-stage centrifugal compressor (SSCC) facility are performed at the Shenyang Blower Works Group Corporation. Numerous experimental studies have been conducted to acquire the knowledge of the compressor blade vibration under rotating stall condition. Since the stall and surge condition will do great harm to the compressor test rig, these operating points are carried out carefully during the experiments. Both transient and quasi-steady operating conditions are tested in detail. Figure 3 presents the operation line of the compressor test rig and shows detailed experimental measurement scheme. The behavior of the compressor test rig near the stall and surge boundary is mostly concerned. The compressor stage is first throttled at 100%Ω_{norm} speed to study the stall behavior from aerodynamic point of view. In order to find the rotating stall induced vibration phenomena, speed ramp (continuous varying speed) testing of the compressor at a low mass flow rate is measured. Two operating speeds (100%Ω_{norm} and 87%Ω_{norm}) are further tested with five selected mass flow rates (OP1-OP10 denoted in Figure 3). These quasi-steady operating conditions are combined to reveal the rotating stall mechanism and quantify the blade vibration. Detailed experimental results and discussion will be given in Section 4. Investigated compressor test rig and different measurement techniques will be elaborated in the following part of Section 3.

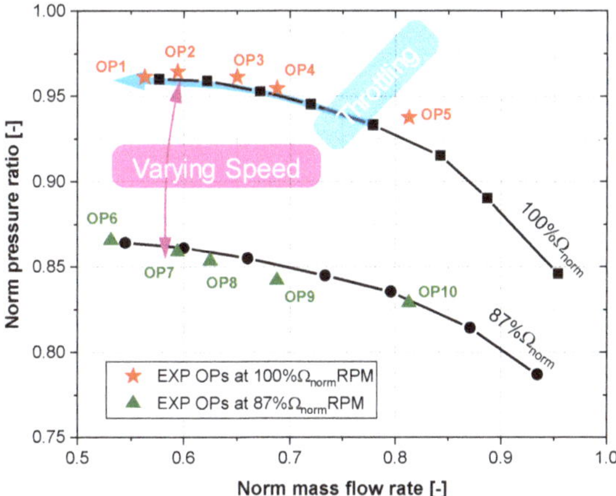

Figure 3. Operation line and experimental measurement scheme of the compressor facility.

3.1. Centrifugal Compressor Test Rig

The structure of the test compressor is shown in Figure 4. It consists of variable inlet guide vanes (VIGVs), an unshrouded backswept centrifugal impeller, a vaned diffuser, and the return channel. The number of blades and the main dimensions of these components are listed in Table 1. IGV blades are fixed in the axial inlet duct and used to deflect the flow in the tangential direction. The designed variable angle range is 40° to 120° (Here, 90° corresponds to the full open of IGV blades). Upstream and downstream of the compressor stage are the horizontal inlet duct and outlet pipe. A 2100 kW Electric motor is installed to drive the compressor with required rotating speeds. The test rig running speed ranges from 500 RPM to 9,000 RPM. A driven gear (drive ratio is 126/43 = 2.93) and a fluid coupling are further used to connect the motor and compressor shaft.

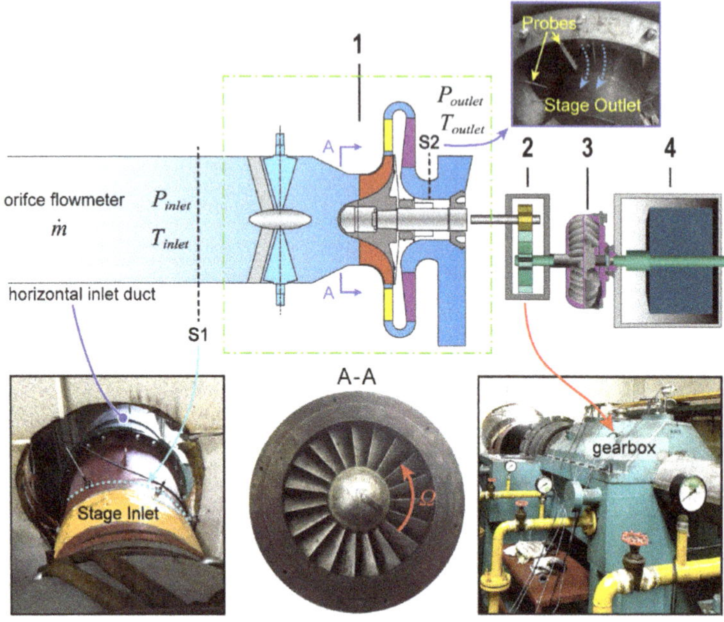

Figure 4. Schematic representation of the experimental test rig.

Table 1. Specifications of the investigated compression stage.

Parameters	Value
Number of inlet guide vanes	11
Number of impeller blades	19
Number of diffuser vanes	20
Number of return channel vanes	18
Impeller outlet diameter D2 (mm)	810
Impeller outlet width b2 (mm)	57.5
Diffuser inlet diameter D3 (mm)	900
Diffuser outlet diameter D4 (mm)	1242

3.2. Data Acquisition

3.2.1. Unsteady Static Pressure Measurement

The unsteady static pressure is measured at several locations within flow passage from compressor inlet to diffuser outlet. Figure 5 presents the meridional and axial view of these transducer positions. In order to determine the flow instability in the streamwise direction, the pressure pulsation at the compressor inlet (M1–C1) and in the diffuser passage (A1–A4) are monitored during the experiment. At the same time, the circumferential pressure pulsation at impeller-diffuser interface (M2) is also monitored for extra 4 points (B2–B5). Monitoring points A1 and B1 are in the same position. Holes used to install the pressure transducer at all these positions are opened at impeller and diffuser shroud casing component. The dynamic probe used here is the Model 106B52 produced by PCB Piezotronics. In order to avoid the disturbance caused by the transducer itself, each sensor is treated carefully and flush-mounted with the casing inner surface. The black dots at measurement plane M2 denotes the same location of A1/B1 at each passage inlet and only two diffuser passages are drawn here for simplicity. The angles in the circumferential direction between B1–B2, B2–B3, and B3–B4 are 72°. Considering non-uniform distribution of the sensors can depict the propagate of rotating stall cells more properly, sensor at location B5 is positioned at the adjacent diffuser passage of B4, such that the angle between B4–B5 is 36 degrees. Upon finishing the installation of these sensors, all signals are transmitted to a signal acquisition card, which is used for multi-channel high-precision measurement. Transient static pressure signals are acquired and recorded in DASP software, provided by Beijing Oriental Vibration and Noise Technology Institute. The sampling rate for the pressure transducers depends on the frequencies of interest. It is set to 20.48 kHz which is enough to resolve the blade passing frequency (BPF) and the rotating stall frequency (RSF).

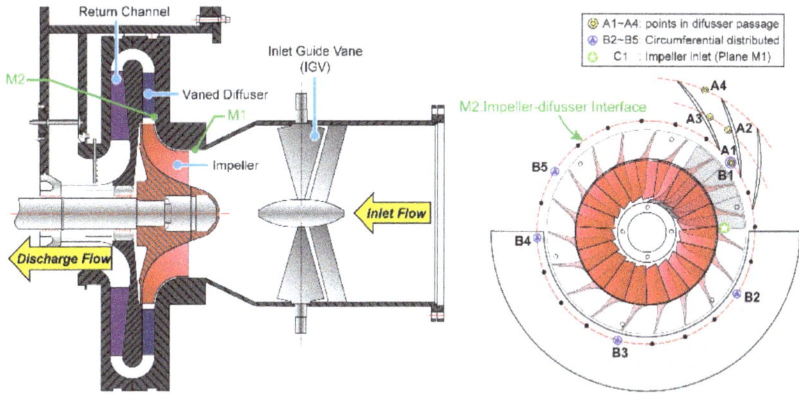

Figure 5. Casing-mounted pressure transducer positions in meridional and axial view.

3.2.2. Strain Gauge and Tip Timing Measurement

In this experiment, the non-synchronous vibration of the blade is monitored by 4 strain gauges in total. The installation position of these strain gauges (G1–G4) is shown in Figure 6b. The strain gauge G3 is positioned at half-span height and near the leading edge of the blade. It is used to capture the blade first bending mode since it is most sensitive to rotating stall disturbance. G1 and G2 are located at the tip of the blade to acquire the response of higher modes considering rotor-stator interactions. G1 is parallel to the blade tip and G2 is perpendicular to G1. At the root of the blade, strain gauge G4 is used to test the strength of the blade. Compact dynamic data acquisition instrument is installed in the nose cone (shown in Figure 6b) of the impeller to continually acquire the signal. A rounded bulb structure is further used to cover the hole. Only the strain gauge is exposed in the airflow environment. Considering the additional mass of the strain acquisition equipment, the balance of the rotor is conducted so that the shaft unbalance vibration is small. Since only one blade vibration is measured in detail by strain gauges, blade tip timing measurement is used to supplement the strain test for detailed vibration quantification. The tip deflection of all blades is measured simultaneously. Figure 6a gives the distribution of the tip timing sensors. The distribution angle between each BTT sensor is 120°. The light source (Red path) of each probe is provided by the laser box. The reflected laser (Blue path) will be transmitted back as each blade passes through the sensor. After receiving the blade triggered pulse, a high-speed counting acquisition card NI6602 is used here to obtain the blade arrival time sequence using 80 MHz internal counter. These data are further combined with the rotating speed and the tip radius to obtain the instantaneous vibration displacement. All pre- and post-processing work of the BTT signal can be finished in the in-house developed measurement system.

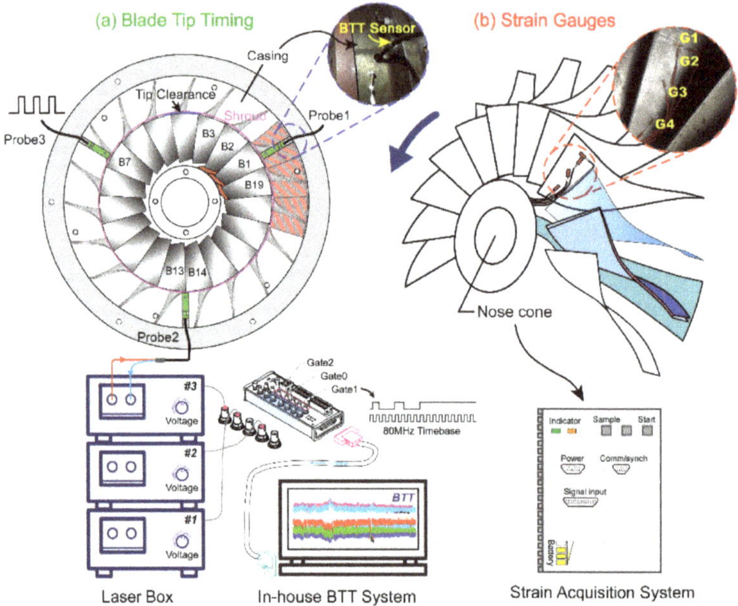

Figure 6. Installation and arrangement of tip timing probes (**a**) on the casing and strain gauges (**b**) on one impeller blade.

4. Results and Discussion

4.1. Stall Induced Flow Instability Behavior during Throttling Process

The raw pressure measurements and corresponding low pass filtered signal with the blade passing frequency f_{BPF} when entering surge are presented in Figure 7. This pressure signal is measured at the inlet of the diffuser passage. Before flow instability starts, the pressure fluctuating frequency is dominated by the jet-wake flow feature at the outlet of the impeller. During the throttling process, the pressure will become unstable and low-frequency variation with large amplitude starts to appear. This change in signal structure means that after these stall cells formed in the flow passage, the jet-wake flow pattern in stable operating condition will be superimposed by the stall dominated fluctuation. And, these stall cells will become stable and propagate in circumference after a short time. The static pressure measured during the throttling process is further analyzed by the wavelet transform. The scalogram values are scaled by the maximum absolute value at each level and frequencies are displayed on a linear scale. The flow behavior described above can also be viewed in time-frequency space. When the instability is further enhanced after 750 revolutions, a stable low-frequency mode occurs, which is consistent with the signal structure evolution in the time domain. If the flow rate is further reduced, the compression system will enter surge and large fluctuation of static pressure occurs.

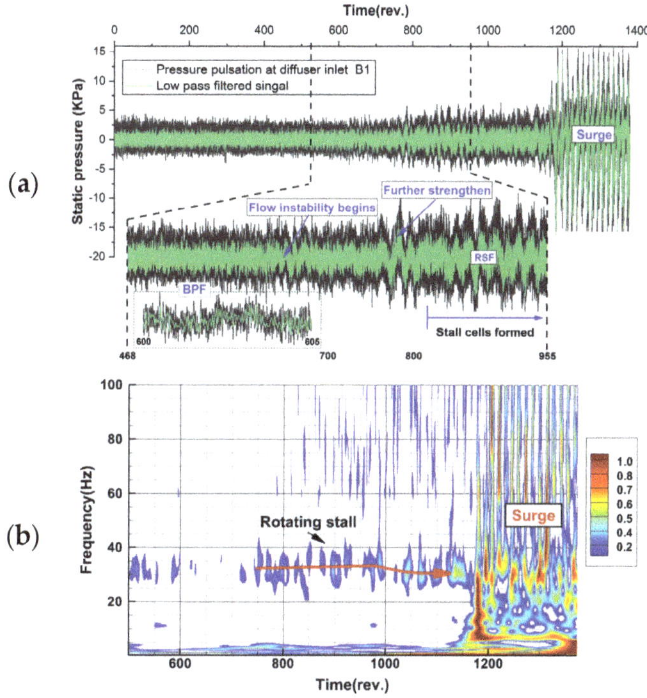

Figure 7. Rotating stall evolution and surge happened during throttling process at $100\%\Omega_{\text{norm}}$: (**a**) pressure traces from 0 to 1370 revolutions; (**b**) time-frequency spectrum based on wavelet transform.

4.2. Rotating Stall Induced Vibration Identification during Speed Ramp

Different rotating stall induced vibration phenomenon is first found by speed ramp testing method. The centrifugal compressor is first set to run at $100\%\Omega_{\text{norm}}$ speed and the mass flow is also adjusted to low flow rate to make the impeller disturbed by the rotating stall. Then, the shaft speed is continuously reduced to about $50\%\Omega_{\text{norm}}$ speed after all the signal acquisition equipment is get prepared. Figure 8a gives the time signal of gauge G3. The envelope of the signal is calculated to show the interaction between impeller mode with engine order excitation. The rotating passing frequency (RPF) of the shaft is also plotted on the right axis. Figure 8b gives the entire short time Fourier transform (STFT) spectrum of the result. It should be mentioned that the frequency value is normalized by the shaft frequency at the beginning. The engine order number is displayed as the left y-axis. These integers are used to show different engine order lines when the speed is continuously reduced. The non-synchronous blade response information is mainly found around the mode 1 of the impeller. It is consistent with the rotating stall induced vibration frequency determined by Equation (2) since the stall cell number is limited. Four different non-integer exciting regions are found and denoted by the dashed line. However, compared to mode 1 response of the impeller, the rotating stall induced vibration is still not so apparent. It is mainly due to the transient and instability nature of rotating stall under variable speed operation condition. The stalled state will change according to impeller speed and mass flow rate. But the stall cell formed rotating aerodynamic force gives us a dangerous warning because it has the same or even larger energy level compared to synchronous exciting. Large amplitude vibrations will appear if the exciting frequency coincides with the impeller modes.

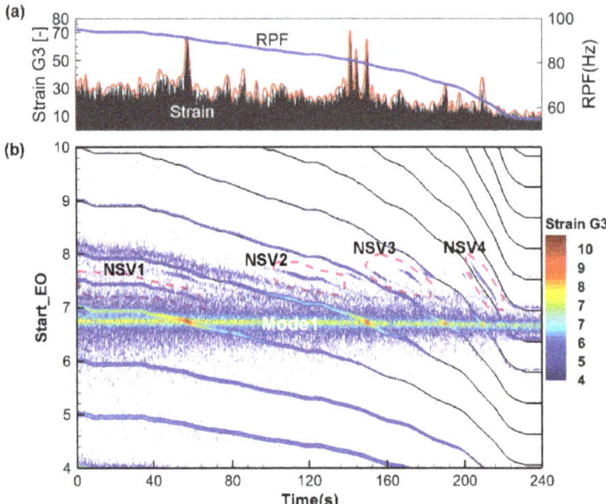

Figure 8. Blade vibration measured by Strain Gauge G3 during Speed Ramp: (**a**) time domain signal; (**b**) time-frequency spectrum based on STFT.

Although the variable speed testing method gives us a quick way to view the rotating stall induced vibration information. Due to the operation is transient and unstable, it is not suitable for stall parameters identification. In addition, in order to evaluate the blade vibration, quasi-steady operation under the rotating stall condition is also needed.

4.3. Stall Parameters Identification Based on Circumferential Pressure Pulsation Signals

In order to characterize the stall cell pattern during different quasi-steady operating conditions, the FFT spectrum combined with circumferential pressure distribution is first used for manual identification of the stall parameters since it is the most used and reliable method for industrial application [28]. Pressure pulsation signal spectrum is first analyzed to identify stall occurrences. Figure 9a gives the frequency spectrum measured at diffuser inlet when the compressor operates at OP1 depicted in Figure 3. A significant low frequency of $f_{PC} = 21.9$ Hz below the shaft frequency occurs, and there exists apparent signal modulation phenomenon on both sides of the BPF, which means the original rotational perturbation of the impeller is modulated by stall cell-induced fluctuation. In Figure 9b, a graphical matter based on bandpass filtered pressure traces is further used to acquire the stall parameters, including the number of stall cells NB_C, rotating speed f_{cell} and propagation direction [26]. Within the graph, pressure signals are plotted from bottom to top according to their circumferential positions in direction of impeller rotation. In order to form one revolution, the pressure signal of the first sensor is plotted once more. As it can be viewed from the graph, the low-frequency disturbance is arising and decaying in different stable periods. However, it will also become weak and disappear for a short time but reformed immediately. The propagation direction is first analyzed based on the dynamic behavior of the pressure signals from closely distributed two sensors, B5 and B4. It can be clearly judged that the stall propagation direction is the same as the impeller rotation. During stall formed period, it can be found that the circumferential 5-channel band-pass filtered pressure signal is almost the same phase at any time, but the fourth channel is in the inverted state. According to the sensor distribution angles, this phenomenon indicates that 36° corresponds to half of the distance between two adjacent stall cells in a cylindrical coordinate system along the direction of propagation. The identified stall mode is a stall group consisting of five cells. If we connect a peak of B1-0° with the corresponds one of B5-108°, which is delayed one and a half cycle, and extend to P1-360°, the stall propagation trajectory which crosses all signal peak positions are extracted.

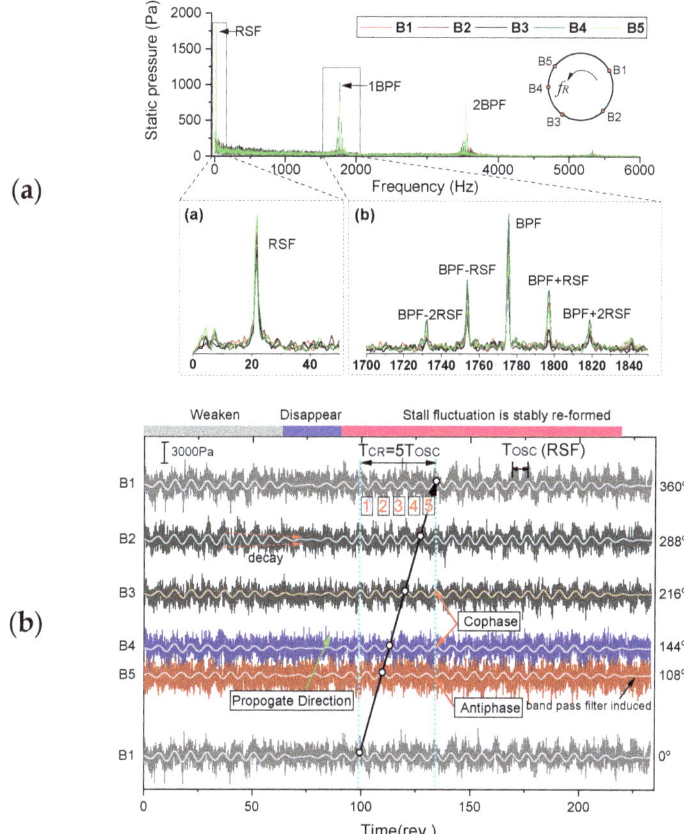

Figure 9. Rotating stall identification and parameter characterization. (**a**) FFT spectrum of diffuser inlet pressure signal; (**b**) spatial pressure distribution at stalled condition.

However, the pressure trace diagram method is a manual approach and can lead to some misunderstandings in some cases. The cross-correlation is further used to confirm the identified stall mode and further automate the identification process.

$$R_{ij}[n] = \frac{1}{N}\sum_{k=1}^{N} P_i(k)P_j(k+n) \tag{16}$$

Equation (16) is the discrete form of the cross-correlation between signal P_i and P_j. Figure 10a shows the cross-correlation results of the 5-channel pressure signals. Before cross-correlation, bandpass filtering is needed to extract the stall wave. The delay time can be calculated by the number of delayed sampling points corresponding to the local maximum of the cross-correlation. As can be seen, the delayed sampling points is around 450–500 which means the delay time is about 0.022–0.024 s based on the sampling frequency. In order to access the long-term behavior of the stall propagation, the delay time between stall cells is further calculated based on 30s' pressure pulsation signal. Figure 10b gives the box plot of the result. This result means upon the stall cell is formed around the impeller it will not change and keep stable. Since the aerodynamic force is not changed, the induced vibration should also be stable which can be further proven in the following content.

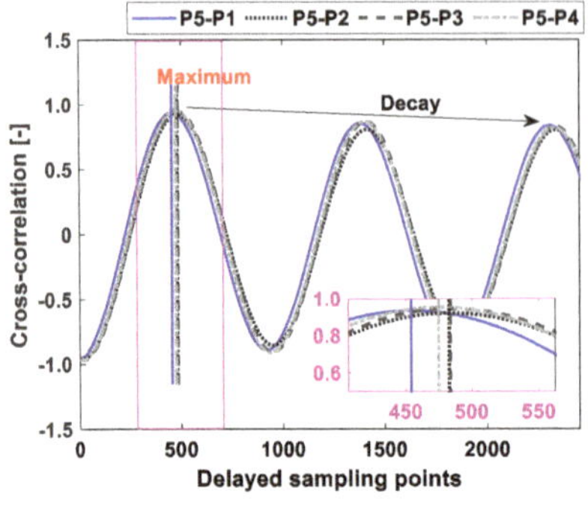

(a) cross-correlation of Pressure signals.

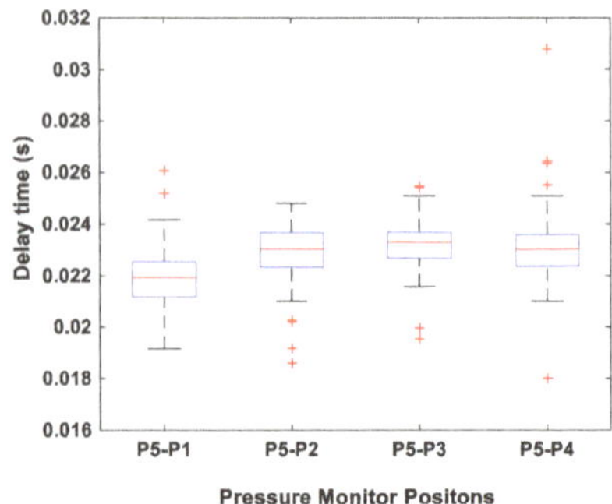

(b) identified delay time for 30 s Pressure signal

Figure 10. Stall mode identification and parameter characterization using cross-correlation analysis.

First, the five stall cells induced pressure traces can be plotted in polar coordinate based on the identified stall parameters, as shown in Figure 11a. These circumferential distributions of the pressure wave give a much better description of the footprint of the rotating stall cells at different monitoring positions. Figure 11b gives the corresponding time-frequency spectrum of blade vibration. During the rotating stall period, the five stall cells mode induced blade vibration is first found. The engine order of the vibration is 4.77. It is not an integer of the shaft rotating passing frequency which means non-synchronous vibration happened during rotating stall operating condition. Based on the identified parameters and Equation (2), the relationship can be verified:

$$\begin{cases} f_e^{Strain} = 4.773 \cdot f_{\text{shaft}} = 446.3\,\text{Hz} \\ f_e^{5cell} = \text{NB}_C \cdot f_{\text{shaft}} - \text{NB}_C \cdot f_{\text{cell}} = 445.625\,\text{Hz} \\ f_{\text{shaft}} = 93.5\,\text{Hz}, \text{NB}_C = 5 \end{cases} \tag{17}$$

So, it has $f_e^{Strain} \approx f_e^{5cell}$, where f_e^{Strain} is the vibration frequency existing in the blade response signal, f_e^{5cell} is the predicted blade vibration frequency calculated by rotating stall parameters and Equation (2). f_{cell} is the rotating frequency of one individual stall cell and has $f_{\text{cell}} = f_{\text{PC}}/\text{NB}_C$. As can be seen, these two frequencies agree quite well. There is only a small difference (less than 1 Hz) between f_e^{Strain} and f_e^{5cell} due to numerical error or spectral resolution. And the propagation direction is the same as the impeller rotation since "$-$" is used in Equation (17) instead of "$+$". The method of judging the direction of propagation through the low-frequency component fluctuation curve is still not very robust. Under some circumstances, the direction information may lead to being misinterpreted or cannot be decided only based on the signal dynamic behaviors. For example, if B1-B3 signals are used, the direction is hard to be decided. As well, it will also cause the cell number be misidentified when the spatial resolution of sensors is very limited. Since the relationship between pressure pulsation and blade vibration provided by Equation (2) is proven by analytical models and experimental analysis. Considering the relationship between the vibration and pressure fluctuating not only contains the cell number information but also includes the propagation direction, the vibration information can be combined to give a more reliable result.

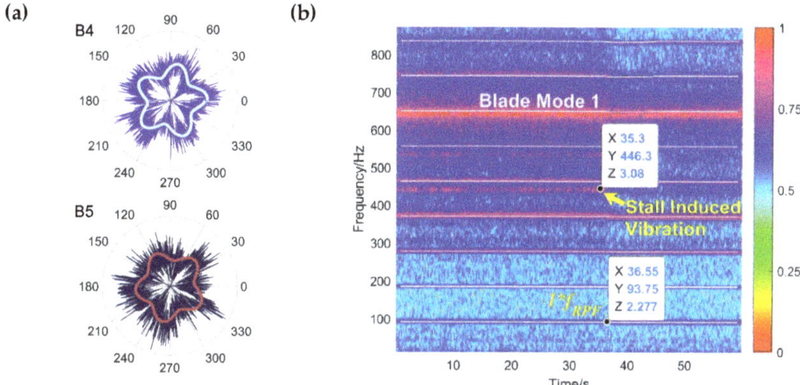

Figure 11. Rotating stall induced impeller blade vibration: (**a**) 5 stall cells dominated aerodynamic exciting force in polar coordinate and (**b**) time-frequency analysis of Strain Gauge G3.

4.4. Vibration Quantification Based on Tip Timing Measurement

The tip displacement signal is measured close to the leading edge of the impeller blade. Detailed monitoring position can be referred to Figure 6a. Based on the blade tip timing system, the tangential vibration of all blades under different operating points can be recorded for further displacement evaluation. Figure 12 gives the calculated blade tip displacement signal. The vibration behavior under low mass flow rate is concerned. For the operating point A, surge happens in this compression system and leads to an extremely large vibration amplitude. When the mass flow rate is adjusted during the experiment to make the compressor return to rotating stall condition (operating point B), the vibration is reduced significantly. Root mean square (RMS) level of the vibration signal is further calculated to show the influence of different mass flow rate. Results are plotted at the right top of Figure 12. From the perspective of the RMS value, the blade vibration is similar for operating point B to E. However, these small differences still show blade vibration trend and give a reasonable result of the vibration monitoring. The vibration is slightly reduced from operating point B to D, which corresponds to the mass flow change varying from stall to design point. However, when it comes to the choke boundary at operating point E, the vibration amplitude increases inversely.

It should also be mentioned that due to the arrangement of the BTT probes and the limited number of sensors, the frequency of the BTT signal cannot be accurately recovered. Non-intrusive detection of stall-vibration can be performed based on the tip timing signal in the future. The layout of the BTT probes and the experimental operating points of the compressor should be carefully designed and selected to capture this phenomenon from the perspective of aerodynamics and signal processing.

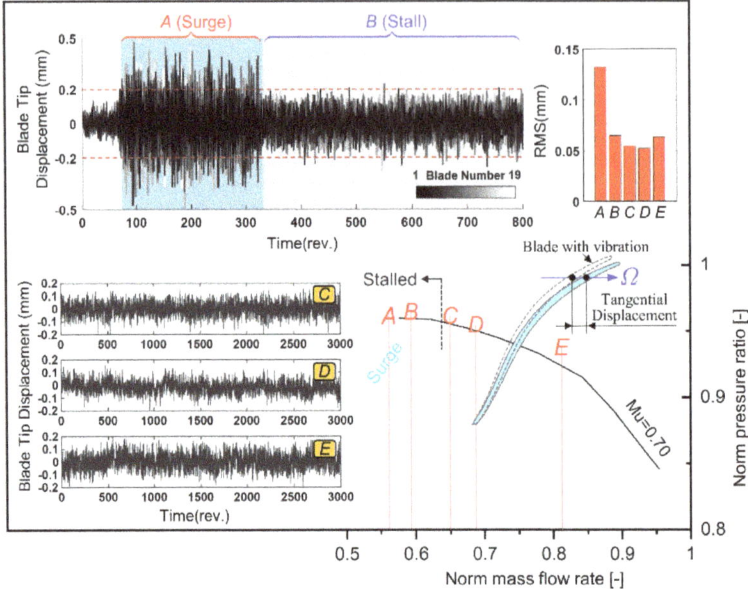

Figure 12. Blade tip timing results under five operating points with different mass flow rate varying from surge to choke.

5. Conclusions

This paper focuses on the non-synchronous blade vibration at stall and near surge operating of the centrifugal compressor. A theoretical model was first built up to describe the impeller vibration. Under the traveling wave excitation, the forced response function of the simplified disc was deduced. The impeller resonance condition was further acquired based on the steady part of the response function. The relationship between stall wave and impeller vibration was also correlated using the parameters which describe the stall mode. The derived relationship is consistent with previous understanding. The stall excitation law of the rotating stall can be viewed from a novel aspect, and at the same time, it can be found that both the rotating stall and its higher harmonics can induce the blade vibration and resonance with corresponding nodal diameter modes.

Experimental work was further conducted based on the industrial compressor test rig using an unshrouded centrifugal impeller equipped with vaned diffuser. During experiments, multiple signals from casing-mounted pressure transducers, strain gauges, and tip timing sensors were simultaneously acquired to provide a detailed insight into this physical phenomenon. Both transient and quasi-steady operating of compressor were designed and tested in detail to reveal the underlying mechanism and verify the relationship between pressure pulsation and blade vibration. The experimental investigation included: (1) detection and parameter characterization of rotating stall cells. (2) clearly interpretation and relationship verification of non-synchronous blade vibration. (3) quantify the impact of stall cells on all blade response amplitude. The throttling test of the compressor shown the dynamical behavior of airflow and a stable low-frequency mode appeared as it approached the stall boundary. The rotating stall was first understood from aerodynamic perspective. Then, the fluid-structure interaction measurements were further conducted using varying speed operating of the compressor. With time-frequency analysis method, the blade non-synchronous vibration regions were quickly found. Further, the quasi-steady operating of compressor was selected and measured for a long time in order to stabilize the stall and identify the specific cells mode. Both circumferential pressure tracking and cross-correlation methods were used to give an appropriate result. The frequency spectrum and spatial distribution characteristics of the pressure signal were also correlated with the blade vibration, which also shown the correctness of the theoretical model. The quantified blade response provides realistic compressor representative vibration data, which is the basis for the aeroelastic design of the impeller blades considering rotating stall. The overall test method in this paper is also important for stall vibration and resonance-free design using experimental verification methods.

Stall cells are extra excitation sources and can cause blade forced vibration at a specific frequency, which should be considered for more reliable operating of the compressor. In addition, the stall cell number is the core parameter which mostly affects the vibration frequency and occurrence of blade resonance. The first-order impeller mode is quite close to the stall excitation frequency and requires more attention in the future design phase of the impeller structure. The theoretical and experimental work of this paper contributes to the basic understanding and quantification of compressor stall and vibration. Further investigations are required to non-intrusive detect and control the rotating cells.

Author Contributions: Conceptualization, H.L.; Data curation, X.Z. and Q.Z.; Formal analysis, X.Z.; Funding acquisition, H.L. and S.Y.; Investigation, X.Z. and Q.Z.; Methodology, X.Z.; Writing—original draft, X.Z.; Writing—review & editing, X.Z. and H.L.

Funding: The work was supported by the Natural Science Foundation of China under Grant No. 51575075 and U1808214 are gratefully acknowledged.

Acknowledgments: The work was supported by the Natural Science Foundation of China under Grant No. 51575075 and U1808214 are gratefully acknowledged. The experimental work has been supported by Shenyang Blower Works Group Corporation. The authors greatly acknowledge the experimental support as well as the helpful technical discussion and the permission to publish.

Conflicts of Interest: The authors declare no conflict of interest.

References

1. Krain, H. Review of Centrifugal Compressor's Application and Development. *J. Turbomach.* **2005**, *127*, 25–34. [CrossRef]
2. Singh, M.P.; Vargo, J.J. Reliability Evaluation of Shrouded Blading Using the SAFE Interference Diagram. *J. Eng. Gas Turbines Power* **1989**, *111*, 601–609. [CrossRef]
3. Haupt, U.; Seidel, U.; Abdel-Hamid, A.N.; Rautenberg, M. Unsteady Flow in a Centrifugal Compressor with Different Types of Vaned Diffusers. *J. Turbomach.* **1988**, *110*, 293–302. [CrossRef]
4. Day, I.J. Stall, Surge, and 75 Years of Research. *J. Turbomach.* **2015**, *138*, 011001. [CrossRef]
5. Emmons, H.W.; Pearson, C.E.; Grant, H.P. Compressor Surge and Stall Propagation. *Trans. ASME* **1955**, *77*, 455–469.
6. Greitzer, E.M. Surge and Rotating Stall in Axial Flow Compressors—Part I: Theoretical Compression System Model. *J. Eng. Power* **1976**, *98*, 190–198. [CrossRef]
7. Greitzer, E.M. Surge and Rotating Stall in Axial Flow Compressors—Part II: Experimental Results and Comparison With Theory. *J. Eng. Power* **1976**, *98*, 199–211. [CrossRef]
8. Cumptsy, N.A.; Greitzer, E.M. A Simple Model for Compressor Stall Cell Propagation. *J. Eng. Power* **1982**, *104*, 170–176. [CrossRef]
9. Jansen, W. Rotating Stall in a Radial Vaneless Diffuser. *J. Basic Eng.* **1964**, *86*, 750–758. [CrossRef]
10. Lennemann, E.; Howard, J.H.G. Unsteady Flow Phenomena in Rotating Centrifugal Impeller Passages. *J. Eng. Gas Turbines Power* **1970**, *92*, 65–72. [CrossRef]
11. Abdelhamid, A.N. Analysis of Rotating Stall in Vaneless Diffusers of Centrifugal Compressors. In Proceedings of the International Gas Turbine Conference and Products Show, New Orleans, LA, USA, 10–13 March 1980; p. V01BT02A089. [CrossRef]
12. Abdelhamid, A.N. Effects of Vaneless Diffuser Geometry on Flow Instability in Centrifugal Compression Systems. In Proceedings of the International Gas Turbine Conference and Products Show, Houston, TX, USA, 9–12 March 1981; p. V001T03A008. [CrossRef]
13. Ferrara, G.; Ferrari, L.; Baldassarre, L. Rotating stall in centrifugal compressor vaneless diffuser: Experimental analysis of geometrical parameters influence on phenomenon evolution. *Int. J. Rotating Mach.* **2004**, *10*, 433–442. [CrossRef]

14. Fujisawa, N.; Hara, S.; Ohta, Y. Unsteady behavior of leading-edge vortex and diffuser stall in a centrifugal compressor with vaned diffuser. *J. Therm. Sci.* **2016**, *25*, 13–21. [CrossRef]
15. Fujisawa, N.; Inui, T.; Ohta, Y. Evolution Process of Diffuser Stall in a Centrifugal Compressor With Vaned Diffuser. *J. Turbomach.* **2018**, *141*, 041009. [CrossRef]
16. Iwamoto, S.; Yoneda, K.; Tokuyama, S.; Higuchi, H. Impeller Stall Induced by Reverse propagation of Non-Uniform Flow Generated at Return Channel. In Proceedings of the 45th Turbomachinery & 32nd Pump Symposia, Houston, TX, USA, 12–15 September 2016; pp. 1–23. [CrossRef]
17. Chen, J.; Hasemann, H.; Seidel, U.; Jin, D.; Huang, X.; Rautenberg, M. The Interpretation of Internal Pressure Patterns of Rotating Stall in Centrifugal Compressor Impellers. In Proceedings of the International Gas Turbine and Aeroengine Congress and Exposition, Cincinnati, OH, USA, 24–27 May 1993; p. V03AT15A043. [CrossRef]
18. Seidel, U.; Chen, J.; Haupt, U.; Hasemann, H.; Jin, D.; Rautenberg, M. Rotating Stall Flow and Dangerous Blade Excitation of Centrifugal Compressor Impeller: Part 1—Phenomenon of Large-Number Stall Cells. In Proceedings of the International Gas Turbine and Aeroengine Congress and Exposition, Orlando, FL, USA, 3–6 June 1991; p. V001T01A044. [CrossRef]
19. Hasemann, H.; Haupt, U.; Jin, D.; Seidel, U.; Chen, J.; Rautenberg, M. Rotating Stall Flow and Dangerous Blade Excitation of Centrifugal Compressor Impeller: Part 2—Case Study of Blade Failure. In Proceedings of the International Gas Turbine and Aeroengine Congress and Exposition, Orlando, FL, USA, 3–6 June 1991; p. V001T01A045. [CrossRef]
20. Jenny, P.; Bidaut, Y. Experimental Determination of Mechanical Stress Induced by Rotating Stall in Unshrouded Impellers of Centrifugal Compressors. *J. Turbomach.* **2016**, *139*, 031011. [CrossRef]
21. Mischo, B.; Jenny, P.; Mauri, S.; Bidaut, Y.; Kramer, M.; Spengler, S. Numerical and Experimental FSI-Study to Determine Mechanical Stresses Induced by Rotating Stall in Unshrouded Centrifugal Compressor Impellers. In Proceedings of the ASME Turbo Expo 2018: Turbomachinery Technical Conference and Exposition, Oslo, Norway, 11–15 June 2018; p. V07CT36A015. [CrossRef]
22. Vahdati, M.; Simpson, G.; Imregun, M. Unsteady Flow and Aeroelasticity Behavior of Aeroengine Core Compressors During Rotating Stall and Surge. *J. Turbomach.* **2008**, *130*, 031017. [CrossRef]
23. Sorokes, J.M.; Marshall, D.F.; Kuzdzal, M.J. A Review of Aerodynamically Induced Forces Acting on Centrifugal Compressors, and Resulting Vibration Characteristics of Rotors. In Proceedings of the 45th Turbomachinery & 32nd Pump Symposia, Houston, TX, USA, 12–15 September 2016; pp. 1–23.
24. Giachi, M.; Ferrara, G.; Tapinassi, L.; Belardini, E.; Bianchini, A.; Vannini, G.; Ferrari, L.; Biliotti, D. A Systematic Approach to Estimate the Impact of the Aerodynamic Force Induced by Rotating Stall in a Vaneless Diffuser on the Rotordynamic Behavior of Centrifugal Compressors. *J. Eng. Gas Turbines Power* **2013**, *135*, 112502. [CrossRef]
25. Biliotti, D.; Bianchini, A.; Vannini, G.; Belardini, E.; Giachi, M.; Tapinassi, L.; Ferrari, L.; Ferrara, G. Analysis of the Rotordynamic Response of a Centrifugal Compressor Subject to Aerodynamic Loads Due to Rotating Stall. *J. Turbomach.* **2015**, *137*, 021002. [CrossRef]
26. Levy, Y.; Pismenny, J. The Number and Speed of Stall Cells During Rotating Stall. In Proceedings of the Volume 6: Turbo Expo 2003, Parts A and B, Atlanta, GA, USA, 16–19 June 2003; pp. 889–899. [CrossRef]
27. Mehdigholi, H. Forced Vibration of Rotating Discs and Interaction with Non-Rotating Structures. Ph.D. Thesis, University of London, London, UK, 1991.
28. Bianchini, A.; Biliotti, D.; Giachi, M.; Belardini, E.; Tapinassi, L.; Ferrari, L.; Ferrara, G. Some Guidelines for the Experimental Characterization of Vaneless Diffuser Rotating Stall in Stages of Industrial Centrifugal Compressors. In Proceedings of the ASME Turbo Expo 2014: Turbine Technical Conference and Exposition, Düsseldorf, Germany, 16–20 June 2014; p. V02DT42A028. [CrossRef]

Article

Applying Deep Learning to Continuous Bridge Deflection Detected by Fiber Optic Gyroscope for Damage Detection

Sheng Li [1,*], Xiang Zuo [2], Zhengying Li [2] and Honghai Wang [1]

[1] National Engineering Laboratory for Fiber Optic Sensing Technology, Wuhan University of Technology, Wuhan 430070, China; wanghh@whut.edu.cn
[2] School of Information Engineering, Wuhan University of Technology, Wuhan 430070, China; zuoxiang@whut.edu.cn (X.Z.); zhyli@whut.edu.cn (Z.L.)
* Correspondence: lisheng@whut.edu.cn

Received: 4 January 2020; Accepted: 6 February 2020; Published: 8 February 2020

Abstract: Improving the accuracy and efficiency of bridge structure damage detection is one of the main challenges in engineering practice. This paper aims to address this issue by monitoring the continuous bridge deflection based on the fiber optic gyroscope and applying the deep-learning algorithm to perform structural damage detection. With a scale-down bridge model, three types of damage scenarios and an intact benchmark were simulated. A supervised learning model based on the deep convolutional neural networks was proposed. After the training process under ten-fold cross-validation, the model accuracy can reach 96.9% and significantly outperform that of other four traditional machine learning methods (random forest, support vector machine, k-nearest neighbor, and decision tree) used for comparison. Further, the proposed model illustrated its decent ability in distinguishing damage from structurally symmetrical locations.

Keywords: bridge damage detection; fiber optic gyroscope; deep learning; convolutional neural network

1. Introduction

Dynamic modal analysis has been the most commonly used approach for structural damage detection in civil engineering [1–3]. The use of wavelet, Hilbert–Huang transform, and other signal processing methods are also the conventional choices for structural damage detection that directly analyze the perturbation of vibration signals [4]. Various structural non-destructive testing approaches [5–7] are also significant means for detecting structural damage. Over the last decade, machine-learning algorithms have been used to address a wide range of vibration-based damage detection problems [8,9]. Although most of these techniques are based on vibration responses and such approaches still dominate the diagnosis and prognosis of structural health monitoring [10], feature extraction processes heavily relying on handcrafted intervention prior to damage classification [11] have often become major challenges that limit the effectiveness of various methods.

With the ability of automatic feature extraction and classification, deep convolutional neural networks (CNN) have been explored to address the range of difficulties in such following areas as computer vision [12,13], speech recognition [14], natural language processing [15], medical image processing [16], pathological signal classification [17,18], mechanical fault diagnosis [19–21], impact evaluation of natural disasters on infrastructure systems [22], and structural damage detection [23–25].

Most of the research efforts on deep CNN-based structural damage detection are essentially associated with the supervised learning processes. In this emerging area, Cha et al. [23] pioneered the deep CNN study of damage detection for cracks in concrete structures, and subsequently, Cha et al. [26]

further expanded the detection objectives of structural damage based on the faster Region-CNN (R-CNN). The recent study based on R-CNN to quantify the identified concrete spalling damage in terms of volume was reported in [27]. Xue and Li [28] established a fully convolutional neural networks model to classify the concrete tunnel lining defects. Other image-based researches on structural damage detection using deep learning were reported in [29–31]. In addition to two-dimensional convolution operations on structural images, one-dimensional convolution operations which usually spend a considerably cheaper computational cost than that of recurrent neural networks [32], are employed by researchers to perform the signal-based structural damage detection. The structural vibration signal as a typical type of one-dimensional time series [33] data is used to perform deep CNN-based structural damage detection. For instance, Abdeljaber et al. [18] proposed a method for detecting structural damage using one-dimensional CNN for multi-nodal vibration testing of steel frames. Lin et al. [25] simulated the vibration response of simply supported beams under various damage scenarios and proposed a procedure for detecting the categories of structural damage using the one-dimensional CNN. Huang et al. [34] analyzed the mechanical operation process through vibration signal by constructing a one-dimensional CNN.

Although deep CNN technology to some extent relieves the heavy pre-processing on the raw data or feature crafting for the damage detection when using vibration signals, the analysis of bridge damage detection is more complex comparing to that of simple structures, which needs more support of structural responses. In other words, compared with the amount of the degree of structural freedom, the scale of available vibration sensors used for bridge structural damage detection are often finite or even insufficient. To obtain as much structural dynamic information as possible in the case of limited measurement points, sensor optimization layout [35] is generally considered, which results in a decrease in damage detectability of complex structures. Therefore, a novel type of structural response which can easily cover the whole and local test requirement and provide enough structural information for the analysis of damage detection by using deep CNN should be attempted and explored. In this paper, we aimed at the multi-dimensional type of signal and chose a test technique for continuous curve mode of deformation based on fiber optic gyroscope (FOG) to produce continuous deflection of bridge. Detailed fundamental principles of the FOG-based testing technique were reported in [36–38]. A corresponding sample set based on supervised learning techniques was established, and a specific one-dimensional CNN model was proposed to automate feature extraction and classification. Specifically, the scheme procedure for the production of structural damage scenarios based on deformation responses was elaborated. The deformation responses and corresponding output labels were established through data augmentation and pre-processing. Furthermore, architectures and algorithms of the proposed one-dimensional CNN, and the partition rules of the dataset used for method verification were discussed. Finally, the performance of the proposed approach was compared with the results of other pattern recognition methods, all of which were conducted under the ten-fold cross-validation [39].

2. Design and Implementation of Structural Damage Scenarios

2.1. Experimental Platform and Instrumentation

A scale-down model of cable-stayed bridge was used as the experimental platform to represent the responses due to the simulation damage. The model shown in Figure 1 with the main span of 9.7 m, tower height of 3.46 m, deck width of 0.55 m, and 56 stay cables, was manufactured at a scale of 1:40. Figure 2 further illustrated a structure diagram and a physical structure of the device dedicated to measuring continuous deflection of the bridge model, which was triggered by a remote controller to perform mobile measurement. When the test device [36] moved along the surface of the bridge model, data of continuous deflection can be collected at sampling rate 150 Hz. Since the measurement period based on the motion carrier was relatively short, the continuous deflection obtained chronologically at each time was regarded as a multi-dimensional variable acquired at the same moment, and the deflection of main span was chosen as the structural deformation response used for subsequent analysis.

Figure 1. Experimental platform for testing continuous deflection.

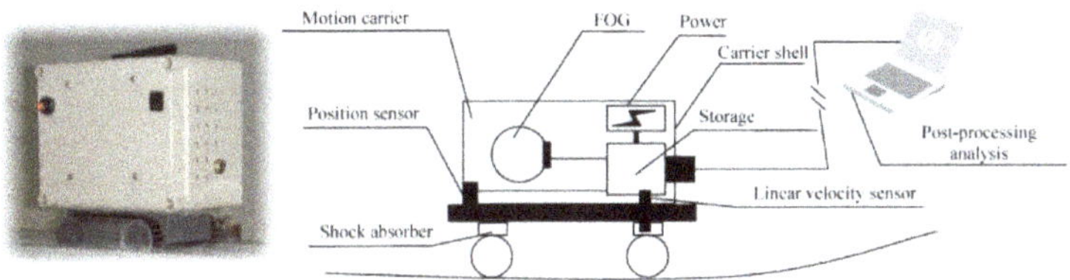

Figure 2. Measuring device integrated in a motion carrier.

2.2. Experimental Design and Procedures

The change in structural geometry can reflect a certain degree of transformation of interior mechanical properties. Further, structural damage is one reason for the change of interior mechanical properties of structure. Therefore, the different damage scenarios of the structure theoretically have corresponding structural deformation states. Continuous deflection can provide the dense deformation information, which can present more abundant structural response information than other finite point-based geometry measurement methods [40–42]. In the context of an experiment based on supervised learning, it was assumed that the change in the continuous deflection of bridge was only due to the result of structural damage. A metal pad (42.8L × 12.8W × 0.2H cm) with slope at both ends was used to simulate structural deformation caused by damage rather than physically destroying the structure [43]. The pad as an obstacle was placed on the movement path of the measuring device to simulate the deformation caused by structural damage. Compared with the situation without the pad, the measuring device can capture responses of the continuous deflection of bridge under the disturbance of the pad. This localized continuous deflection caused by the influence of the pad was clearly the most important of the continuous deflection of the entire bridge. Using such local responses instead of the global deflections can undoubtedly simplify the training process of the following supervised learning algorithm.

By this way, as shown in Figure 3, when the pad was not placed, the corresponding continuous deflection of the bridge was defined as $\mathbf{U}_0$. For each of the three damage scenarios, one pad was placed at a position each time, and therefore, $\mathbf{U}_1$, $\mathbf{U}_2$, and $\mathbf{U}_3$ can be obtained. Here, $\mathbf{U}_0$, $\mathbf{U}_1$, $\mathbf{U}_2$, and $\mathbf{U}_3$ as raw data of continuous deflection represented four types of simulated structure states, respectively. To improve the training efficiency and save the computational overhead of the supervised learning, $\mathbf{U}_1$, $\mathbf{U}_2$, and $\mathbf{U}_3$ were truncated to $\mathbf{u}_1$, $\mathbf{u}_2$, and $\mathbf{u}_3$. Such truncated selection in the areas affected by the pad can be estimated through both the original testing curves and the dimension of pad in the context of experiment based on supervised learning. In the intact and three damage scenarios, the actual benchmarks of $\mathbf{u}_1$, $\mathbf{u}_2$, and $\mathbf{u}_3$ were $\mathbf{u}_{01}$, $\mathbf{u}_{02}$, and $\mathbf{u}_{03}$, respectively. A common baseline for the three damage scenarios was defined to facilitate analysis. The weights of $\mathbf{u}_{01}$, $\mathbf{u}_{02}$, and $\mathbf{u}_{03}$ were regarded as equal and their average $\mathbf{u}_0$ was designated as the nominal benchmark of $\mathbf{u}_1$, $\mathbf{u}_2$ and $\mathbf{u}_3$. The following work utilized $\mathbf{u}_0$, $\mathbf{u}_1$, $\mathbf{u}_2$, and $\mathbf{u}_3$ to conduct the damage detection based on deep CNN algorithm.

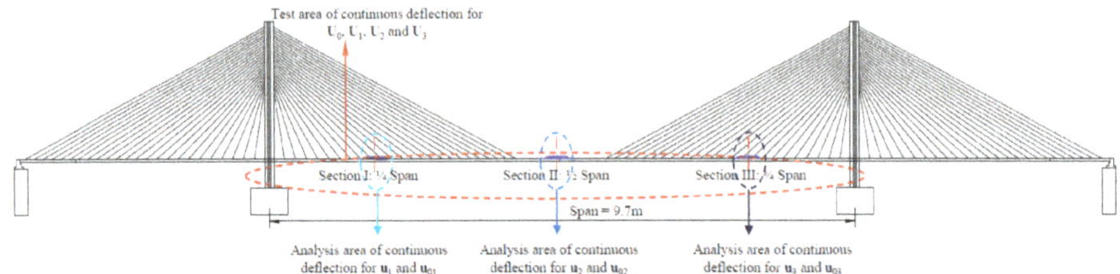

Figure 3. Locations of the metal pad used to assist in simulated damage scenarios.

2.3. *Raw Samples of Continuous Deflection of Bridge*

The spatial resolution in motion direction refers to the adjacent sampling interval of the device in Figure 2. This parameter is determined by the wheel diameter and the reticle of the rotational speed code wheel, and is approximately 1.48 mm. For the continuous deflection of the main span, taking intact scenario for an example shown in Figure 4, the deformation response of $\mathbf{U}_0$ consisted of sequence data of 6554 dimensions which depicted the length of main span of 9.7 m. Due to the high spatial resolution, the continuous curve clearly reflected a certain degree of pre-camber applied at the main span. Moreover, the continuous curve revealed that the experiment platform did not exhibit completely symmetrical structural deformation owing to the handcrafted control for the cable force.

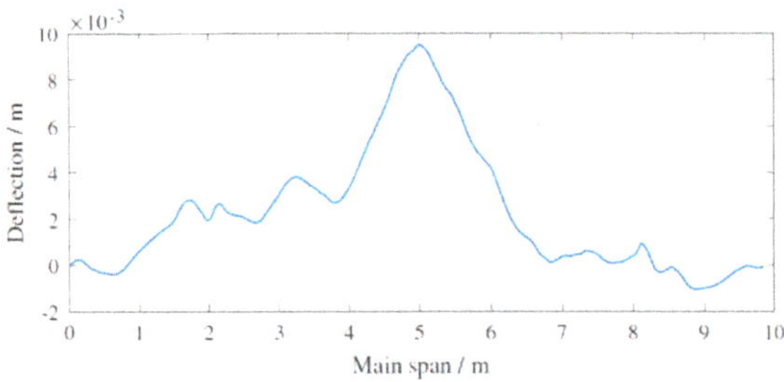

Figure 4. Deformation of main span depicted by continuous deflection.

The local continuous deflection curves of $\mathbf{u}_0$, $\mathbf{u}_1$, $\mathbf{u}_2$ and $\mathbf{u}_3$ were shown in Figure 5. Each type of the local continuous deflection curve contained 390 dimensions of sequence data. The coverage length of the area affected by the pad was considered to be the primary basis for determining the length of the local continuous deflection. Moreover, through preliminary data observation, the length of the region having the largest influence range among the three disturbance positions of the pad was selected, rounded, and defined as the final truncated length, which guaranteed the consistency of multiple sets of sample dimensions. The continuous curve mode test technique was used to separately collect the structural response of the scale-down bridge model under intact and simulated structural damage, and five groups of $\mathbf{U}_0$, $\mathbf{U}_1$, $\mathbf{U}_2$, and $\mathbf{U}_3$, were collected, respectively. Therefore, five groups of $\mathbf{u}_0$, $\mathbf{u}_1$, $\mathbf{u}_2$, and $\mathbf{u}_3$ corresponded to four types of structural conditions, namely intact, damage$_{1/4}$, damage$_{1/2}$, and damage$_{3/4}$, and these were used as raw samples to conduct the following study.

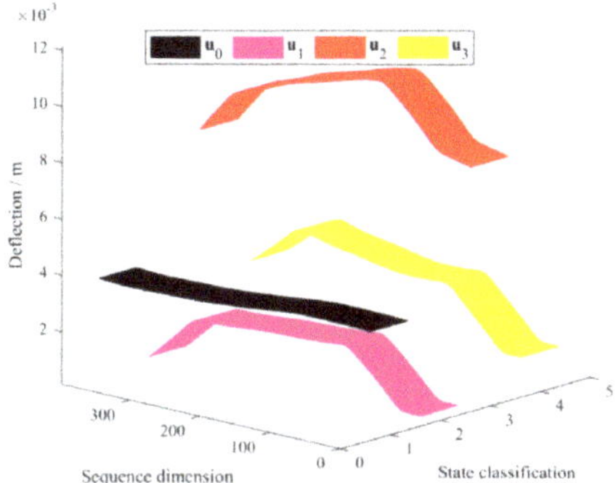

Figure 5. Truncated deformations depicted by continuous deflections.

3. Detection Methodology Based on Deep CNN

3.1. Data Augmentation and Pre-processing

Data augmentation and pre-processing are two essential tasks before carrying out deep learning. The former is always the first choice to boost the performance of a deep network. For image recognition based on deep CNN, there are a wide range of ways to perform data augmentation [12,44,45]. However, the above approaches are not suitable for signal-based pattern classification when using deep CNN algorithm. As shown in Figure 6a, dividing the raw acquisition signals into the same sub-fragment directly is common means of data augmentation [24,25]. It can be seen from Figure 6b that, for fragments of the same length as that in Figure 6a, the overlapping zone set in the adjacent fragments causes the amount of the fragment m to be larger than n shown in Figure 6a, which effectively increases the amount of data size.

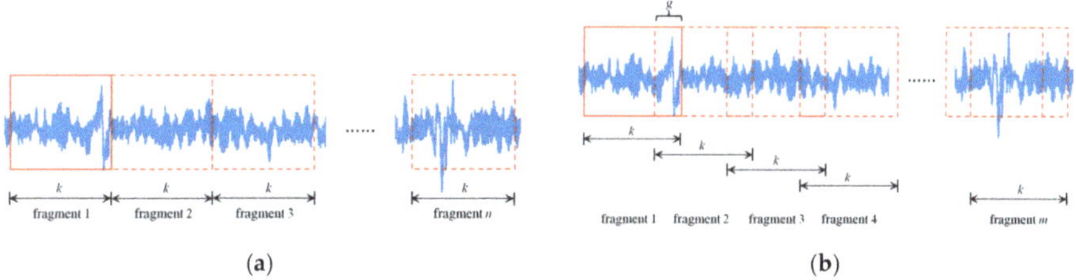

Figure 6. Data augmentation of (**a**) common means and (**b**) adopted operation.

Since the original experimental samples were small, the overlapping zone was taken as $g = 1$. It was obvious that the larger the value of k in Figure 6, the smaller the number of fragments after data augmentation and vice versa, which also indicated the greater number of fragments needed more computational overhead of model training. With the consideration of a tradeoff result between the training objective of model and the computational overhead, the length of fragment was set as $k = 50$, followed by the 390-dimensional original sequence becoming 341 50-dimensional sequence samples. For the five raw groups of $\mathbf{u}_0$, $\mathbf{u}_1$, $\mathbf{u}_2$, and $\mathbf{u}_3$, after data augmentation, the sample set of $\mathbf{u}_0'$, $\mathbf{u}_1'$, $\mathbf{u}_2'$, and $\mathbf{u}_3'$, each including 1705 samples, corresponding to Figure 7a–d, were shown by mesh graphics.

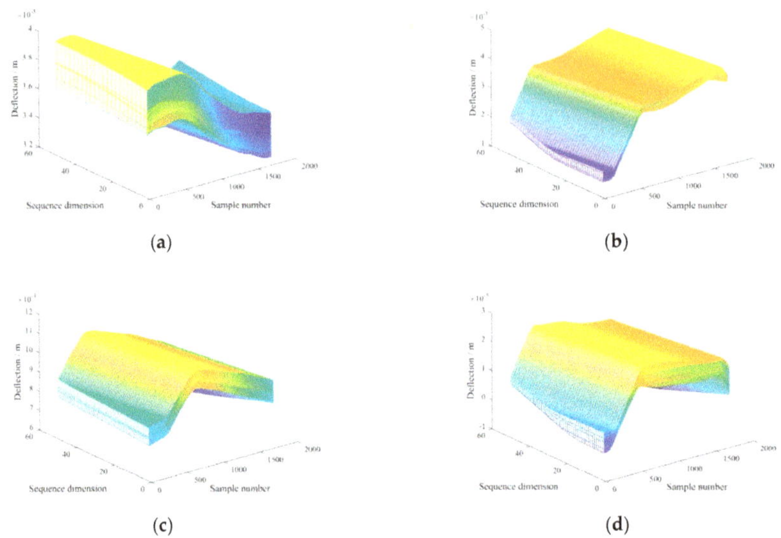

Figure 7. Sample set represented by mesh graphics through data augmentation for (**a**) intact, (**b**) damage$_{1/4}$, (**c**) damage$_{1/2}$, and (**d**) damage$_{3/4}$.

To eliminate the difference in the deflection amplitudes of four types of samples in Figure 7 and boost a better classification effect [46,47], a type of min-max normalization [48] expressed in Equation (1) is used to normalize all the amplitudes to the range of 0~1.

$$\mathbf{u}_i'' = \frac{\mathbf{u}_i' - \mathbf{min}(\mathbf{u}_i')}{\mathbf{max}(\mathbf{u}_i') - \mathbf{min}(\mathbf{u}_i')}, (i = 0, 1, 2, 3 \ldots) \tag{1}$$

As shown in Table 1, raw and truncated represented the continuous deflections of the test area and analysis area shown in Figure 3, respectively. After data augmentation and normalization based on the truncated stage, each category of the four-dataset including intact and three types of simulated damage contained 1705 samples. The four types of state, namely, $\mathbf{u}_0''$, $\mathbf{u}_1''$, $\mathbf{u}_2''$, and $\mathbf{u}_3''$ were used as input data, in which $\mathbf{u}_0''$ represented the intact baseline and the rest three represented different damage scenarios. One-hot form was used to describe the output labels corresponding to the four categories, meaning that the label vector was generated by the rule that the vector had all zero elements except the position *j*, where *j* was the type number of structural state.

Table 1. The dataset details of training and test.

Processing Stage	Variable				Dimension of Each Variable	Total Samples
Raw	$\mathbf{U}_0$	$\mathbf{U}_1$	$\mathbf{U}_2$	$\mathbf{U}_3$	6554	4 × 5 = 20
Truncated	$\mathbf{u}_0$	$\mathbf{u}_1$	$\mathbf{u}_2$	$\mathbf{u}_3$	390	
Augmentation	$\mathbf{u}_0'$	$\mathbf{u}_1'$	$\mathbf{u}_2'$	$\mathbf{u}_3'$	50	4 × 1705 = 6820
Normalization	$\mathbf{u}_0''$	$\mathbf{u}_1''$	$\mathbf{u}_2''$	$\mathbf{u}_3''$		

The entire measurement process was performed under stable temperature field, and the data of this study came from actual measurements, which already contained noise disturbances existing in the indoor environment. Therefore, extra interferences of simulated noise and temperature effect were not further considered here.

3.2. Descriptions of the Proposed CNN Architecture

Table 2 gave the details of the proposed CNN structure through the trial-and-error under the current computing resource configuration. The model structure was inspired by Cifar-10 [49], in which operations of convolution and pooling were not pairwise used. Figure 8 showed the graphical representation of CNN structure with 50 input sample lengths where the green, blue, and yellow referred to the kernel size, max-pooling, and fully connected layer, respectively.

Table 2. The details of CNN structure.

Layers	Type	No. of Neurons (Output Layers)	Kernel Size	Stride	Padding	Activation
0-1	Convolution	25 × 20	2	2	Same	PReLU
1-2	Convolution	25 × 20	2	1	Same	PReLU
2-3	Max-pooling	24 × 20	2	1	Valid	——
3-4	Convolution	12 × 32	2	2	Same	PReLU
4-5	Convolution	12 × 32	2	1	Same	PReLU
5-6	Max-pooling	11 × 32	2	1	Valid	——
6-7	Convolution	11 × 20	2	1	Same	PReLU
7-8	Convolution	11 × 20	2	1	Same	PReLU
8-9	Max-pooling	10 × 20	2	1	Valid	——
9-10	Flatten	200	——	——	——	——
10-11	Dense	128	——	——	——	PReLU
11-12	Dense with dropout	64	——	——	——	PReLU
12-13	Dense	4	——	——	——	Softmax

Note: PReLU—parametric rectified linear unit.

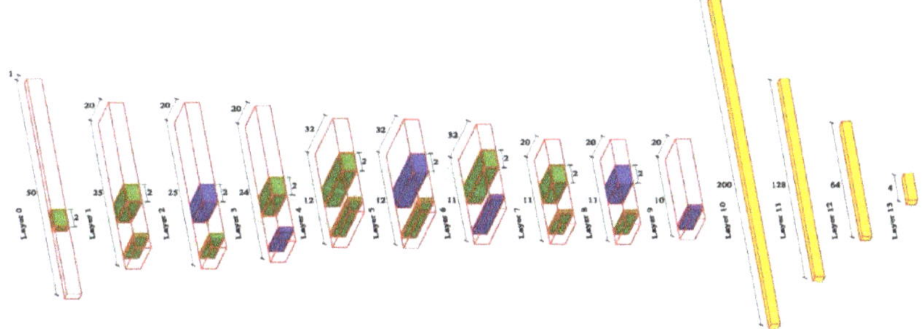

Figure 8. The proposed deep CNN architecture.

Layer 0 as the input layer in Figure 8 was convolved with a kernel of size 2 to produce Layer 1. The convolution and cross-correlation were used interchangeably in deep learning [50], which can be described as:

$$\mathrm{f}(i) = \sum_{n=1}^{N} \mathrm{s}(i+n)\mathrm{k}(n) \tag{2}$$

where s is input signal, k is filter, and N is the number of elements in s. The output vector f is the cross-correlation of s and k. Next, Layer 1 was convolved with a same kernel size to produce Layer 2. After two times of convolution, a max-pooling of size 2 was applied to every feature map (Layer 3). By repeating the above operations two times, other four convolutional layers and two max-pooling layers were created. In Layer 9, the neurons were then fully connected to 200 neurons in Layer 10 by flatten. Eventually, Layer 10 was fully connected to 128 neurons in Layer 11 and Layer 11 was fully connected to 64 neurons in Layer 12. Finally, Layer 12 was connected to the last layer (Layer 13) with 4 output neurons which represented intact, $\text{damage}_{1/4}$, $\text{damage}_{1/2}$ and $\text{damage}_{3/4}$.

Because the gradient of the left side of rectified linear unit (ReLU) [51] as shown in (3) is always zero, the activation operation may become invalid during training process if the weights updated by a large gradient become zero after being activated.

$$\mathrm{f}(x) = \begin{cases} x, & x \geq 0 \\ 0, & x < 0 \end{cases} \tag{3}$$

The Leaky ReLU method [52] is a good alternative to address such problem by considering a parameter α in (4),

$$\mathrm{f}(x) = \begin{cases} x, & x \geq 0 \\ \alpha x, & x < 0 \end{cases} \tag{4}$$

where α is usually set to a small number, and once α is set, its value will keep constant. This allows a small, non-zero gradient when the unit is not active. The parametric rectified linear unit (PReLU) [53] which has the same mathematical expression to Leaky ReLU, takes this idea further by making the coefficient α into a parameter that is learnt along with the other neural network parameters. Since it was not necessary to consider how to specify α, PReLU was used to take the place of Leaky ReLU in this work as an activation function for the convolutional layers (1, 2, 4, 5, 7 and 8) and two fully connected layers (11 and 12).

Further, the Softmax function was used to compute the probability distribution of the four output classes, which can be expressed as follows:

$$p_k = \frac{e^{x_k}}{\sum\limits_{i=1}^{n} e^{x_i}} \tag{5}$$

where x_k is the input of last layer, n is the number of output nodes and output values of p_k are between 0 and 1 and their sum equals to one. Equation (5) was used for Layer 13 in Figure 8 to predict which category the input signals (intact, $\text{damage}_{1/4}$, $\text{damage}_{1/2}$, or $\text{damage}_{3/4}$) belonged to.

Compared with shallow neural networks, deep CNN as a more complicated model contains more hidden layers and more weights, and is particularly prone to overfitting. In the proposed deep CNN, a dropout rate of 0.35 was used before the classification layer (Layer13) as shown in Table 2, which together with early stopping [54] mentioned in the following, effectively suppressed incidence of overfitting during all training processes.

3.3. Training Setting

Ten percent of the total dataset was used for test, while the rest 90% was divided into two parts, namely, training (80%) and validation (20%). The reason for validation was to evaluate the performance of the model for each epoch and prevent overfitting.

Because the cross entropy function is much more sensitive to the error, the learning rules derived from the cross entropy function generally yield better performance. Here, categorical cross entropy was used as the objective function to estimate the difference between original and predicted damage types, expressed as follows:

$$J = \sum_{i=1}^{k} \left[-d_i \ln(y_i) - (1 - d_i) \ln(1 - y_i) \right] \tag{6}$$

where J is the cross entropy, y_i is the output of prediction class, d_i is the original class in the training data, and k is the number of output nodes.

To minimize the above objective function, adaptive moment estimation (Adam) was selected as the optimization algorithm. It calculated an adaptive learning rate for each parameter and stored both an exponentially decaying average of past squared gradients and an exponentially decaying average of past gradients [55]. Details about the training parameters in this work are given in Table 3, in which the early stopping technique was used to control training epochs and further avoid overfitting, and the parameters set in Adam were based on the suggestion in [56].

Table 3. The training parameters of CNN structure in this work.

Batch Size	Epoch	Patience in Earlystopping	Adam			
			Initial Learning Rate	β_1	β_2	ε
128	5000	500	0.001	0.9	0.009	1.0×10^{-8}

Moreover, a ten-fold cross-validation approach was used in this study, the purpose of which was to reduce the sensitivity of algorithm performance to data partitioning and to obtain as much valid information as possible from the enhanced data. First, all the prepared dataset was randomly divided into ten equal parts. Nine out of ten parts of the total were used to train the proposed deep CNN while the remaining one-tenth dataset were used to test the performance of the model. This strategy was repeated ten times by shifting the training and test dataset. The accuracies reported in the paper were the average values obtained from ten evaluations.

4. Results and Discussion

The proposed CNN model was implemented by Python package Tensorflow and Keras [57]. The average training runtime of each fold for the proposed model was approximately 15 minutes, which was run on a GPU core (GTX 1080 Ti) with twelve 2.20 GHz processors (Intel Xeon E5-2650 v4). According to the setting in Table 3, the training processes showed that in the initial 500 epochs, the convergence speed was rather quickly for all of the dataset from the ten-fold cross-validation, but it still needed approximately 3000 to 4500 epochs to reach the best performance based on the patience rule set in early stopping. The typical training process regarding accuracy and loss represented by fold 2 is shown in Figure 9, which stopped at the epochs of 3036.

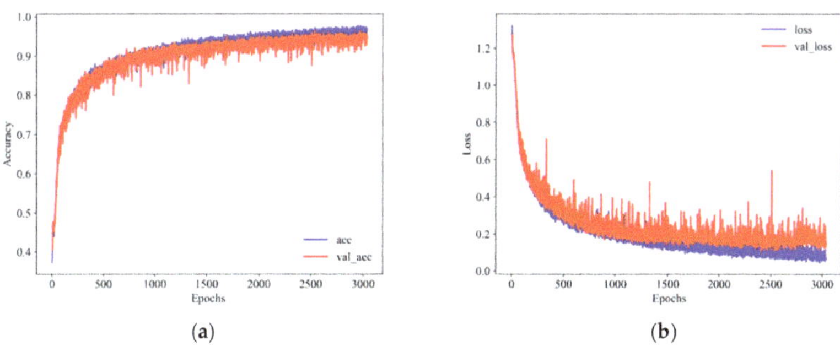

(**a**) (**b**)

Figure 9. Typical training processes about (**a**) accuracy and (**b**) loss.

The confusion matrix cross all ten-fold was presented in Figure 10a. It was observed that 98.3% of $\mathbf{u}_0''$ signals were correctly classified as intact. Moreover, 1.7% of $\mathbf{u}_0''$ were erroneously classified as other damage categories. Further, a high percentage of 98.4% of $\mathbf{u}_1''$ signals were correctly classified as damage$_{1/4}$ with 1.3% of $\mathbf{u}_1''$ wrongly classified as damage$_{3/4}$. For $\mathbf{u}_2''$ the accuracy rate for damage$_{1/2}$ reached 96.8% with 2.9% of $\mathbf{u}_2''$ wrongly predicted as damage$_{3/4}$. Similarly, 94.2% of $\mathbf{u}_3''$ signals were correctly classified as damage$_{3/4}$ with 5.8% wrongly classified as intact (1.1%), damage$_{1/4}$ (1.9%), and damage$_{1/2}$ (2.8%).

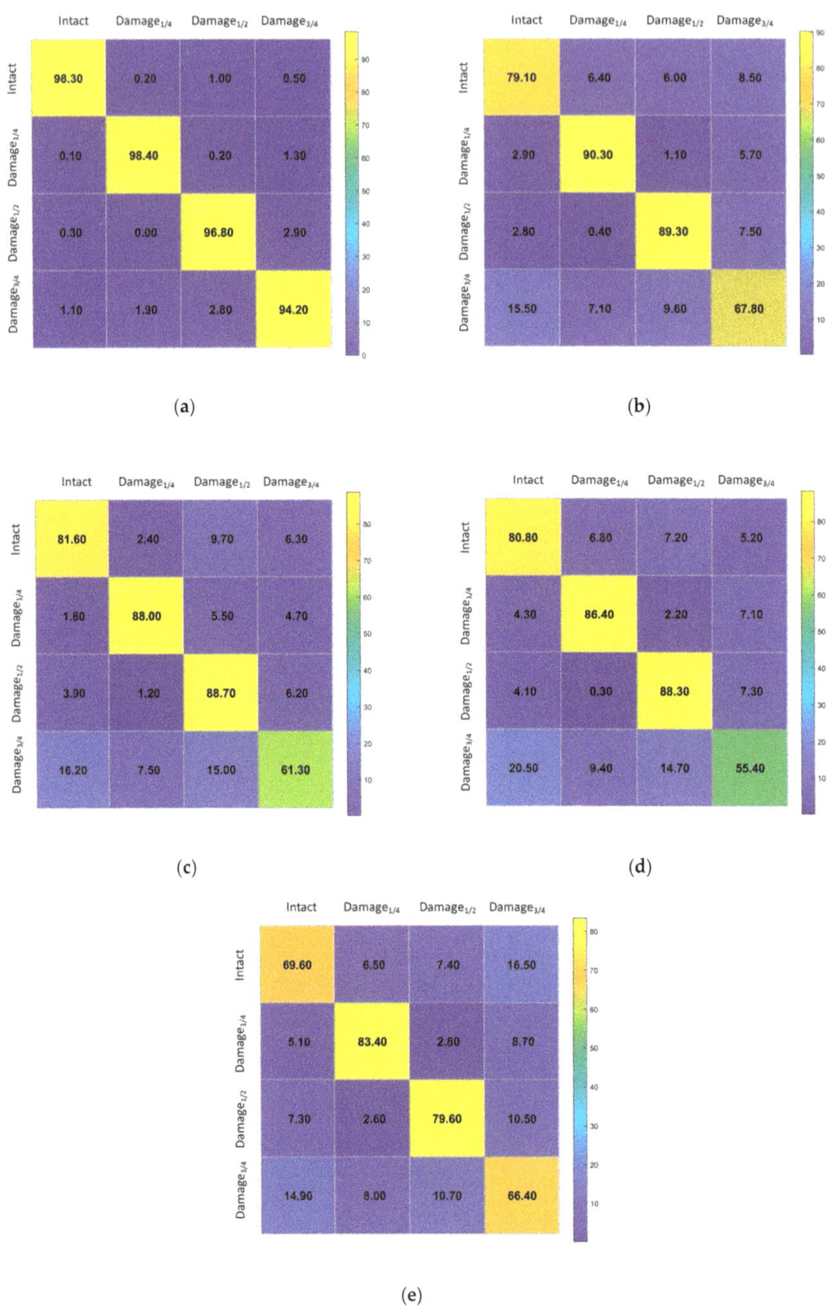

Figure 10. The confusion matrices of (**a**) CNN, (**b**) random forest, (**c**) support vector machine, (**d**) k-nearest neighbor, and (**e**) decision trees.

Furthermore, to evaluate the capability in each fold of cross-validation, average accuracy results shown in Figure 11 for different classes were compared between the proposed model and other four pattern recognition methods. When the samples were directly used to classify without heavy consideration regarding features extraction, the accuracy of automatic detection for proposed CNN model (96.9%) was obviously better than that of random forest (RF) (81.6%), support vector machine (SVM) (79.9%), k-nearest neighbor (KNN) (77.7%), and decision trees (DT) (74.8%). Here, the allocation of dataset of the four comparison methods was consistent with the proposed deep CNN algorithm. To fully compete with the proposed model, the most decent key hyperparameters set in sklearn [58] for RF, SVM, KNN and DT were derived through trial-and-error. To further quantify the effect of classifiers, Figure 10b–e show confusion matrices of the other four methods, respectively. It was observed that the best accuracy in various comparison methods can reach to 90.3% as shown in Figure 10b, which was still inferior to the lowest accuracy 94.2% as shown in Figure 10a.

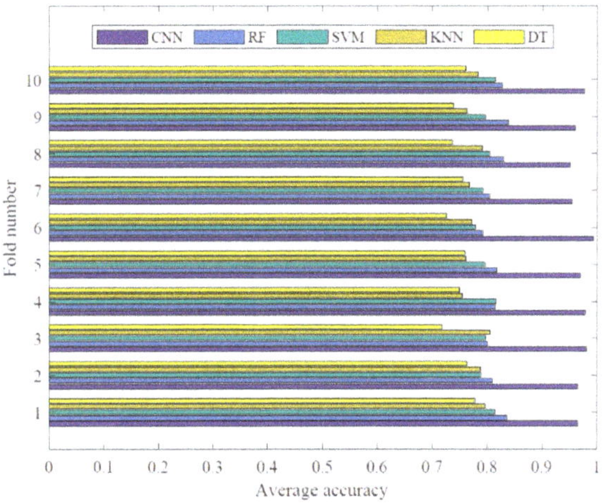

Figure 11. Comparisons of average accuracy in each fold of cross-validation.

Next, as shown in Figure 12a, for all five methods, the classification effects on $damage_{1/4}$ obviously outperformed the results of the other three categories. Moreover, the detection results of $damage_{3/4}$ were the worst in all methods, having a direct influence on the average accuracy of various ways. Further, as shown in Figure 12b, the classification imbalance presented in the confusion matrix was most severe when KNN was used as the classifier. This phenomenon may be related to the relatively lower algorithm complexity of KNN [59,60] compared with other methods mentioned in this work. Only the proposed approach based on deep CNN effectively mitigated this imbalance, although the accuracy of $damage_{3/4}$ in Figure 10a was still slightly less than the other three classes. The limited data samples should be a major aspect for such imbalance. In addition, current tests were all from the one-way results and lack of the data from the opposite direction. This may introduce a cumulative system error to the results of the structural response. Further, only slight pre-processing was carried out for the original dataset, which reduced the learning ability of each method mentioned in this paper. Actually, as shown in Table 4, the other four machine learning methods for comparison had fewer key parameters to consider in the process of balancing training accuracy and training error than the proposed method. This weak complexity, determined by the principles of the algorithm, resulted in a poor predictive effect on training and validation. Therefore, under the same circumstance, the proposed approach clearly demonstrated better overall performance in automatic feature extraction than other comparison means.

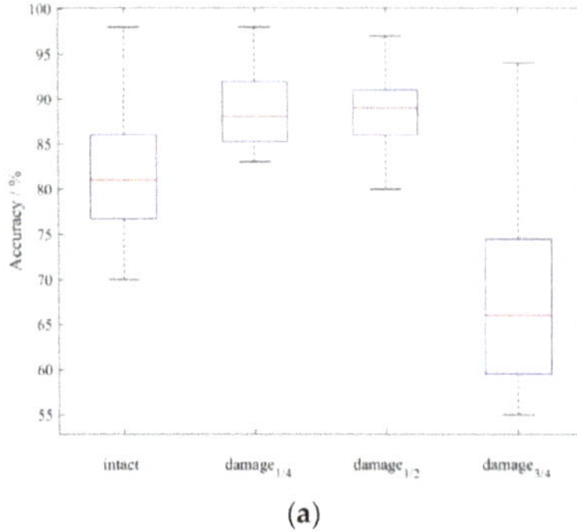

(a)

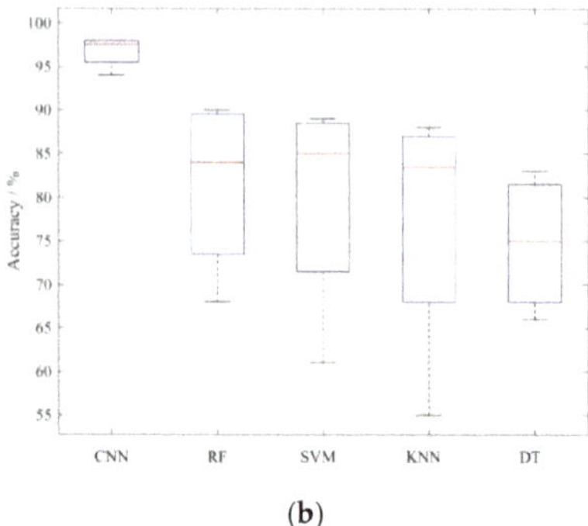

(b)

Figure 12. Accuracy distribution based on (**a**) damage category and (**b**) detection method.

Table 4. Key parameters set in the comparison methods.

RF	SVM	KNN	DT
FS criteria = gini Number of DT = 150	Kernel = rbf gamma = 10 C = 10	$K = 6$ DM = euclidean	FS criteria = entropy

Note: FS—Feature selection, gamma—The influence of kernel radius, C—The penalty parameter, K—k value defined in KNN, DM—Distance metric.

5. Conclusions

A deep learning CNN model with 11 trainable hidden layers was proposed to automatically extract and classify the bridge damage represented by the continuous deflection of bridge. Although current research on the use of FOG-based test technique to detect the damage of a scale-down bridge model through deep learning is just a pilot study, the following conclusions can be drawn:

(1) In the case where it is easy to measure the FOG-based continuous deflection of the target structure, it is convenient to build structural deformation database that can provide sufficient training samples for deep learning-based damage detection.
(2) Based on the data preparation strategies adopted in this work, one-dimensional convolution operation can effectively extract the detailed features of bridge deflection after a slight data pre-processing.
(3) The deep CNN-based method as a classifier has at least 15.3% accuracy advantage over other traditional methods mentioned in this paper in distinguishing different types of bridge deformation modes.
(4) Even if the same level structural damage occurs at a symmetrical position, the proposed method can still achieve satisfactory results with a deviation of only 4.2% for the recognition accuracy of damage at the symmetrical position.
(5) For an actual bridge with a complete deformation monitoring database, the advantage of deep learning on automatic extracting of features of large-scale database can be exploited to search the damage or provide the preliminary diagnostic findings. Moreover, since the FOG-based measurement system has higher test accuracy for larger distributed deflections [61], the proposed method should be more suitable for long-span bridges.

Author Contributions: Data curation, X.Z.; Funding acquisition, S.L.; Methodology, S.L.; Project administration, Z.L.; Resources, H.W.; Supervision, H.W.; Writing—original draft, X.Z.; Writing—review & editing, S.L. All authors have read and agreed to the published version of the manuscript

Funding: This research was funded by the National Natural Science Foundation of China grant number 61875155.

Acknowledgments: The research work reported in this paper was supported by the National Engineering Laboratory for Fiber Optic Sensing Technology, Wuhan University of Technology.

Conflicts of Interest: The authors declare no conflict of interest.

References

1. Moughty, J.J.; Casas, J.R. A state of the art review of modal-based damage detection in bridges: Development, challenges, and solutions. *Appl. Sci.* **2017**, *7*, 510. [CrossRef]
2. Sarah, F.; Hejazi, F.; Rashid, R.S.M.; Ostovar, N. A review of dynamic analysis in frequency domain for structural health monitoring. *IOP Conf. Ser.: Earth Environ. Sci.* **2019**, *357*, 012007. [CrossRef]
3. Sabato, A.; Niezrecki, C.; Fortino, G. Wireless MEMS-based accelerometer sensor boards for structural vibration monitoring: A review. *IEEE Sens. J.* **2017**, *17*, 226–235. [CrossRef]
4. Amezquita-Sanchez, J.P.; Adeli, H. Signal processing techniques for vibration-based health monitoring of smart structures. *Arch. Comput. Method Eng.* **2016**, *23*, 1–15. [CrossRef]
5. Tao, Y.H.; Fitzgerald, A.J.; Wallace, V.P. Non-contact, Non-destructive testing in various industrial sectors with terahertz technology. *Sensors* **2020**, *20*, 712. [CrossRef]
6. Tashakori, S.; Baghalian, A.; Unal, M.; Fekrmandi, H.; Senyurek, V.Y.; McDaniel, D.; Tansel, I.N. Contact and non-contact approaches in load monitoring applications using surface response to excitation method. *Measurement* **2016**, *89*, 197–203. [CrossRef]
7. Tashakori, S.; Baghalian, A.; Senyurek, V.Y.; Unal, M.; McDaniel, D.; Tansel, I.N. Implementation of heterodyning effect for monitoring the health of adhesively bonded and fastened composite joints. *Appl. Ocean Res.* **2018**, *72*, 51–59. [CrossRef]
8. Farrar, C.R.; Worden, K. *Structural Health Monitoring: A Machine Learning Perspective*; John Wiley and Sons: New York, NY, USA, 2013; pp. 17–43.
9. Burgos, D.A.T.; Vargas, R.C.G.; Pedraza, C.; Agis, D.; Pozo, F. Damage identification in structural health monitoring: A brief review from its implementation to the use of data-driven applications. *Sensors* **2020**, *20*, 733. [CrossRef]
10. Webb, G.T.; Vardanega, P.J.; Middleton, C.R. Categories of SHM deployments: Technologies and capabilities. *J. Bridge Eng.* **2015**, *20*, 04014118. [CrossRef]
11. Pan, H.; Azimi, M.; Yan, F.; Lin, Z. Time-frequency-based data-driven structural diagnosis and damage detection for cable-stayed bridges. *J. Bridge Eng.* **2018**, *23*, 04018033. [CrossRef]
12. Krizhevsky, A.; Sutskever, I.; Hinton, G.E. ImageNet classification with deep convolutional neural networks. In Proceedings of the 26th Annual Conference on Neural Information Processing Systems (NIPS 2000), Lake Tahoe, NV, USA, 3–6 December 2012.
13. Koziarski, M.; Cyganek, B. Image recognition with deep neural networks in presence of noise-dealing with and taking advantage of distortions. *Integr. Comput. Aided Eng.* **2017**, *24*, 337–349. [CrossRef]
14. Abdel-Hamid, O.; Mohamed, A.; Jiang, H.; Penn, G. Applying convolutional neural networks concepts to hybrid NN-HMM model for speech recognition. In Proceedings of the IEEE International Conference on Acoustics, Speech and Signal Processing, Kyoto, Japan, 25–30 March 2012.
15. Hu, B.; Lu, Z.; Li, H.; Chen, Q. Convolutional neural network architectures for matching natural language sentences. In Proceedings of the 28th Annual Conference on Neural Information Processing Systems (NIPS 2014), Montreal, QC, Canada, 8–13 December 2014.
16. Wang, G. A perspective on deep imaging. *IEEE Access* **2016**, *4*, 8914–8924. [CrossRef]
17. Acharya, U.R.; Fujita, F.; Oh, S.L.; Hagiwara, Y.; Tan, J.H.; Adam, M. Application of deep convolutional neural network for automated detection of myocardial infarction using ECG signals. *Inf. Sci.* **2017**, *415–416*, 190–198. [CrossRef]
18. Acharya, U.R.; Oh, S.L.; Hagiwara, Y.; Tan, J.H.; Adeli, H. Deep convolutional neural network for the automated detection and diagnosis of seizure using EEG signals. *Comput. Biol. Med.* **2018**, *100*, 270–278. [CrossRef] [PubMed]
19. Chen, Z.; Li, C.; Sanchez, R. Gearbox fault identification and classification with convolutional neural networks. *Shock Vib.* **2015**, *2015*. [CrossRef]
20. Guo, X.; Chen, L.; Shen, C. Hierarchical adaptive deep convolution neural network and its application to bearing fault diagnosis. *Measurement* **2016**, *93*, 490–502. [CrossRef]

21. Janssens, O.; Slavkovikj, V.; Vervisch, B.; Stockman, K.; Loccufier, M.; Verstockt, S.; Van de Walle, R.; Van Hoecke, S. Convolutional neural network based fault detection for rotating machinery. *J. Sound Vib.* **2016**, *377*, 331–345. [CrossRef]
22. Nabian, M.A.; Meidani, H. Deep learning for accelerated reliability analysis of transportation networks. *Comput. Aided Civ. Infrastruct. Eng.* **2018**, *33*, 443–458. [CrossRef]
23. Cha, Y.; Choi, W.; Büyüköztürk, O. Deep learning-based crack damage detection using convolutional neural networks. *Comput. Aided Civ. Infrastruct. Eng.* **2017**, *32*, 361–378. [CrossRef]
24. Abdeljaber, O.; Avci, O.; Kiranyaz, M.S.; Gabbouj, M.; Inman, D.J. Real-time vibration-based structural damage detection using one-dimensional convolutional neural networks. *J. Sound Vib.* **2017**, *388*, 154–170. [CrossRef]
25. Lin, Y.; Nie, Z.; Ma, H. Structural damage detection with automatic feature-extraction through deep learning. *Comput. Aided Civ. Infrastruct. Eng.* **2017**, *32*, 1025–1046. [CrossRef]
26. Cha, Y.; Choi, W.; Suh, G.; Mahmoudkhani, S.; Büyüköztürk, O. Autonomous structural visual inspection using region-based deep learning for detecting multiple damage types. *Comput. Aided Civ. Infrastruct. Eng.* **2018**, *33*, 731–747. [CrossRef]
27. Beckman, G.H.; Polyzois, D.; Cha, Y. Deep learning-based automatic volumetric damage quantification using depth camera. *Autom. Constr.* **2019**, *99*, 114–124. [CrossRef]
28. Xue, Y.; Li, Y. A fast detection method via region-based fully convolutional neural networks for shield tunnel lining defects. *Comput. Aided Civ. Infrastruct. Eng.* **2018**, *33*, 638–654. [CrossRef]
29. Gao, Y.; Mosalam, K.M. Deep transfer learning for image-based structural damage recognition. *Comput. Aided Civ. Infrastruct. Eng.* **2018**, *33*, 748–768. [CrossRef]
30. Kang, D.; Cha, Y.J. Autonomous UAVs for structural health monitoring using deep learning and an ultrasonic beacon system with geo-tagging. *Comput. Aided Civ. Infrastruct. Eng.* **2018**, *33*, 885–902. [CrossRef]
31. Zhang, L.; Zhou, G.; Han, Y.; Lin, H.; Wu, Y. Application of Internet of Things technology and convolutional neural network model in bridge crack detection. *IEEE Access* **2018**, *6*, 39442–39451. [CrossRef]
32. Lipton, Z.C.; Berkowitz, J.; Elkan, C. A Critical Review of Recurrent Neural Networks for Sequence Learning. Available online: https://arxiv.org/abs/1506.00019 (accessed on 10 December 2019).
33. Bagnall, A.; Lines, J.; Bostrom, A.; Large, J.; Keogh, E. The great time series classification bake off: A review and experimental evaluation of recent algorithmic advances. *Data Min. Knowl. Discov.* **2017**, *31*, 606–660. [CrossRef]
34. Huang, S.; Tang, J.; Dai, J.; Wang, Y. Signal status recognition based on 1DCNN and its feature extraction mechanism analysis. *Sensors* **2019**, *19*, 2018. [CrossRef]
35. Chang, M.; Pakzad, S. Optimal sensor placement for modal identification of bridge systems considering number of sensing nodes. *J. Bridge Eng.* **2014**, *19*, 04014019. [CrossRef]
36. Li, S.; Hu, W. A novel bridge curve mode measurement technique based on FOG. *Optik* **2015**, *126*, 3442–3445. [CrossRef]
37. Yang, D.; Wang, L.; Hu, W.; Zhang, Z.; Fu, J.; Gan, W. Singularity detection in FOG-based pavement data by wavelet transform. In Proceedings of the 25th International Conference on Optical Fiber Sensors, Jeju, Korea, 24–28 April 2017.
38. Li, L.; Tang, J.; Gan, W.; Hu, W.; Yang, M. The continuous line-shape measurement of bridge based on tri-axis fiber optic gyro. In Proceedings of the 25th International Conference on Optical Fiber Sensors, Jeju, Korea, 24–28 April 2017.
39. Zemmour, E.; Kurtser, P.; Edan, Y. Automatic parameter tuning for adaptive thresholding in fruit detection. *Sensors* **2019**, *19*, 2130. [CrossRef] [PubMed]
40. Liu, Y.; Deng, Y.; Cai, C.S. Deflection monitoring and assessment for a suspension bridge using a connected pipe system: A case study in China. *Struct. Control Health Monit.* **2015**, *22*, 1408–1425. [CrossRef]
41. Bogusz, J.; Figurski, M.; Grzegorz, N.; Marcin, S.; Wrona, M. GNSS-based multi-sensor system for structural monitoring applications. Proceedings of the Institute of Navigation International Technical Meeting 2012 (ITM 2012), Newport Beach, CA, USA, 30 January–1 February 2012.
42. Hou, X.; Yang, X.; Huang, Q. Using inclinometers to measure bridge deflection. *J. Bridge Eng.* **2005**, *10*, 564–569. [CrossRef]
43. Yang, D.; Wang, L.; Hu, W.; Ding, C.; Gan, W.; Liu, F. Trajectory optimization by using EMD- and ICA-based processing method. *Measurement* **2019**, *140*, 334–341. [CrossRef]

44. Szegedy, C.; Liu, W.; Jia, Y.; Sermanet, P.; Reed, S.; Anguelov, D.; Erhan, D.; Vanhoucke, V.; Rabinovich, A. Going deeper with convolutions. In Proceedings of the IEEE Conference on Computer Vision and Pattern Recognition (CVPR 2015), Boston, MA, USA, 7–12 June 2015.
45. Deep Image: Scaling up Image Recognition. Available online: http://www.academia.edu/download/55673602/97789723e45650442cfaedff17e1c11a5080.pdf (accessed on 16 September 2019).
46. Han, J.; Kamber, M.; Pei, J. *Data Mining: Concepts and Techniques*, 3rd ed.; Morgan Kaufmann: Waltham, MA, USA, 2011; pp. 83–124.
47. Jahan, A.; Edwards, K.L. A state-of-the-art survey on the influence of normalization techniques in ranking: Improving the materials selection process in engineering design. *Mater. Des.* **2015**, *65*, 335–342. [CrossRef]
48. Patro, S.G.K.; Sahu, K.K. Normalization: A Preprocessing Stage. Available online: https://arxiv.org/abs/1503.06462 (accessed on 16 September 2019).
49. Krizhevsky, A. Learning Multiple Layers of Features from Tiny Images. Available online: http://www.cs.toronto.edu/~{}kriz/lea-rning-features-2009-TR.pdf (accessed on 10 December 2019).
50. Ketkar, N. *Deep Learning with Python: A Hands-On Introduction*, 1st ed.; Apress: New York, NY, USA, 2017; pp. 63–66.
51. Nair, V.; Hinton, G.E. Rectified linear units improve restricted Boltzmann machines. In Proceedings of the 27th International Conference on Machine Learning (ICML 2010), Haifa, Israel, 21–25 June 2010.
52. Maas, A.L.; Hannun, A.Y.; Ng, A.Y. Rectifier nonlinearities improve neural network acoustic models. In Proceedings of the 30th International Conference on Machine Learning (ICML'13), Atlanta, GA, USA, 16–21 June 2013.
53. He, K.; Zhang, X.; Ren, S.; Sun, J. Delving deep into rectifiers: Surpassing human-level performance on ImageNet classification. *arXiv* **2015**, arXiv:1502.01852.
54. Caruana, R.; Lawrence, S.; Giles, L. Overfitting in neural nets: Backpropagation, conjugate gradient, and early stopping. In Proceedings of the 14th Annual Neural Information Processing Systems Conference (NIPS 2000), Denver, CO, USA, 27 November–2 December 2000.
55. Ruder, S. An Overview of Gradient Descent Optimization Algorithms. Available online: https://arxiv.org/abs/1609.04747 (accessed on 16 September 2019).
56. Kingma, D.P.; Ba, J. Adam: A Method for Stochastic Optimization. Available online: https://arxiv.org/abs/1412.6980 (accessed on 10 December 2019).
57. Keras. Available online: http://github.com/keras-team/keras (accessed on 10 December 2019).
58. Hackeling, G. *Mastering Machine Learning with Scikit-Learn*, 1st ed.; Packt Publishing: Birmingham, UK, 2014; pp. 97–185.
59. Kubat, M. *An Introduction to Machine Learning*, 2nd ed.; Springer: New York, NY, USA, 2017; pp. 43–56.
60. Noi, P.T.; Kappas, M. Comparison of random forest, k-nearest neighbor, and support vector machine classifiers for land cover classification using sentinel-2 imagery. *Sensors* **2018**, *18*, 18. [CrossRef]
61. Li, S.; Hu, W.; Yang, Y.; Liu, F.; Gan, W. Research of FOG-based measurement technique for continuous curve modes of long span bridge. *Bridge Constr.* **2014**, *44*, 69–74.

Article

Combining SDAE Network with Improved DTW Algorithm for Similarity Measure of Ultra-Weak FBG Vibration Responses in Underground Structures

Sheng Li [1,*], Xiang Zuo [2], Zhengying Li [2], Honghai Wang [1] and Lizhi Sun [3]

1 National Engineering Laboratory for Fiber Optic Sensing Technology, Wuhan University of Technology, Wuhan 430070, China; wanghh@whut.edu.cn
2 School of Information Engineering, Wuhan University of Technology, Wuhan 430070, China; zuoxiang@whut.edu.cn (X.Z.); zhyli@whut.edu.cn (Z.L.)
3 Department of Civil and Environmental Engineering, University of California, Irvine, CA 92697-2175, USA; lsun@uci.edu
* Correspondence: lisheng@whut.edu.cn

Received: 23 March 2020; Accepted: 11 April 2020; Published: 12 April 2020

Abstract: Quantifying structural status and locating structural anomalies are critical to tracking and safeguarding the safety of long-distance underground structures. Given the dynamic and distributed monitoring capabilities of an ultra-weak fiber Bragg grating (FBG) array, this paper proposes a method combining the stacked denoising autoencoder (SDAE) network and the improved dynamic time wrapping (DTW) algorithm to quantify the similarity of vibration responses. To obtain the dimensionality reduction features that were conducive to distance measurement, the silhouette coefficient was adopted to evaluate the training efficacy of the SDAE network under different hyperparameter settings. To measure the distance based on the improved DTW algorithm, the one nearest neighbor (1-NN) classifier was utilized to search the best constraint bandwidth. Moreover, the study proposed that the performance of different distance metrics used to quantify similarity can be evaluated through the 1-NN classifier. Based on two one-dimensional time-series datasets from the University of California, Riverside (UCR) archives, the detailed implementation process for similarity measure was illustrated. In terms of feature extraction and distance measure of UCR datasets, the proposed integrated approach of similarity measure showed improved performance over other existing algorithms. Finally, the field-vibration responses of the track bed in the subway detected by the ultra-weak FBG array were collected to determine the similarity characteristics of structural vibration among different monitoring zones. The quantitative results indicated that the proposed method can effectively quantify and distinguish the vibration similarity related to the physical location of structures.

Keywords: similarity measure; subway tunnel; distributed vibration; feature extraction; autoencoder; ultra-weak FBG

---

1. Introduction

Over the past decades, with the rapid development of rail transit infrastructure in China, the operation safety and security of subway systems have attracted much attention. According to the recent research progress of distributed optical fiber-sensing technology [1–7], the requirement for time- and space-continuous monitoring for the geotechnical underground structures [8] has gradually become feasible. Comparisons between various commonly used sensors for underground structure monitoring were reported in [9,10], which revealed that the ultra-weak fiber optic Bragg grating (FBG) array [11] can be used for both static and dynamic measurements [12–14]. In the field of dynamic measurement, it was reported that the distributed vibration detected by the ultra-weak FBG array can be applied to track train and identify incursion [10,15]. Moreover, the change of the structural vibration responses usually reflects the evolution of the structure state to a certain extent. A wide range of research reports concerning the vibration-based structural condition assessment can be found in [16–19]. Compared with ground transportation, the daily operation of underground trains is of obvious regularity. For example, the speed of trains in each travel zone always follows the operation schedule, and the number of passengers does not change suddenly within a certain period due to commuting habits. Moreover, the temperature and humidity fields of underground infrastructure are relatively stable due to the management measures of tunnel ventilation. Therefore, it can be assumed that the structural vibration responses corresponding to the excitation of multiple passing trains in a certain structural state should be stable and similar. With the support of distributed vibration monitoring adapted to the long-distance underground structures, it is possible to quantify the structural status by measuring the similarity of structural vibration responses for a specified monitoring area under different stages and this is the research motivation of the paper.

The vibration responses of subway tunnel structures can be regarded as a collection of typical one-dimensional time-series signals. The similarity measure between time series can often be converted to measure the distance between vectors. The Euclidean distance (ED) [20] and its variants based on common L_p-norm [21] are the most straightforward methods for similarity measures of such one-dimensional time-series. However, there is a slight difference in the length of duration in the vibration responses excited by each train passing through the monitoring area, making the ED and its variants unable to directly perform the similarity measure for unequal-length sequences. Even when dealing with equal-length vibration signals, these methods are susceptible to noise and time misalignment and are unable to deal with local time-shifting. Dynamic time warping (DTW) [22] is an option to overcome time-shifting, which allows a time series to be either stretched or compressed to provide a better match with another time series. Therefore, it can be used to handle similarity measures between inconsistent length sequences. Another group of similarity measures suitable for processing unequal-length time series is developed based on the concept of the edit distance for strings [23]. Compared with DTW which only considers the constrain bandwidth, the similarity measure based on the edit distance requires tuning more parameters [24–26] to find the most similar set of matching patterns. It is reported [15] that the data amount is huge for vibration responses detected by ultra-weak FBG of each monitoring area under the excitation of passing trains. This often results in high time complexity and is expensive in terms of processing and storage costs to directly use the above methods to perform a similarity measure on the raw format of high-dimensional vibration responses of underground structures. Furthermore, it is difficult to completely avoid random outlier interference during data collection and transmission. Therefore, the results of the similarity measure based on any algorithm may significantly deviate from expectations if the raw signals are not carefully wrangled.

Feature extraction should be the most intuitive idea to solve the above problems. It can improve the effectiveness and efficiency of the similarity measure by maintaining the characteristics of the original signal in a smaller dimensionality. Compared with principal component analysis (PCA) [27], linear discriminant analysis (LDA) [28] and other linear feature extraction methods, manifold learning [29], restricted Boltzmann machine (RBM) [30], autoencoder (AE) [31], as typical representatives of non-linear feature extraction methods, can retain much richer sample features of high-dimensional vibration signals. High computational complexity is the bottleneck of manifold learning based on local domain classification and its feature extraction process is sensitive to noise [32]. Therefore, this method is not suitable for extracting the characteristics of the vibration responses of underground structures that cannot avoid noise interference. RBM and its derivative deep belief network [33] use the probability distribution rather than the real-valued sequence to express the characteristics of the hidden layer. These two methods for dimensionality reduction are not suitable for the similarity measure of real-valued sequences. The training of AE resembles that of the RBM. However, models of AE can be easier to train than that of RBM with contrastive divergence and are thus preferred in contexts where RBM training is less effective [34]. Adding a denoising process makes

AE models substantially more robust to input variations or distortion, causing the deep network formed by a stacked denoising autoencoder (SDAE) with higher accuracy than that of the stacked autoencoder (SAE) [35,36]. Thus, the SDAE network is used to achieve feature extraction before the similarity measure in this paper. In the following second section, the implementation process of the proposed similarity measure is introduced in combination with typical one-dimensional datasets in the public UCR (University of California, Riverside) time-series data archives [37]. This part also illustrates the metrics used to evaluate the effectiveness of feature extraction and similarity measure. After that, based on the ultra-weak FBG vibration response of the actual underground structure, the feasibility and significance of the proposed similarity measure method in engineering are discussed.

2. Methodology and Implementation of Signal Similarity Measure

2.1. Overview of the Procedure for Signal Similarity Measure

As shown in Figure 1, the proposed method and performance test constituted the processing flow for the similarity measurement of one-dimensional time-series signals. The method in the left part of Figure 1 contained dimensionality reduction of the original sequence through feature extraction based on the SDAE network and distance measurement for the extracted feature sequences of equal length through the improved DTW algorithm. The silhouette coefficients and one nearest neighbor (1-NN) classifier in the right part of Figure 1 were used to evaluate the performances of feature extraction and distance measure, respectively.

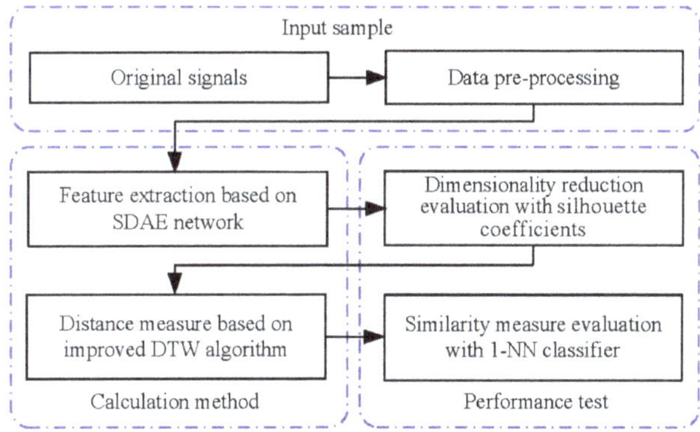

Figure 1. Similarity measurement and evaluation process.

For the general time series involving the similarity measure, the need for dataset partitioning and the purpose of each divided dataset are shown in Figure 2, which primarily included two parts, namely, the unsupervised learning through the SDAE network and the supervised learning through the 1-NN classifier. Since the cost of collecting and processing the distributed high-dimensional vibration responses is often expensive or even prohibitive, two datasets with data labels (CinCECGTorso and SemgHandMovementCh2 [38]) were selected from the UCR time series archives to help explain the implementation process. The two selected datasets have moderate sample sizes and relatively long sequence lengths, ensuring the operation feasibility of dimensionality reduction based on the SDAE network under acceptable computing overhead. Moreover, both the selected datasets and the vibration of interest face some negatives in common, such as redundant information and outliers, which should be overcome when measuring the similarity of sequences, although their appearance and type are varied. The default training set and test set ratio of each dataset in UCR databases are different. For each of the selected raw datasets used for the subsequent research, we first merged the training and test sets, then shuffled the samples, and finally set a uniform split ratio to form datasets A and B. Table 1 shows the final processing results, in which the ratio of datasets A to B was three. Note that other partitioning ratios were also acceptable as long as the dataset to be partitioned had a sufficient sample size to ensure corresponding algorithm training. The Python packages Tensorflow and Keras, as well as their libraries [39], were utilized to establish the SDAE network and calculate different distance metrics, in which the operation of the 1-NN classifier that can choose different distance measurement methods referred to in the work by Regan [40].

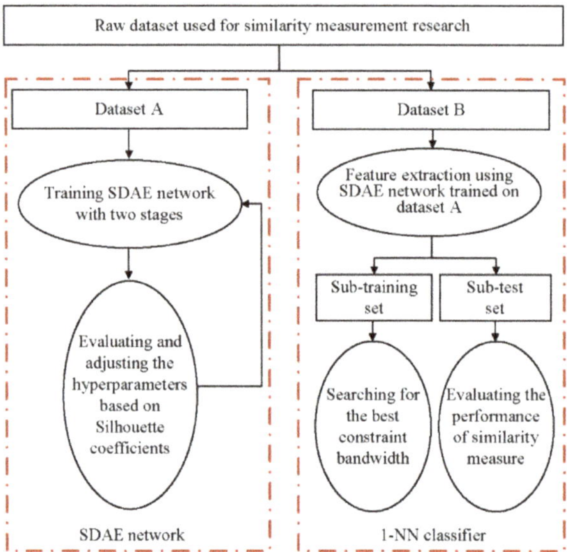

Figure 2. Functions of the dataset used for similarity measurement during implementation.

Table 1. Experimental datasets for validation method.

Name of Datasets	Sequence Length	Size of the Dataset A	Size of the Dataset B	Number of Labels
CinCECGTorso	1639	1050	350	4
SemgHandMovementCh2	1500	675	225	6

2.2. Feature Extraction Based on Stacked Denoising Autoencoder (SDAE) Network

As shown in the left part of Figure 2, feature extraction was primarily based on unsupervised training to shorten the sequence length in the second column of Table 1. Here, labels were just a supplement to fine-tune in the second training stage of the SDAE network, which was generally stacked by multiple three-layer DAE models. Figure 3 shows the structure of a typical SDAE network, which was formed by stacking three sub-DAE networks. Because the noise was actively added to the input data, hidden layers in such networks can retain more robust sample features during the learning process [41]. Here, greedy layer-wise training [42] that can boost the network learning efficiency was a preferred solution to conduct the pre-training process. In the first stage of feature extraction, the initial features of the input sample can be forcibly extracted through the unsupervised learning network. To obtain a better feature extraction effect, labels of the input sample were used to establish a classification output layer to perform a supervised training. Thereafter, a feature extraction model based on the SDAE network can be obtained through training the dataset A in Table 1. When a new sample dataset B was fed into the trained model, the feature representation of the last hidden layer can be regarded as reduced-dimensional features of the original input.

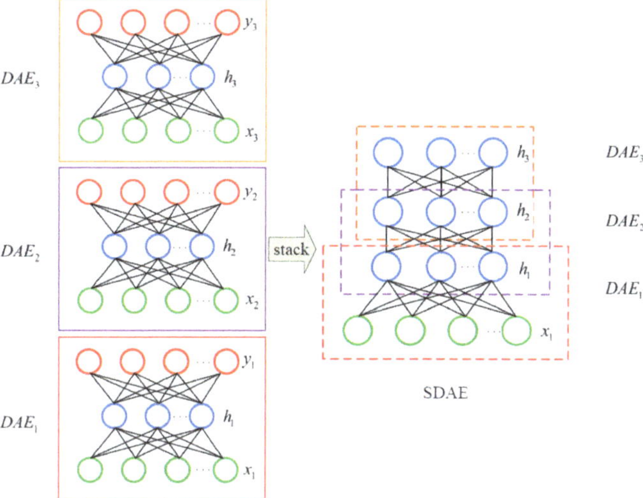

Figure 3. Schematic diagram of the stacked denoising autoencoder (SDAE) network stacking process.

2.3. Dimensionality Reduction Evaluation with Silhouette Coefficients

Through the above processing, a sequence having a length shorter than that of the original sequence listed in the second column in Table 1 can be obtained. It was straightforward that the effect and rationality of dimensionality reduction needed to be evaluated, indicating that the hyperparameter settings of the SDAE network should be assessed. Ideally, the feature vector generated due to dimensionality reduction should be able to represent the category information of the original sample to the greatest extent, namely, good feature extraction results should make the dimensionality reduction sequences belong to the same category closer, and the distance between the dimensionality reduction sequences belong to different categories farther. Silhouette coefficients [43] described as Equation (1) provide a single value measuring both the above two traits.

$$s_i = \frac{b_i - a_i}{\max(a_i, b_i)} \quad (1)$$

where s_i is the silhouette coefficient for observation i, a_i is the mean distance between i and all observations of the same class, and b_i is the mean distance between i and all observations from the different classes. Silhouette coefficients range between −1 and 1, with 1 indicating dense, well-separated different categories. Therefore, the mean silhouette coefficient for all observations can be used to evaluate the impact of the selection of various key hyperparameters on the performance of feature extraction based on the SDAE network for the two UCR datasets in Table 1. The operation based on grid search [44] combined with cross-validation [45] can guarantee to find the most accurate set of hyperparameter settings within a specified range, but it required iterating through all possible parameter combinations, which was very time-consuming in the face of large datasets and multiple parameters of interest. Another feasible option was to optimize the hyperparameter set step by step. Considering the characteristics of the SDAE network, the key hyperparameters that are usually concerned are the number of network layers, the number of hidden layer nodes, and the noise level [41]. As shown in Figure 4, the number of hidden layers of the SDAE network can be firstly determined by the mean silhouette coefficient. Here, in the training process, the adaptive moment estimation optimizer [46] was used to search the right learning rate automatically, and the maximum number of training epochs can be controlled based on the early stopping [47] technique. Other hyperparameters under the different number of hidden layers were derived through trial-and-error under the control of maximum number of training epochs and minimum reconstruction errors. As shown in Figure 4, when the number of hidden layers of the datasets CinCECGTorso and SemgHandMovementCh2 were set to two and three, respectively, the node number of the last hidden layer that reflected the dimensionality reduction effect can be further analyzed. Here, the hyperparameter configurations determined through trial and error in the previous step were used as the initial settings for the next tuning step and the key hyperparameter determined in the previous step remained constant in the subsequent tuning step. As shown in Figure 5, the number of nodes in the last hidden layer was expressed as a percentage of the original sequence length. After the number of hidden layers and the number of nodes in the last hidden layer were determined in turn, the reasonable value of the denoising coefficient [48] in the input layer of the SDAE network can be discussed. Figure 6 gave the relationship between different denoising coefficient and corresponding silhouette coefficient based on the tuning strategy of hyperparameters mentioned above. Hence, based on the hyperparameters determined by the maximum mean silhouette coefficients, the network structures of the SDAE for the two selected datasets used to obtain the dimensionality reduction sequences can be established.

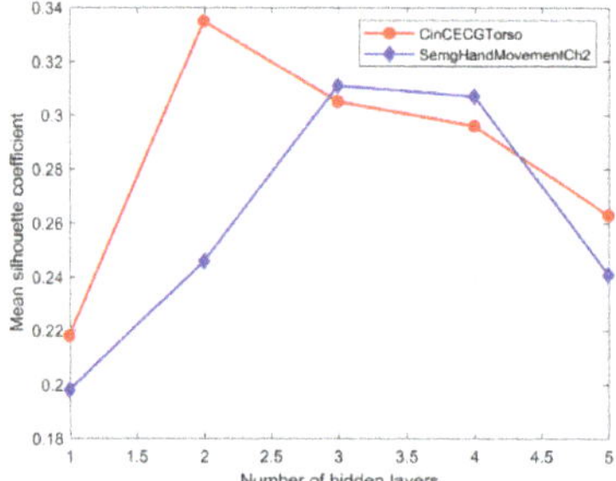

Figure 4. Mean silhouette coefficients at different number of hidden layers.

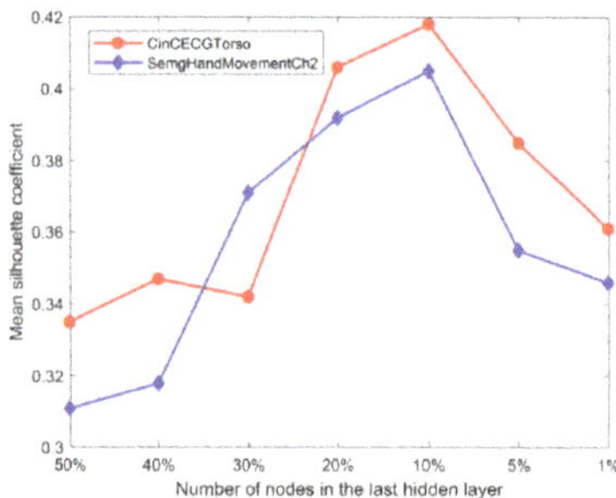

Figure 5. Mean silhouette coefficients at different reduction dimensionality.

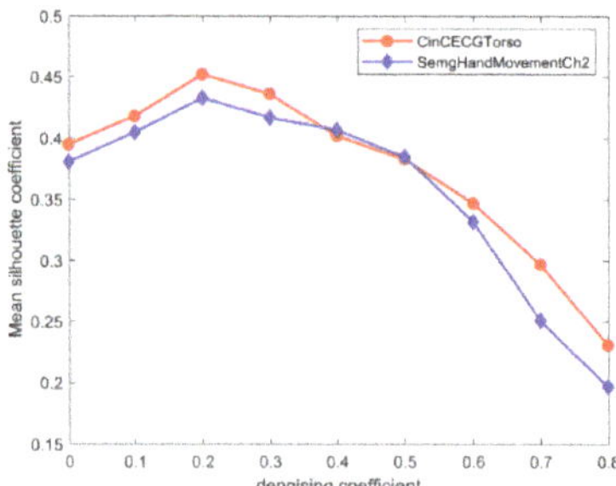

Figure 6. Mean silhouette coefficients at different denoising coefficients.

After the above step-by-step tuning of hyperparameters, the optimal mean silhouette coefficients of the datasets CinCECGTorso and SemgHandMovementCh2 with the feature extraction dimensions reduced to 20% of the original sequence length were 0.455 and 0.432, respectively. Figure 7 further shows that the capability of SDAE-based feature extraction was significantly better than that of other methods, although all the silhouette coefficient results did not exceed 0.5. Here, various comparison methods maintained a unified feature dimension reduction scale. Therefore, the SDAE network training process used for feature extraction for dataset B listed in Table 1 in this section was the premise for the subsequent similarity measure of reducing dimensionality sequences.

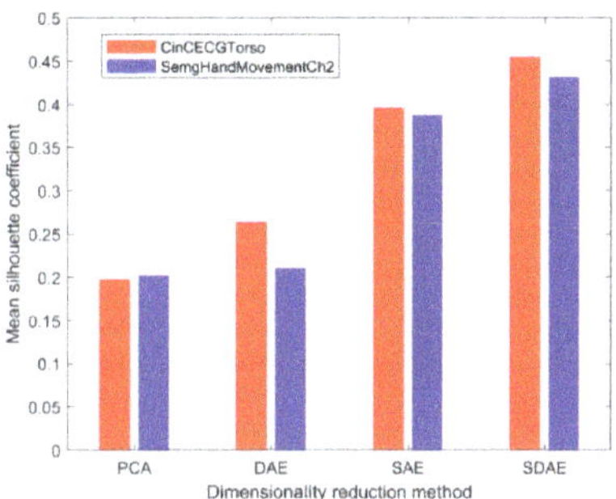

Figure 7. Comparison of different dimensionality reduction methods.

2.4. Distance Measure Based on Improved Dynamic Time-Warping (DTW) Algorithm

Since the features were extracted as dimensionality reduction sequences of equal length, the impact of high time complexity and low calculation efficiency can be effectively avoided when measuring distance based on the DTW algorithm. Although the reported window-based constraint methods have some positive effects on avoiding the DTW's matching path from falling into the suboptimum under certain circumstances, improvements against the influences of undesired warping [49] still deserve attention. Based on the DTW with a constraint of Sakoe–Chubaband [50] (hereinafter abbreviated as SDTW), warping offset distance (d_{WOD}) was defined in the proposed improved DTW algorithm to further mitigate the effects of undesired warping. The defined d_{WOD} was the area between the optimal matching path and the diagonal path under the SDTW algorithm. As shown in Figure 8, these two paths were derived from the distance matrix D of two equal-length sequences after feature extraction, and the d_{WOD} described in Equation (2) can be shown as the cumulative sum of the differences between each point on the optimal matching path and each corresponding point to the unbiased state. By aligning the feature points of two sequences processed by the SDAE network, this method not only ensured that the matching path can recognize the slight warping of the time axis but also realized the constraint on the length of the matching path. Detailed definition of the distance matrix of DTW and the searching method of the optimal matching sequence based on dynamic programming can be found in [51]:

$$d_{\text{WOD}} = \sum_{i=1}^{m} \left| w_i - dia(i) \right| \tag{2}$$

where w_i and $dia(i)$ represent the i-th point in the optimal matching path and the i-th point in the diagonal of the distance matrix D, respectively. The sum of d_{WOD} and the distance based on the SDTW (d_{SDTW}) was used as the distance metric of the improved DTW algorithm in Equation (3) and therefore $d_{\text{similarity}}$ was regarded as the result of the similarity measure:

$$d_{\text{similarity}} = d_{\text{SDTW}} + d_{\text{WOD}} \tag{3}$$

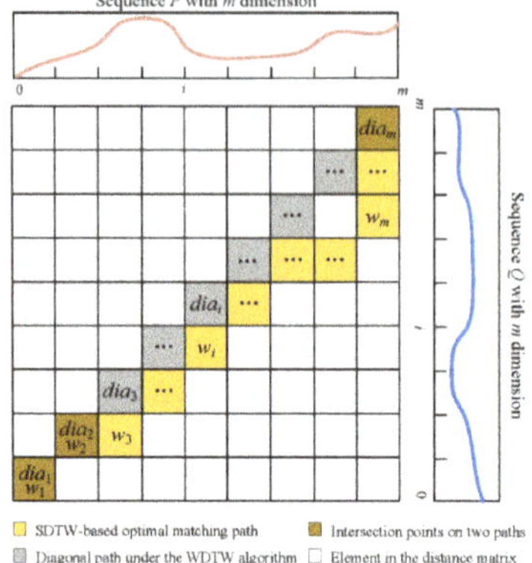

Figure 8. Warping offset distance expressed by the diagram of the DTW distance matrix.

2.5. Similarity Measure Evaluation with One Nearest Neighbor (1-NN) Classifier

The bandwidth r defines the constraint range of the matching path in the distance matrix and suppresses the influence of undesired convergence in the matching path [52]. Because there was a correlation between the defined warping offset distance and the SDTW algorithm, as well as the SDTW-based distance and the constraint bandwidth r, different r not only affected the optimal matching path of the SDTW but also led to the change of $d_{\text{similarity}}$. The r determined the efficacy of the proposed similarity measurement method. It was reported that the 1-NN classifier on labeled data was a feasible way to evaluate the efficacy of the selected distance metric and its classification accuracy directly reflected the effectiveness of the similarity measure [53]. Moreover, the 1-NN classifier can be used to search for a proper r and the idea was to train a labeled dataset with different bandwidth constraints based on two distance metrics d_{SDTW}, d_{WOD}, respectively. Then, two sets of classification error rates $E_{\text{SDTW}}(r)$ and $E_{\text{WOD}}(r)$ at different r through the 1-NN classifier model can be derived. We defined E_{SUM} as the sum of $E_{\text{SDTW}}(r)$ and $E_{\text{WOD}}(r)$ and the constraint bandwidth r that minimized E_{SUM} was considered to be the appropriate choice for calculating $d_{\text{similarity}}$.

Figure 9 depicted the possibly typical variation of E_{SUM} at different r. For cases I and IV, it was easy to determine the appropriate r based on the minimum E_{SUM}. For case II, it can be considered that the constraint bandwidth did not affect the distance measured by the SDTW algorithm, and the first r corresponding to the minimum can be seen as the candidate. For the situation in case III that multiple candidate values within the convergence region corresponded to the same minimum value E_{SUM}, the median of these candidate values was selected as r. Here, the general rules for determining and adjusting the preset range for r can refer to [52].

According to the data-processing procedure in the right part of Figure 2, the dataset B listed in Table 1 was further divided into sub-training and sub-test sets after the dimension reduction through the SDAE network. The dataset information used for the supervised learning of the 1-NN classifier was given in Table 2. Here, the sample size of the test set was made significantly larger than that of the training set according to the ratio commonly adopted in the dataset sheet of UCR archives [38]. First, the best r was searched based on the training results of the 1-NN classifier under two distance metrics. Figure 10 showed the variation of the classification error rate E_{SUM} of two datasets with respect to r after the dimensionality reduction in Table 2, and r for datasets of CinCECGTorso and SemgHandMovementCh2 should choose 2 and 3, respectively. Next, the defined $d_{\text{similarity}}$ under the specified r was used as the distance metric of the 1-NN classifier to perform supervised training on the sub-training set. Also, other distance metrics can be applied in the 1-NN classifier to train the sub-training set. Furthermore, the performance evaluation of the similarity measure can be transformed into a comparison of the classification error rate of the 1-NN classifier under different distance measures.

The generalization capacities of the 1-NN classifier with different distance metrics were compared in Figure 11 for the sub-test set by the classification error rate. The bar distribution reflected that the distance based on the improved DTW had lower classification error rates for the two sub-test datasets than that of the other distance measure functions, which also meant that the proposed distance metric was more suitable for similarity evaluation.

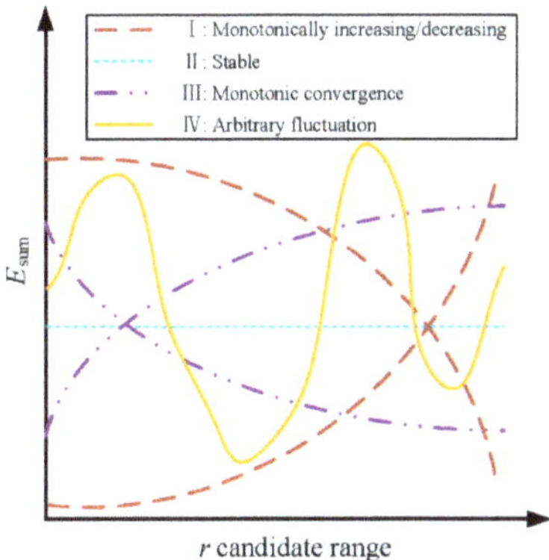

Figure 9. Typical variation of the sum of classification error rates at different constrains.

Table 2. Datasets for searching the constraint bandwidth and evaluating the distance metrics.

Name of Datasets	Sequence Length	Size of the Sub-Training Set	Size of the Sub-Test Set	Number of Labels
CinCECGTorso	164	50	300	4
SemgHandMovementCh2	150	25	200	6

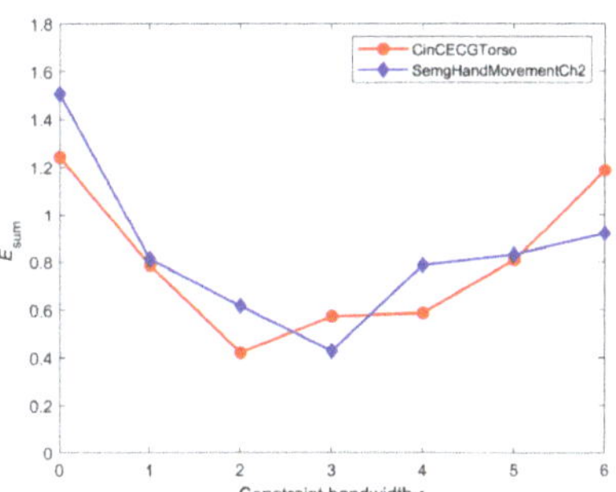

Figure 10. Variation of the defined error of two datasets with different constraint bandwidths.

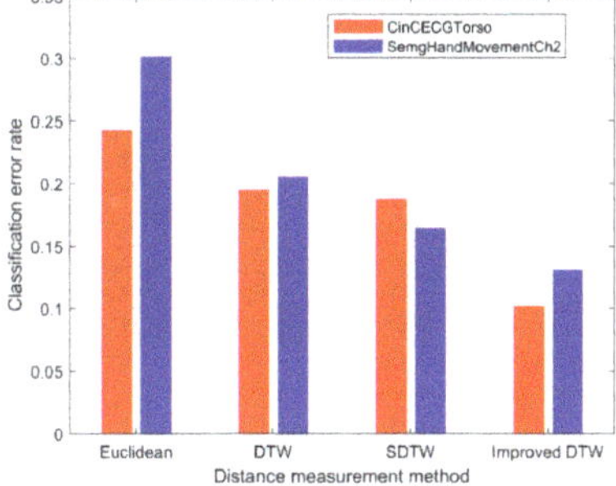

Figure 11. Performance comparison of 1-NN classifier under different distance metrics.

3. Similarity Measurement for On-Site Vibration Monitoring

3.1. Vibration Sequences Acquisition and Preparation

As shown in Figure 12, the regular moving loads caused by subway trains can be regarded as a vibration source. Owing to such excitation, the surface waves propagate omni-directionally on the ground. Because the surface wave couples to the track bed and subway rail, a distributed sensing optic fiber mounted beside the rail along the on-site monitoring area can detect the vibration and can be used to establish the vibration database for each monitoring zone. Figure 13 showed part of the actual engineering scenario in the subway tunnel. The monitoring area of interest covered three underground stations with a total length of nearly three kilometers. According to the spatial resolution of the sensing optic fiber and the on-spot layout of the tunnel structure, more than 500 vibration regions along the track bed can be distinguished based on the interrogated address of the light interference [54]. Here, the repeatability of the demodulator was revealed in [55] and the layout of the monitoring system can refer to [15]. When a train passed, the real-time vibration response triggered in each monitoring zone was fully transmitted back to the platform monitoring center at a sampling rate of 1 kHz and processed by the demodulator and servers. Therefore, the database of vibration response caused by passing train can be established for each monitoring zone and the location code of the monitoring area can be used as a unique label of each vibration sequence database.

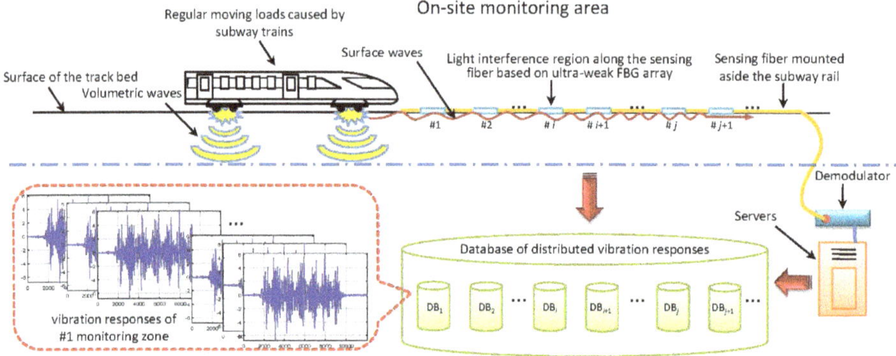

Figure 12. The processing sketch from ultra-weak fiber Bragg grating (FBG) array to distributed vibration.

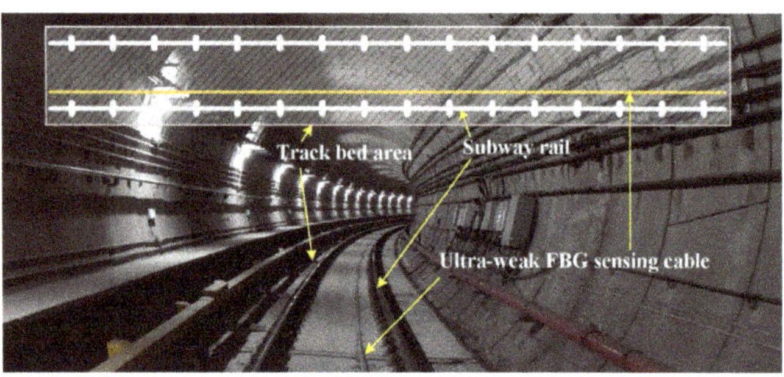

Figure 13. Field layout of ultra-weak FBG sensing cable used for detecting distributed vibration.

Figure 14 demonstrates the typical vibration responses of a track bed area due to a passing subway train. The triggered vibration responses of each monitoring area automatically recorded due to the passing of the train were basically within 12 s [15]. The characteristics of the vibration response were mainly composed of pulses with a duration of about 9 s caused by the axle weight. To meet the requirement that the node number of the input samples in the SDAE network must be consistent, the main vibration characteristics caused by the action of the train axle in each sample were retained. The sampling points at both ends of the vibration response were then truncated to match the minimum sequence length of the vibration response. Finally, min-max normalization [56] was used to normalize all the vibration amplitudes to the range of 0~1, which can boost a better learning efficiency for the SDAE network.

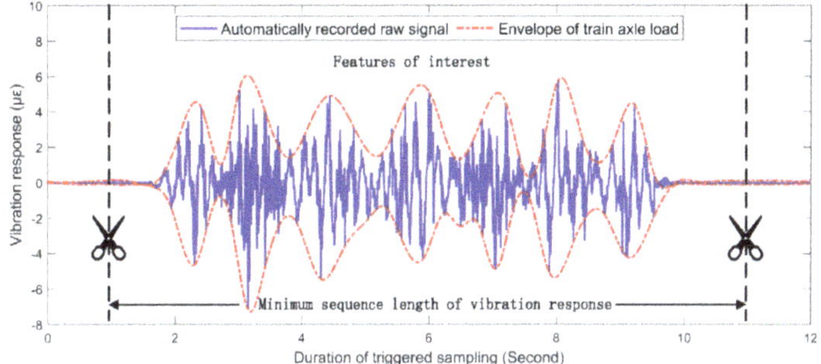

Figure 14. Typical vibration response of a monitoring zone caused by a passing train.

3.2. Result Analysis and Discussion

Considering the processing power of current experimental hardware that was composed of a graphics processing unit (GPU) core (GTX 1080 Ti) with twelve 2.20 GHz processors (Intel Xeon E5-2650 v4), we collected the distributed vibration responses caused by 100 passing trains within 2 h in the subway, aiming at three randomly selected monitoring zones labeled #130, #135 and #145 to perform similarity measurement through training the SDAE network and searching the optimal constraint bandwidth. The three selected zones belong to the common track bed in the same traveling area and the ultra-weak FBG sensors used to detect vibration were installed with the consistent craft [15]. First of all, all samples were truncated into the sequence with 10,000 dimensionalities and processed by the min-max normalization. Based on the step-by-step parameter tuning, the most appropriate SDAE network structure assessed by the silhouette coefficient is shown in Figure 15, which set the denoising coefficient to 0.2, contained 5 hidden layers and reduced the input length from 10,000 to 600. The constraint bandwidth was then set to 13 by the 1-NN classifier training. For each set of candidate hyperparameters, it took approximately 2–2.5 h to perform the task on feature extraction and bandwidth search of the 100 groups of datasets of the three monitoring zones.

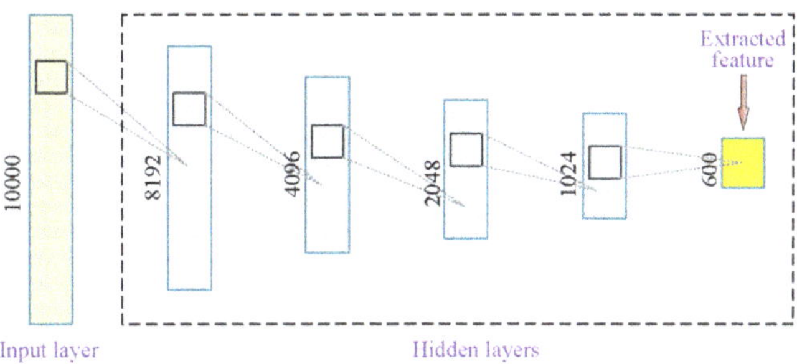

Figure 15. Hidden layers schematic of the SDAE network used for processing vibration responses.

After completing the two-step training of feature extraction of the SDAE network and distance measurement of the improved DTW algorithm, the established model can be applied to calculate the similarity of new samples. In another subway operation period, three groups of vibration responses of the #130, #135, and #145 monitoring zones caused by passing trains were collected and used to verify the proposed method of measurement similarity.

Figure 16 shows similarity measures between two vibration sequences related to the monitoring zone and the passing train. The left side of the dotted boundary in the bar graph displays the similarities between each pair of vibration responses of the same monitoring zone under different passing trains, while the right side of the boundary shows the similarities of different monitoring areas passed by the same train. Here, the subscripts A, B, and C of the monitoring area numbers in each bar column indicate the different passing trains, and two related monitoring area labels for measuring distance are connected by the symbol '&'. Obviously, the threshold of the two comparison types represented by the distance derived from the improved DTW algorithm can be identified, and it was about 800 με. The distance unit here depended on the vibration signal denoted by the strain-induced phase variation between two ultra-weak FBGs [10]. Because the results in the left part of Figure 16 are all below the threshold, it is quantitatively revealed that the similarity of the vibration response at the same physical location in the underground structure is significantly higher than the measurement results between different locations. Moreover, by using the mean distance based on the improved DTW algorithm, the similarities for monitoring zones #130, #135 and #145 can be determined as 700 με, 656 με and 756 με, respectively. The quantitative outcomes based on distance measure not only indicated that the similarity of the structural vibration changed with the location of the underground structure but also revealed that the proposed method can effectively quantify and distinguish such similarity difference. Hence, the condition change of the surrounding structure and environment can be tracked by similarities of structural vibration detected by distributed ultra-weak FBG array.

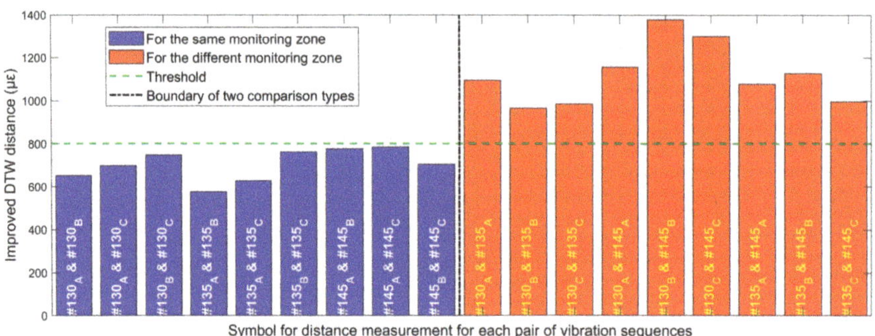

Figure 16. Similarity comparison based on the improved DTW distance.

4. Conclusions

This study proposed a similarity measure method to quantify the distributed vibration responses of underground structures, which involved feature extraction by the SDAE network and distance measurement by the improved DTW algorithm. Combining two datasets of one-dimensional time series from UCR archives, the detailed implementation processes for the similarity measure were introduced, and the advantages of feature extraction and distance measure in the proposed method were revealed according to algorithm comparisons. Considering the current processing capabilities of the experimental hardware, the size of the field dataset used to train the SDAE network was limited, but the subsequent experimental outcomes on distance measure still agreed well with the expected cognition. The prediction results of similarity based on the modeling of 100 groups of vibration sequences in three monitoring zones on the subway site demonstrated that the vibration similarity of the same monitoring zone was significantly higher than that from different ones. Moreover, the similarities of distributed vibration closely related to the physical location of the underground structure can be distinguished effectively by the improved DTW distance, demonstrating that the proposed method assisted with the distributed vibration detected by the ultra-weak FBG array is promising for quantifying structural status and locating structural anomalies.

Author Contributions: Conceptualization, S.L.; Data curation, X.Z.; Funding acquisition, S.L.; Investigation, S.L.; Methodology, X.Z.; Project administration, H.W.; Resources, Z.L.; Software, H.W.; Supervision, L.S.; Validation, Z.L.; Writing—original draft, X.Z.; Writing—review and editing, S.L. and L.S. All authors have read and agreed to the published version of the manuscript.

Funding: This research was funded by the National Natural Science Foundation of China grant numbers 61875155 and 61735013.

Acknowledgments: The research work reported in this paper was supported by the National Engineering Laboratory for Fiber Optic Sensing Technology, Wuhan University of Technology, and the Smart Nanocomposites Laboratory, University of California, Irvine.

Conflicts of Interest: The authors declare no conflict of interest.

References

1. Hong, C.; Zhang, Y.; Li, G.; Zhang, M.; Liu, Z. Recent progress of using Brillouin distributed fiber optic sensors for geotechnical health monitoring. *Sens. Actuators A Phys.* **2017**, *258*, 131–145. [CrossRef]
2. Lu, P.; Lalam, N.; Badar, M.; Liu, B.; Chorpening, B.T.; Buric, M.P.; Ohodnicki, P.R. Distributed optical fiber sensing: Review and perspective. *Appl. Phys. Rev.* **2019**, *6*, 041302. [CrossRef]
3. Thévenaz, L. Review and progress in distributed fiber sensing. In Proceedings of the Optical Fiber Sensors, OFS 2006, Cancun, Mexico, 23 October 2006.
4. Yang, M.; Li, C.; Mei, Z.; Tang, J.; Guo, H.; Jiang, D. Thousand of fiber grating sensor array based on draw tower: A new platform for fiber-optic sensing. In Proceedings of the Optical Fiber Sensors, OFS 2018, Lausanne, Switzerland, 24–28 September 2018.
5. Muanenda, Y. Recent advances in distributed acoustic sensing based on phase-sensitive optical time domain reflectometry. *J. Sens.* **2018**, *2018*, 3897873. [CrossRef]
6. Liu, X.; Jin, B.; Bai, Q.; Wang, Y.; Wang, D.; Wang, Y. Distributed fiber-optic sensors for vibration detection. *Sensors* **2016**, *16*, 1164. [CrossRef] [PubMed]
7. Lai, J.; Qiu, J.; Fan, H.; Zhang, Q.; Hu, Z.; Wang, J.; Chen, J. Fiber Bragg grating sensors-based in situ monitoring and safety assessment of loess tunnel. *J. Sens.* **2016**, *2016*, 8658290. [CrossRef]
8. Pamukcu, S.; Cheng, L.; Pervizpour, M. Chapter 1—Introduction and overview of underground sensing for sustainable response. In *Underground Sensing. Monitoring and Hazard Detection for Environment and Infrastructure*, 1st ed.; Pamukcu, S., Cheng, L., Eds.; Academic Press: Cambridge, MA, USA, 2018; pp. 1–42.
9. Gong, H.; Kizil, M.S.; Chen, Z.; Amanzadeh, M.; Yang, B.; Aminossadati, S.M. Advances in fibre optic based geotechnical monitoring systems for underground excavations. *Int. J. Min. Sci. Technol.* **2017**, *29*, 229–238. [CrossRef]
10. Gan, W.; Li, S.; Li, Z.; Sun, L. Identification of ground intrusion in underground structures based on distributed structural vibration detected by ultra-weak FBG sensing technology. *Sensors* **2019**, *19*, 2160. [CrossRef]
11. Yang, M.; Bai, W.; Guo, H.; Wen, H.; Yu, H.; Jiang, D. Huge capacity fiber-optic sensing network based on ultra-weak draw tower gratings. *Photonic Sens.* **2016**, *6*, 26–41. [CrossRef]
12. Zhou, L.; Li, Z.; Xiang, N.; Bao, X. High-speed demodulation of weak fiber Bragg gratings based on microwave photonics and chromatic dispersion. *Opt. Lett.* **2018**, *43*, 2430–2433. [CrossRef]
13. Li, Z.; Tong, Y.; Fu, X.; Wang, J.; Guo, Q.; Yu, H.; Bao, X. Simultaneous distributed static and dynamic sensing based on ultra-short fiber Bragg gratings. *Opt. Express* **2018**, *26*, 17437–17446. [CrossRef]
14. Bai, W.; Yang, M.; Hu, C.; Dai, J.; Zhong, X.; Huang, S.; Wang, G. Ultra-weak fiber Bragg grating sensing network coated with sensitive material for multi-parameter measurements. *Sensors* **2017**, *17*, 1509. [CrossRef] [PubMed]
15. Nan, Q.; Li, S.; Yao, Y.; Li, Z.; Wang, H.; Wang, L.; Sun, L. A novel monitoring approach for train tracking and incursion detection in underground structures based on ultra-weak FBG sensing array. *Sensors* **2019**, *19*, 2666. [CrossRef] [PubMed]

16. Moughty, J.J.; Casas, J.R. A state of the art review of modal-based damage detection in bridges: Development, challenges, and solutions. *Appl. Sci.* **2017**, *7*, 510. [CrossRef]
17. Amezquita-Sanchez, J.P.; Adeli, H. Signal processing techniques for vibration-based health monitoring of smart structures. *Arch. Comput. Method Eng.* **2016**, *23*, 1–15. [CrossRef]
18. Abdeljaber, O.; Avci, O.; Kiranyaz, M.S.; Gabbouj, M.; Inman, D.J. Real-time vibration-based structural damage detection using one-dimensional convolutional neural networks. *J. Sound Vib.* **2017**, *388*, 154–170. [CrossRef]
19. Wang, T.; Han, Q.; Chu, F.; Feng, Z. Vibration based condition monitoring and fault diagnosis of wind turbine planetary gearbox: A review. *Mech. Syst. Signal Process.* **2019**, *126*, 662–685. [CrossRef]
20. Faloutsos, C.; Ranganathan, M.; Manolopoulos, Y. Fast subsequence matching in time-series databases. In Proceedings of the 1994 ACM SIGMOD International Conference on Management of Data, Minneapolis, MN, USA, 24–27 May 1994.
21. Yi, B.K.; Faloutsos, C. Fast time sequence indexing for arbitrary L_p norms. In Proceedings of the 26th International Conference on Very Large Data Bases, VLDB 2000, Cairo, Egypt, 10–14 September 2000.
22. Liu, L.; Li, W.; Jia, H. Method of time series similarity measurement based on dynamic time warping. *Comput. Mater. Contin.* **2018**, *57*, 97–106. [CrossRef]
23. Riesen, K.; Schmidt, R. Online signature verification based on string edit distance. *Int. J. Doc. Anal. Recognit.* **2019**, *22*, 41–54. [CrossRef]
24. Gold, O.; Sharir, M. Dynamic time warping and geometric edit distance: Breaking the quadratic barrier. *ACM Trans. Algorithms* **2018**, *14*, 50. [CrossRef]
25. Yu, M.; Wang, J.; Li, G.; Zhang, Y.; Deng, D.; Feng, J. A unified framework for string similarity search with edit-distance constraint. *VLDB J.* **2017**, *26*, 249–274. [CrossRef]
26. Chen, X.; Huo, H.; Huan, J.; Vitter, J.S. An efficient algorithm for graph edit distance computation. *Knowl. Based Syst.* **2019**, *163*, 762–775. [CrossRef]
27. Shlens, J. A Tutorial on Principal Component Analysis. Available online: https://arxiv.org/abs/1404.1100 (accessed on 17 March 2020).
28. Izenman, A.J. Chapter 8–Linear Discriminant Analysis. In *Modern Multivariate Statistical Techniques*, 2nd ed.; Springer: New York, NY, USA, 2013; pp. 237–280.
29. Yan, J.; Sun, H.; Chen, H.; Junejo, N.U.R.; Cheng, E. Resonance-based time-frequency manifold for feature extraction of ship-radiated noise. *Sensors* **2018**, *18*, 936. [CrossRef] [PubMed]
30. Liu, J.; Chen, F.; Wang, D. Data compression based on stacked RBM-AE model for wireless sensor networks. *Sensors* **2018**, *18*, 4273. [CrossRef] [PubMed]
31. Pathirage, C.S.N.; Li, J.; Li, L.; Hao, H.; Liu, W.; Ni, P. Structural damage identification based on autoencoder neural networks and deep learning. *Eng. Struct.* **2018**, *172*, 13–28. [CrossRef]
32. Seyyedsalehi, S.Z.; Seyyedsalehi, S.A. Simultaneous learning of nonlinear manifolds based on the bottleneck neural network. *Neural Process. Lett.* **2014**, *40*, 191–209. [CrossRef]
33. Yang, H.; Shen, S.; Yao, X.; Sheng, M.; Wang, C. Competitive deep-belief networks for underwater acoustic target recognition. *Sensors* **2018**, *18*, 952. [CrossRef]
34. Hearty, J. *Advanced Machine Learning with Python*; Packt Publishing: Birmingham, UK, 2016; pp. 57–61.
35. Fan, Z.; Bi, D.; He, L.; Shiping, M.; Gao, S.; Li, C. Low-level structure feature extraction for image processing via stacked sparse denoising autoencoder. *Neurocomputing* **2017**, *243*, 12–20. [CrossRef]
36. Xia, M.; Li, T.; Liu, L.; Xu, L.; de Silva, C.W. Intelligent fault diagnosis approach with unsupervised feature learning by stacked denoising autoencoder. *IET Sci. Meas. Technol.* **2017**, *11*, 687–695. [CrossRef]
37. Dau, H.A.; Bagnall, A.; Kamgar, K.; Yeh, C.M.; Zhu, Y.; Gharghabi, S.; Ratanamahatana, C.A.; Keogh, E. The UCR time series archive. *IEEE/CAA J. Autom. Sin.* **2019**, *6*, 1293–1305. [CrossRef]
38. Chen, Y.; Keogh, E.; Hu, B.; Begum, N.; Bagnall, A.; Mueen, A.; Batista, G. The UCR Time Series Classification Archive. Available online: https://www.cs.ucr.edu/~{}eamonn/time_series_data/ (accessed on 10 February 2020).
39. Keras. Available online: http://github.com/keras-team/keras (accessed on 4 April 2020).
40. Regan, M. K Nearest Neighbors with Dynamic Time Warping. Available online: https://github.com/markdregan/K-Nearest-Neighbors-with-Dynamic-Time-Warping (accessed on 4 April 2020).

41. Vincent, P.; Larochelle, H.; Lajoie, I.; Bengio, Y.; Manzagol, P. Stacked denoising autoencoders: Learning useful representations in a deep network with a local denoising criterion. *J. Mach. Learn. Res.* **2010**, *11*, 3371–3408.
42. Bengio, Y.; Lamblin, P.; Popovici, D.; Larochelle, H. Greedy layer-wise training of deep networks. In Proceedings of the 20th Annual Conference on Neural Information Processing Systems, NIPS 2006, Vancouver, BC, Canada, 4–7 December 2006.
43. Aranganayagi, S.; Thangavel, K. Clustering categorical data using silhouette coefficient as a relocating measure. In Proceedings of the International Conference on Computational Intelligence and Multimedia Applications, ICCIMA 2007, Sivakasi, Tamil Nadu, India, 13–15 December 2007.
44. Hackeling, G. *Mastering Machine Learning with Scikit-Learn*, 1st ed.; Packt Publishing: Birmingham, UK, 2014; pp. 84–86.
45. Browne, M.W. Cross-validation methods. *J. Math. Psychol.* **2000**, *44*, 108–132. [CrossRef] [PubMed]
46. Kingma, D.P.; Ba, J. Adam: A Method for Stochastic Optimization. Available online: https://arxiv.org/abs/1412.6980 (accessed on 18 March 2020).
47. Caruana, R.; Lawrence, S.; Giles, L. Overfitting in neural nets: Backpropagation, conjugate gradient, and early stopping. In Proceedings of the 14th Annual Neural Information Processing Systems Conference (NIPS 2000), Denver, CO, USA, 27 November–2 December 2000.
48. Poole, B.; Sohl-Dickstein, J.; Ganguli, S. Analyzing Noise in Autoencoder and Deep Networks. Available online: https://arxiv.org/abs/1406.1831 (accessed on 18 March 2020).
49. Chen, Q.; Hu, G.; Gu, F.; Xiang, P. Learning optimal warping window size of DTW for time series classification. In Proceedings of the 2012 11th International Conference on Information Science, Signal Processing and their Application, ISSPA 2012, Montreal, QC, Canada, 2–5 July 2012.
50. Sakoe, H.; Chiba, S. Dynamic programming algorithm optimization for spoken word recognition. *IEEE Trans. Acoust. Speech Signal Process.* **1978**, *26*, 43–49. [CrossRef]
51. Senin, P. Dynamic Time Warping Algorithm Review. Available online: http://seninp.github.io/assets/pubs/senin_dtw_litreview_2008.pdf (accessed on 18 March 2020).
52. Dau, H.A.; Silva, D.F.; Petitjean, F.; Forestier, G.; Bagnall, A.; Mueen, A.; Keogh, E. Optimizing dynamic time warping's window width for time series data mining applications. *Data Min. Knowl. Discov.* **2018**, *32*, 1074–1120. [CrossRef]
53. Ding, H.; Trajcevski, G.; Scheuermann, P.; Wang, X.; Keogh, E. Querying and mining of time series data: Experimental comparison of representations and distance measures. *Proc. VLDB Endow.* **2008**, *1*, 1542–1552. [CrossRef]
54. Luo, Z.; Wen, H.; Guo, H.; Yang, M. A time- and wavelength-division multiplexing sensor network with ultra-weak fiber Bragg gratings. *Opt. Express* **2013**, *21*, 22799–22807. [CrossRef]
55. Tong, Y.; Li, Z.; Wang, J.; Wang, H.; Yu, H. High-speed Mach-Zehnder-OTDR distributed optical fiber vibration sensor using medium-coherence laser. *Photonic Sens.* **2018**, *8*, 203–212. [CrossRef]
56. Patro, S.G.K.; Sahu, K.K. Normalization: A Preprocessing Stage. Available online: https://arxiv.org/abs/1503.06462 (accessed on 18 March 2020).

Article

Distributed Optical Fiber-Based Approach for Soil–Structure Interaction

Nissrine Boujia [1], Franziska Schmidt [1,*], Christophe Chevalier [1] and Dominique Siegert [1] and Damien Pham Van Bang [2]

[1] Université Paris Est, Ifsttar, 77447 Champs sur Marne, France; nissrine.boujia@ifsttar.fr (N.B.); christophe.chevalier@ifsttar.fr (C.C.); dominique.siegert@ifsttar.fr (D.S.)

[2] INRS, Centre Eau Terre Environnement, Québec, QC G1K 9A9, Canada; damien.pham_van_bang@ete.inrs.ca

* Correspondence: franziska.schmidt@ifsttar.fr

Received: 16 December 2019; Accepted: 3 January 2020; Published: 6 January 2020

Abstract: Scour is a hydraulic risk threatening the stability of bridges in fluvial and coastal areas. Therefore, developing permanent and real-time monitoring techniques is crucial. Recent advances in strain measurements using fiber optic sensors allow new opportunities for scour monitoring. In this study, the innovative optical frequency domain reflectometry (OFDR) was used to evaluate the effect of scour by performing distributed strain measurements along a rod under static lateral loads. An analytical analysis based on the Winkler model of the soil was carefully established and used to evaluate the accuracy of the fiber optic sensors and helped interpret the measurements results. Dynamic tests were also performed and results from static and dynamic tests were compared using an equivalent cantilever model.

Keywords: distributed measurements; fiber optic sensors; scour; soil-structure interaction; winkler model; equivalent length

---

1. Introduction

Bridge scour occurs when flowing water erodes sediments around bridge supports, more precisely piers and abutments. When scour depth reaches a critical value, the stability of the bridge is threatened which may lead to its collapse. Therefore, it is crucial to develop monitoring techniques capable of assessing scour depth in order to anticipate this hydraulic risk. Over the last twenty years, scour monitoring technologies have evolved significantly. Beginning with the use of traditional geophysical instruments such as radar [1] and sonar [2] to developing new sensors [3] and the use of more sophisticated tools such as fiber Bragg grating sensors [4], scour monitoring is an ongoing topic of research.

Over the past few years, the effect of scour on the static response of single piles has gained interest and has been reported by several studies: Lin et al. [5] studied the effect of scour on the response of laterally loaded piles considering the change of stress of the remaining sand. Qi et al. [6] investigated the effect of local and global scour on the p-y curves of piles in sand using various centrifuge model tests. These studies showed that scour induces changes in the response of laterally loaded piles. However, the use of those changes to determine scour depth is still limited. An example of those uses is provided in [7]: A scour sensor was equipped with fiber Bragg grating (FBG) sensors. The monitoring system consists of a cantilever beam equipped with FBG sensors having different wavelengths and placed at various heights. The sensor operates on the principle that as long as a FBG sensor is embedded in the soil, the registered value of strain at the sensor location is negligible. Once scour occurs, the FBG sensor emerges and is subjected to water flow. Consequently, the measured value of strain increases. Thus, knowing the height of the sensors having registered the variation of strain, scour depth can be determined. A major limitation of this scour sensor is that FBG sensors can only perform quasi-distributed monitoring and their number along the fiber is limited [8].

The dynamic monitoring of scour has also become increasingly important in recent years. Many authors suggest monitoring the modal parameters of the structure itself (i.e., spans or piers) [9–11]. Other authors suggest monitoring the vibration frequency of sensors rods embedded in the riverbed [4,12]. A numerical model, based on the Winkler theory of the soil, is then usually used to establish a relationship between the measured frequency and scour depth. One of the main challenges of the dynamic monitoring technique is the difficulty in evaluating the modulus of subgrade reaction K_s through a rational and methodical approach. The value of K_s depends not only on the Young modulus of the soil E_s, but also on various geometric and mechanical parameters of the structure itself.

In this study, the effect of scour on both the static and dynamic responses of a rod/sensor is studied. The optical frequency domain reflectometry (OFDR) technique is used to measure strain to overcome the limitations of FBG sensors. OFDR technology enables measuring strain along structures with millimeter level spacial resolution. A rectangular rod was instrumented along its length with a distributed fiber optic strain sensors (OFDR). The rod-sensor was then tested under static lateral loads for different scour depths. These tests were conducted under tension and compression to make sure that the results are independent of the testing configuration. An analytical model, based on the Winkler soil model was then developed to help in static results interpretations. The key parameter of the proposed model is the modulus of subgrade reaction K_s. In this study, its value was determined from Ménard tests and was further confirmed with the fiber optic measurements. Dynamic tests were also conducted in the same testing conditions. Finally, an equivalent cantilever model is proposed in order to compare the static and dynamic approaches used in this study.

The paper starts with a description of the experimental protocols for static and dynamic tests. A second part highlights the main results of this study and presents the analytical model for the static experiments. A third part introduces introduces a simplified cantilever based model which allows modeling the soil–structure interaction for static and dynamic experiments.

2. Theoretical Formulation

The Winkler approach [13–15] was used to model the static experimental tests. According to this approach, it is assumed that the beam is supported by a series of infinitely closed independent and elastic springs. The governing equation of a laterally loaded beam, partially embedded in the soil, is expressed by Equation (1):

$$\begin{cases} E_b I_b \dfrac{\mathrm{d}^4 w(z)}{\mathrm{d}^4 z} = 0 & \text{for } z \in [-a, 0], \\ E_b I_b \dfrac{\mathrm{d}^4 w(z)}{\mathrm{d}^4 z} + K_s w(z) = 0 & \text{for } z \in [0, D], \end{cases} \tag{1}$$

where E_b and I_b are the Young modulus and the cross section moment of inertia of the beam respectively, $w(z)$ the lateral deflection of the beam, K_s the modulus of subgrade reaction of the soil, a the eccentricity of the load F and D the embedded length of the beam (see Figure 1).

If K_s is constant along the depth, the general solution of this set of equations is given by:

$$\begin{cases} w_1(z) & = a_1 z^3 + a_2 z^2 + a_3 z + a_4 & \text{for } z \in [-a, 0], \\ w_2(z) & = e^{(-z/l_0)} \left(a_5 cos(z/l_0) + a_6 sin(z/l_0)\right) & \text{for } z \in [0, D] \\ & +e^{(z/l_0)} \left(a_7 cos(z/l_0) + a_8 sin(z/l_0)\right), & \end{cases} \tag{2}$$

where the characteristic length of the beam is $l_0 = \left(\frac{4E_bI_b}{K_s}\right)^{\frac{1}{4}}$. It is worth noting that l_0 combines mechanical properties of both the soil and the pile.

For the case of flexible or long piles, as the rod used in this study, positive exponential terms in Equation (2) are negligible. Consequently $a_7 = a_8 = 0$ and only six parameters remain to be determined. The following notations are used to refer to the bending moment $M_i = E_bI_bw_i''$, the shear force $T_i = E_bI_bw_i'''$ and the slope $\theta_i = w_i'$.

By assuming that the displacements are small and that the applied force is strictly perpendicaular to the beam, boundary conditions on moment and shear force for $z = -a$ can be written: $M(-a) = E_bI_bw_1''(-a) = 0$ and $T(-a) = E_bI_bw_1'''(-a) = F$ which gives simple relation of a_1 and a_2: $a_1 = \frac{F}{6E_bI_b}$ and $a_2 = \frac{-Fa}{2E_bI_b}$. Finally, adding four equations expressing the continuity of displacement w, slope θ, bending moment M and shear force T at the ground surface $z = 0$ gives enough constraints:

$$\begin{cases} w_1(0) = w_2(0) \\ \theta_1(0) = \theta_2(0) \\ M_1(0) = M_2(0) \\ T_1(0) = T_2(0). \end{cases} \tag{3}$$

These conditions can be written as $[M]\{X\} = \{A\}$ where:

$$A = \begin{pmatrix} 0 \\ 0 \\ \frac{Fa}{E_bI_b} \\ \frac{F}{E_bI_b} \end{pmatrix}, X = \begin{pmatrix} a_3 \\ a_4 \\ a_5 \\ a_6 \end{pmatrix}$$

$$M = \begin{pmatrix} 0 & -1 & 1 & 0 \\ -1 & 0 & -\frac{1}{l_0} & \frac{1}{l_0} \\ 0 & 0 & 0 & -\frac{2}{l_0{}^2} \\ 0 & 0 & \frac{2}{l_0{}^3} & \frac{2}{l_0{}^3} \end{pmatrix}.$$

The solution of the system is:

$$a_6 = -\frac{Fl_0{}^2}{2E_bI_b}a, a_3 = \frac{F(2a + l_0)}{2E_bI_b}l_0, a_4 = a_5 = \frac{Fl_0{}^2}{2E_bI_b}(a + l_0). \tag{4}$$

The value of the bending moment along the rod is therefore given by Equation (5):

$$\begin{cases} M_1(z) = E_bI_bw_1''(z) = E_bI_b(6a_1z + 2a_2) = F(z + a) & z \in [-a, 0], \\ M_2(z) = E_bI_bw_2''(z) = F\exp\left(\frac{-z}{l_0}\right)\left(l_0\sin(\frac{z}{l_0}) + a(\cos(\frac{z}{l_0}) + \sin(\frac{z}{l_0})\right) & z \in [0, D]. \end{cases} \tag{5}$$

The normal stress induced by the bending moment in the rod at the outermost fibers is given by the known formulation:

$$\sigma = \pm\frac{M}{I_b} \times \frac{h}{2}, \tag{6}$$

where $\frac{h}{2}$ is the distance from the neutral axis to the outermost fibers. The linear elastic strain along the rod is then deduced using the Hooke's law:

$$\epsilon = \pm \frac{M}{E_b I_b} \times \frac{h}{2}. \tag{7}$$

Moreover, the expression of the strain along the beam is given by the following equation:

$$\begin{aligned} \epsilon &= \frac{M}{E_b I_b} z, \\ &= \frac{E_b I_b (6a_1 z + 2a_2)}{E_b I_b} z = 6a_1 z + 2a_2, \text{ for } z \in [-a, 0] \\ &= \frac{F\exp\left(\frac{-z}{l_0}\right)\left(l_0 \sin(\frac{z}{l_0}) + a(\cos(\frac{z}{l_0}) + \sin(\frac{z}{l_0})\right)}{E_b I_b} z, \text{ for } z \in [0, D]. \end{aligned} \tag{8}$$

The positions of extreme strain values verify $M'(z) = 0$. Therefore, the position of the maximum strain in the embedded part of the rod z_{max} is given with Equation (9) and varies with the eccentricity of the load a and the characteristic length l_0:

$$z_{max} = l_0 \arctan\left(\frac{l_0}{l_0 + 2a}\right). \tag{9}$$

3. Fiber Optic Sensing Technology

The sensing technology used to measure strains along the rod is optical frequency domain reflectometry. OFDR enables measurement along a fiber up to 2 km long, with millimetre level spacial resolution [16]. The light emitted from a highly tunable laser source undergoes a coupler and is then divided between two branches: the reference branch and the fiber under test branch. Backscattered lights from both branches are then combined to create an interference signal. This signal is detected by an optical detector. The Rayleigh backscattering induced by the random fluctuations in the refractive index along the fiber length can be modelled as a Bragg grating with random period [17]. As long as the fiber is in a stable state, the Rayleigh backscattering spectrum remains constant. When the surrounding environment of the fiber changes due to external stimulus (as strain and temperature), a spectrum shift occurs. This spectrum shift is expressed using Equation (10):

$$\Delta\nu = C_\epsilon \times \epsilon + C_T \times \Delta T, \tag{10}$$

where $\Delta\nu$ is the Rayleigh spectral shift, ϵ the fiber strain, ΔT the temperature variation of the fiber, C_ϵ and C_T are calibration constants. The typical values of the latter parameters for a standard single-mode fiber, at 1550 nm, are respectively: -0.15 GHz/$\mu\epsilon$ and -1.25 GHz/C°. This strain/temperature dependent spectrum shift can be determined by means of cross correlation between reference scan (meaning the scan performed at ambient temperature and null strain state) and measurement scan (when a temperature perturbation or a strain is applied). Fundamental principles of Rayleigh systems are fully detailed in [18].

In the experimental testing, a fiber optic sensor was glued along the length of the rod (Figure 1). A two-component Methyl Methacrylate paste was used as an adhesive. The commercially available optoelectronic optical backscatter reflectometer (OBR) from Luna Technology [19] was used. Th spatial resolution used during the experiment is 5 mm leading to detailed strain profiles along the rod. The acquisition time of the fiber optic sensing signal is 5 s, which limits its use to static testing.

3.1. Experimental Setup

The experimental setup used to perform the static and dynamic tests on the rod is shown in Figure 1. A rigid tank of dimensions 1 m × 1 m × 1 m was progressively filled with dry sand reaching a final height of 0.7 m. The static lateral loads were applied by a set of dead weights connected to the rod described in Section 4.1.1 with a thread passing through a pulley. The static lateral loads were applied with various eccentricities a. To control the direction, and therefore the value of the lateral loads, the height of the pulley was adjustable in order to keep the part of the thread connected to the rod always in horizontal position.

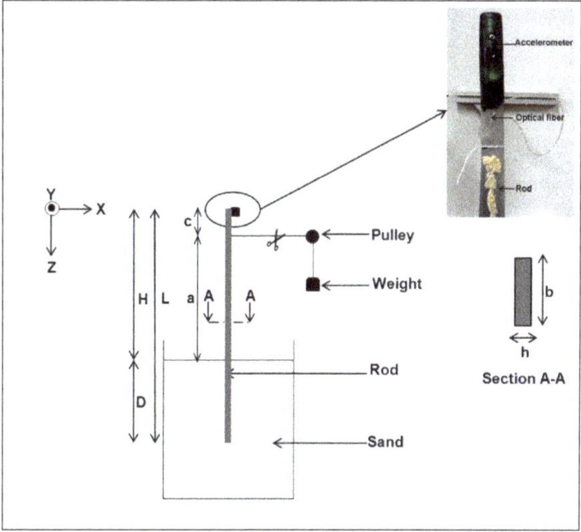

Figure 1. Experimental setup of the static and dynamic tests and close up look to fiber optic installation. L is the length of the beam, with D the embedded length and H the exposed length. a is the exccentricity of the load to the soil level. The scissors show where the thread has been cut to apply the lateral force.

4. Materials and Methods

4.1. Materials Used

4.1.1. Rod Characteristics

An aluminum rectangular rod having width $b = 25$ mm, thickness $h = 5$ mm and length $L = 1170$ mm was used. Its bulk density and Young modulus were respectively $\rho_b = 2700$ kg/m ^{3} and $E_b = 62.2$ GPa. The rod was instrumented using a fiber optic along its length to measure the strain. Following [20], an accelerometer was placed at the head of the rod, to record its transient response.

4.1.2. Soil Characteristics

The static and dynamic tests were conducted using a dry sand of Seine. The mean size of sand grains was $D_{50} = 0.70$ mm and the dry density was $\rho_s = 1700$ kg/m^3. For soil characterization, mini-pressuremeter tests were performed. The average Ménard modulus measured in these tests was $E_m = 0.5$ MPa. The subgrade modulus K_s of the tested soil could then be calculated using the empirical Equation (11) [21]:

$$\begin{cases} K_s = \dfrac{3E_m}{\frac{2}{3}(\frac{B}{B_0})(\frac{\alpha}{2})(2.65\frac{B}{B_0})^{\alpha}} & B > B_0, \\ K_s = \dfrac{18E_m}{4(2.65)^{\alpha} + 3\alpha} & B < B_0, \end{cases} \tag{11}$$

where K_s is the modulus of subgrade reaction, E_m is the Ménard modulus, B is the diameter of the tested pier or rod, $B_0 = 0.6$ m is the reference diameter and α is a rheological parameter depending on the tested soil with $\alpha = \frac{1}{3}$ for sand. Under these assumptions of parameter values, and given the geometry of the rod, the measured modulus of subgrade reaction of the tested soil is given by Equation (12):

$$E_s = 1.3\ MPa. \tag{12}$$

4.2. Test Procedures

4.2.1. Static Tests

The rod was tested under static lateral loads. The following section provides a description of the testing protocol.

Before applying the lateral loads, a reference scan of the fiber optic was performed. A lateral load F was then applied by adding a set of dead weights. A second scan was performed to measure the resulting strain along the rod. Scour was generated by the excavation of a 100 mm thick layer of soil. The various tested configurations are summarized in Table 1. Two loads $F_1 = 2$ N and $F_2 = 4$ N were applied. It is worthy to mention that even if the applied loads were low, the range of generated strains is similar to the usual values of strains along piles.

Table 1. List of static tests.

Test Name	Embedment Length D (cm)	Load Distance from the Tip of the Rod c (cm)	Load
$D40 - c5 - F_1$	40	5	F_1
$D40 - c5 - F_2$	40	5	F_2
$D30 - c5 - F_1$	30	5	F_1
$D30 - c5 - F_2$	30	5	F_2
$D40 - c25 - F_1$	40	25	F_1
$D40 - c25 - F_2$	40	25	F_2
$D30 - c25 - F_1$	30	25	F_1
$D30 - c25 - F_2$	30	25	F_2

The tests presented in Table 1 were first performed with the fiber optic under tension, as it was glued on the side of the rod undergoing experiencing a positive bending moment. But the performance of strain sensors may vary due to the testing configuration, e.g., when tested under tension or under compression [22]. The rod was therefore flipped and similar tests were carried out with the fiber optic under compression to evaluate its performance in both configurations.

4.2.2. Dynamic Tests

Free vibration tests were conducted to measure the frequency of the rod for various scour depth. The following section provides a description of the testing protocol.

The rod was partially embedded in sand. Scour was generated by the progressive excavation of 50 mm thick layers of soil. The embedded length D of the rod varied from 400 mm to 150 mm. For each scour depth, the thread connecting the rod head to the weight was cut inducing the vibration of the rod in the X direction (horizontal). The signal recorded by the accelerometer was then post-processed using a fast Fourier transform (FFT) to measure the first frequency for each scour depth.

5. Results

5.1. Static Testing

5.1.1. Fiber Optic Sensor Performance

The results of the lateral loading tests, for the various configurations presented in Table 1, are shown in Figures 2 and 3. The strains obtained under tension are referred to with 'T' and the ones obtained under compression are referred to with 'C'. Figure 2a,b shows the results of tests conducted with the force F applied at a distance $c = 5$ cm from the tip of the rod. Figure 3a,b shows the tests for a distance $c = 25$ cm from the tip of the rod.

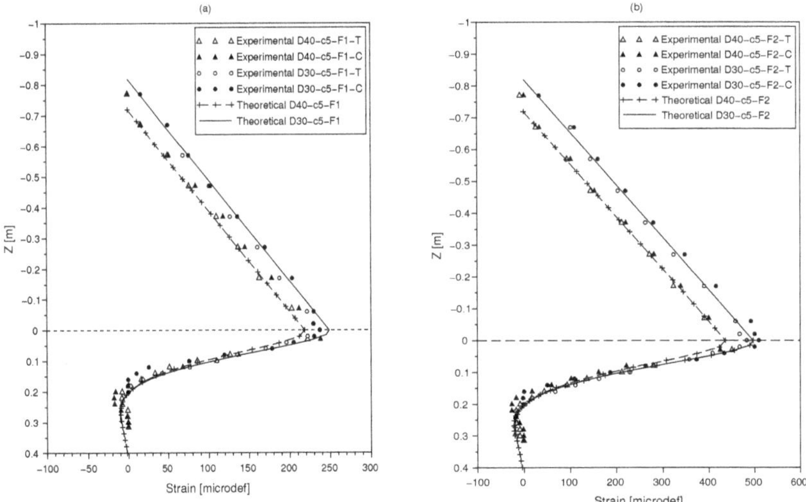

Figure 2. Strain profiles along the rod under tension 'T' and compression 'C' provided by the fiber optic for the load F [(**a**) $F = F1$ (**b**) $F = F2$] applied at $c = 5$ cm from the tip of the rod for an embedded length of $D = 40$ cm and $D = 30$ cm.

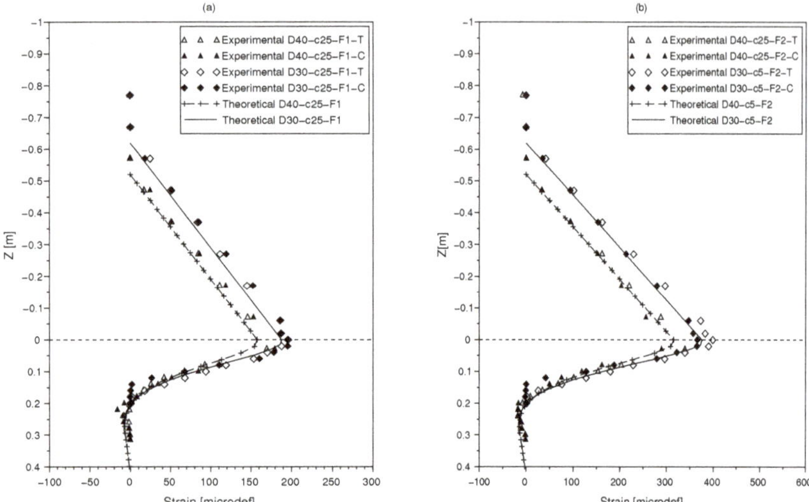

Figure 3. Strain profiles along the rod under tension 'T' and compression 'C' provided by the fiber optic for the load F [(**a**) $F = F1$ (**b**) $F = F2$] applied at $c = 25$ cm from the tip of the rod for an embedded length of $D = 40$ cm and $D = 30$ cm.

It is found that the strain curves obtained with the fiber under compression and tension are similar, proving that the fiber optic performance is not affected by its configuration here.

The strain profile along the exposed part of the rod (i.e., for $z \in [-a, 0]$) was proven to be independent of the soil properties. Therefore, the theoretical strain profile is used to evaluate the accuracy of the fiber optic measurement. Theoretical strain profiles for $z \in [-a, 0]$ are computed using Equations (5) and (7), and compared to the strain measurements with the fiber optic.

As observed in Figures 2 and 3, the experimental and theoretical results are in good agreement. It can be noted that the measurement error does not exceed 7% which highlights the accuracy of the fiber optic sensing technology.

5.1.2. Effect of Scour

As shown in Figures 2 and 3, the strain profiles have a turning point near the ground level plotted with a dashed line. A similar observation was made by [23] who suggested monitoring the maximum bending moment to estimate scour depth.

The static test results also indicate that as scour increases, the bending moment and strain values increase as a consequence of a greater eccentricity of the applied force F (due to the higher exposed length of the beam). However, it can be noted that the effect of scour on the strain profiles is only noticeable near the ground level. As the depth increases, no variation of strain values with scour is noticed.

5.1.3. Experimental Strain Profile Versus Theoretical Prediction

The soil layer used during the experiments was modeled using the Winkler model presented previously. The strain along the rod was then computed using Equations (5) and (7). The value of the modulus of subgrade reaction K_s was determined from mini pressuremeter tests and its value was previously established in Equation (12).

Figures 2 and 3 show the comparison between the measured and the theoretical strains along the rod. The results show a very good agreement between the theoretical strain profile and the experimental results. These results confirm, on the one hand that its is legitimate to model the soil used in this study as a single layer having a constant subgrade modulus K_s with depth, and on the other hand the measured value of the subgrade modulus was also validated as a good agreement was found between the theoretical and experimental strains.

The theoretical positions of the maximum strain are computed using Equation (9) and summarized in Table 2. As it can be seen, the theoretical results confirm that the maximum bending moment is near the ground level for all testing configurations and gives an insight of the parameters that influence its location which are: the eccentricity of the lateral load a and the characteristic length of the rod l_0.

Table 2. Theoretical position of the maximum strain.

Test Configuration	z_{max} (cm)
$D40-c5$	0.46
$D30-c5$	0.40
$D40-c25$	0.62
$D30-c25$	0.53

For a given eccentricity a of the lateral load, the variation of z_{max} position with the characteristic length l_0 can be evaluated by deriving Equation (9):

$$\frac{\partial z_{max}}{\partial l_0} = \frac{2al_0}{(l_0+2a)^2+{l_0}^2} + \arctan\left(\frac{l_0}{l_0+2a}\right) \geq 0. \tag{13}$$

Therefore, as shown by Equation (13), the value of z_{max} increases with the increase of the characteristic length l_0. For this reason, monitoring scour depth using the position of the maximum bending moment/strain can not be generalized for all types of soil and rods. To successfully implement this monitoring technique, it is therefore crucial to carefully design the sensor. The material and geometry in particular should be chosen according to the soil stiffness in order to decrease the characteristic length l_0 and therefore the value of z_{max}.

6. Discussion

In this section, the equivalent cantilever model is introduced using a static approach.

6.1. Static Equivalent Length

This cantilever has a length L_{es} and a similar deflection to that of the rod partially embedded in sand. The equivalent length L_{es} therefore corresponds to the free length of the rod in sand H, increased with a "adjustment static length" d corresponding to the distance between the soil level and the equivalent cantilever base as shown in Figure 4.

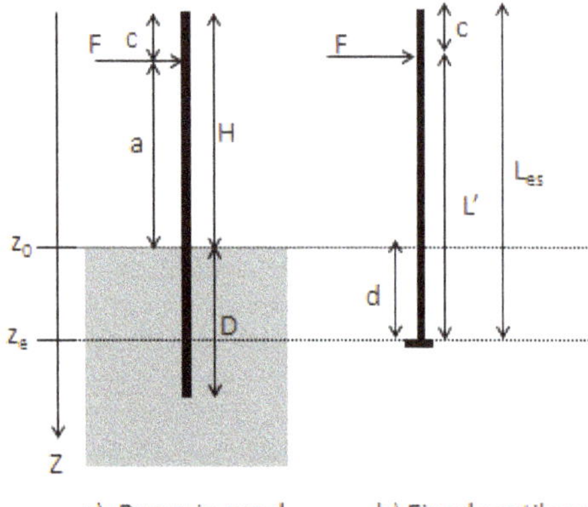

Figure 4. Definition sketch of the static equivalent length. The subfigure (**a**) shows the physical problem, whose static behavior is equivalent to the one of the cantilevered beam of the subfigure (**b**). L is the length of the beam, with D the embedded length and H the exposed length. a is the excentricity of the load to the soil level. z_0 is the soil level, and z_e the level of the fixed of the cantilevered, equivalent beam.

The methodology of identifying the "adjustment static length" d is detailed hereafter.

First, for each tested configuration, the theoretical model is used to determine the deflection along the rod in the sand. To this end, Equation (2) is used to compute the deflection $w(z = -a)$ at the point of force application.

Second, L' is calculated using Equation (14) derived from the Euler-Bernoulli beam theory.

$$L' = \sqrt[3]{\frac{3E_bI_b}{F} \times w(-a)}. \tag{14}$$

Finally, the position of the base z_e of the equivalent cantilever (Figure 4) corresponding to the "adjustment static length" can then be determined using the following equation:

$$z_e = d = L' - a. \tag{15}$$

The previous methodology was applied to the tested configurations and the results are summarized in Table 3. The results show that for all tested configurations, the "adjustment static length" is $z_e = d = 8.4$ cm. Therefore, the rod in the sand is equivalent to a cantilever beam having a total length $L_{es} = H + d$ ($d = 8.4$ cm), where H the exposed length of the rod and the d the "adjustment static length". This result means that for a given range a embedded length, the partially embedded beam can be considered statically as a cantilevered beam of a higher length, to take into account the embedded length that is needed to support the beam.

It is worthy to highlight that in the case of a fixed cantilever, the point $z = z_e = d$ will also have the maximum bending moment which is not the case for the rod partially embedded in sand. Previous results showed that the maximum bending moment is at the ground level.

Table 3. Position Z of the base of equivalent cantilever.

Test Configuration	Deflection w at $z = -a$ (cm)	L' (cm)	d (cm)
$D40 - c5 - (F_1\&F_2)$	4.2	80.4	8.4
$D30 - c5 - (F_1\&F_2)$	6.0	90.4	8.4
$D40 - c25 - (F_1\&F_2)$	1.8	60.4	8.4
$D30 - c25 - (F_1\&F_2)$	2.8	70.4	8.4

6.2. Dynamic Testing of the Effect of Scour

The variation of the first frequency with the embedded length D of the rod is shown is Figure 5. As scour increases, the embedded length D of the rod decreases leading to a decrease of the first frequency of the rod. A 10 cm scour from $D = 40$ cm to $D = 30$ cm caused a 20% variation of the frequency.

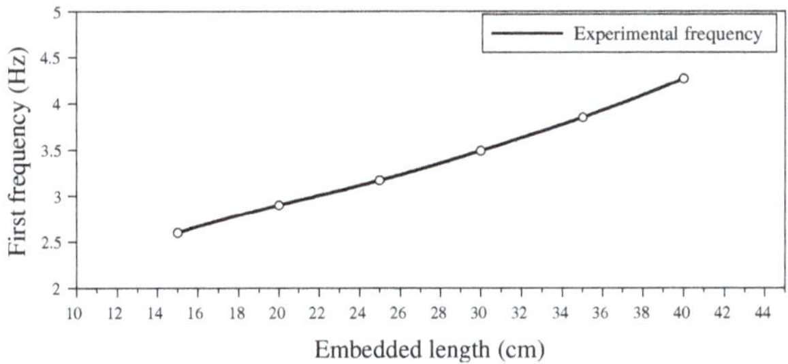

Figure 5. Variation of the first frequency with the embedded length D of the rod.

6.3. Dynamic Equivalent Length

The variation of the first frequency with the exposed length H of the rod was compared to the the frequencies computed using Equation (16) of a cantilever carrying a tip mass m modelling the accelerometer [24,25].

$$f = \frac{1}{2\pi} \times \sqrt{\frac{3E_b I_b}{{L_{ed}}^3(0.24M + m)}}, \quad (16)$$

where m the mass of the accelerometer and M the total mass on the cantilever.

Figure 6 shows that the experimental frequencies are translated against the analytical frequencies of an equivalent cantilever with a free length $L_{ed} = H + d'$ where L_{ed} is the length of the equivalent cantilever, H the exposed length of the rod and d' the "adjustment dynamic length" [26]. The latter, H_c, is determined graphically by shifting the experimental frequency curve to fit the theoretical one (the dotted curve in Figure 6). A good agreement with the theoretical frequencies is obtained for $H_c = 8.4$ cm.

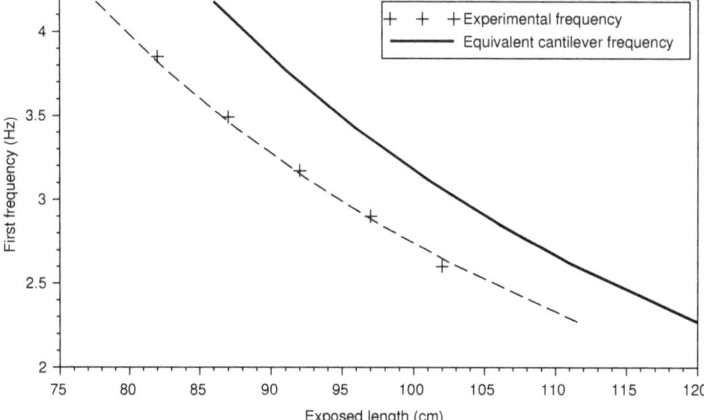

Figure 6. Variation of the first frequency with the exposed length H of the rod.

The comparison between the two adjustment lengths d and d', determined with the static and dynamic approaches, shows that its value is the same for both methods. Therefore, in our testing conditions, the soil-structure interaction can be simplified by the proposed cantilever model.

7. Conclusions

The present study focused on the effect of scour on the static and dynamic responses of a rod partially embedded in sand. Distributed strain measurement using OFDR technique provided a detailed strain profile along the rod. A theoretical formulation was developed using the Winkler model of the soil and compared to the measured experimental values. The errors did not exceed 7% highlighting the accuracy of the OFDR. The static tests results also showed that the fiber optic sensor performed identically under tension and compression which is crucial when the rod will be deformed by the flow.

In order to monitor scour, the turning point of the strain profile was used to identify the ground level. The theoretical model provides insight of the parameters influencing the maximum stain position along the rod, which are: the lateral stiffness of the soil, the Young modulus and the inertia of the tested rod.

The results also showed that the effect of scour on the strain level is only noticeable near the ground. As scour increases, the value of the strain increases along the first layer of the soil. However, no significant variation was detected for greater depth.

Regarding the dynamic tests, the results showed that the first frequency of the rod decreases significantly with scour depth. Finally, an equivalent cantilever model was proposed for both static and dynamic tests. This model correlates both the natural frequency and the deflection of the rod to its exposed length.

Author Contributions: This research article is based on the PhD work of the first author N.B., who made the experimental and numerical investigation. The methodology has been provided and the work has been supervised by the co-authors, depending on their domains of expertise: F.S. for structures, C.C. for geotechnics, D.S. for experimental and sensor issues and D.P.V.B. for hydrological issues. The original draft has been prepared by the first author, reviewing has been done by all co-authors, editing and submitting by the second author. All authors have read and agreed to the published version of the manuscript.

Funding: This research was funded by the ANR French Research Agency within the project SSHEAR No 2014-CE03-0011.

Acknowledgments: The present work benefits from the financial support of the ANR French Research Agency within the project SSHEAR No 2014-CE03-0011. For further information on the project [http://sshear.ifsttar.fr].

Conflicts of Interest: The authors declare no conflict of interest.

Abbreviations

The following abbreviations are used in this manuscript:

a	Load eccentricity from the soil (m);
b	Width of the rod (m);
c	Distance of the lateral force from the tip of the rod (m);
d	Adjustment static length (m);
d'	Adjustment dynamic length (m);
D	Embedded length of the rod (m);
D_{50}	Average grain diameter (mm);
E_b	Young modulus of the rod (MPa);
E_m	Ménard modulus of the soil (MPa);
K_s	Modulus of subgrade reaction (MPa);
f	First frequency (Hz);
h	Thickness of the rod (m);
H	Exposed length of the rod (m);
I_b	Inertia of the rod in the vibration direction (m^4);
L	Total length of the rod (m);
L_{es}	Equivalent static length (m);
L_{ed}	Equivalent dynamic length (m);
M	Mass of the rod (kg);
m	Mass of the accelerometer (kg);
S	Section of the rod (m^2);
α	Rheological parameter of the soil (-);
ρ_s	Bulk density of the soil ($kg.m^{-3}$);
ρ_b	Bulk density of the rod ($kg.m^{-3}$);

References

1. Millard, S.; Bungey, J.; Thomas, C.; Soutsos, M.; Shaw, M.; Patterson, A. Assessing bridge pier scour by radar. *NDT E Int.* **1998**, *31*, 251–258. [CrossRef]
2. Falco, F.D.; Mele, R. The monitoring of bridges for scour by sonar and sedimetri. *NDT E Int.* **2002**, *35*, 117–123. [CrossRef]
3. Ding, Y.; Yan, T.; Yao, Q.; Dong, X.; Wang, X. A new type of temperature-based sensor for monitoring of bridge scour. *Measurement* **2016**, *78*, 245–252. [CrossRef]
4. Zarafshan, A.; Iranmanesh, A.; Ansari, F. Vibration-Based Method and Sensor for Monitoring of Bridge Scour. *J. Bridge Eng.* **2012**, *17*, 829–838. [CrossRef]
5. Lin, C.; Bennett, C.; Han, J.; Parsons, R.L. Scour effects on the response of laterally loaded piles considering stress history of sand. *Comput. Geotech.* **2010**, *37*, 1008–1014. [CrossRef]
6. Qi, W.; Gao, F.; Randolph, M.; Lehane, B. Scour effects on p–y curves for shallowly embedded piles in sand. *Géotechnique* **2016**, *66*, 648–660. [CrossRef]
7. Lin, Y.B.; Chen, J.C.; Chang, K.C.; Chern, J.C.; Lai, J.S. Real-time monitoring of local scour by using fiber Bragg grating sensors. *Smart Mater. Struct.* **2005**, *14*, 664. [CrossRef]
8. Zhu, H.H.; Shi, B.; Zhang, C.C. FBG-Based Monitoring of Geohazards: Current Status and Trends. *Sensors* **2017**, *17*, 452. [CrossRef] [PubMed]

9. Foti, S.; Sabia, D. Influence of Foundation Scour on the Dynamic Response of an Existing Bridge. *J. Bridge Eng.* **2011**, *16*, 295–304. [CrossRef]
10. Prendergast, L.; Hester, D.; Gavin, K.; O'Sullivan, J. An investigation of the changes in the natural frequency of a pile affected by scour. *J. Sound Vib.* **2013**, *332*, 6685–6702. [CrossRef]
11. Elsaid, A.; Seracino, R. Rapid assessment of foundation scour using the dynamic features of bridge superstructure. *Constr. Build. Mater.* **2014**, *50*, 42–49. [CrossRef]
12. Azhari, F.; Loh, K.J. Laboratory validation of buried piezoelectric scour sensing rods. *Struct. Control Health Monit.* **2017**, *24*, e1969. [CrossRef]
13. Winkler, E. *Theory of Elasticity and Strength of Materials*; Dominicus: Prague, Czechoslovakia, 1867.
14. Melville, B.W.; Coleman, S.E. *Bridge Scour*; Water Resources Publication; Springer: Berlin, Germany, 2000.
15. Prendergast, L.; Reale, C.; Gavin, K. Probabilistic examination of the change in eigenfrequencies of an offshore wind turbine under progressive scour incorporating soil spatial variability. *Mar. Struct.* **2018**, *57*, 87–104. [CrossRef]
16. Gifford, D.K.; Froggatt, M.E.; Wolfe, M.S.; Kreger, S.T.; Soller, B.J. Millimeter Resolution Reflectometry Over Two Kilometers. In Proceedings of the 33rd European Conference and Exhibition of Optical Communication, Berlin, Germany, 16–20 September 2007; pp. 1–3.
17. Kreger, S.T.; Gifford, D.K.; Froggatt, M.E.; Soller, B.J.; Wolfe, M.S. High resolution distributed strain or temperature measurements in single-and multi-mode fiber using swept-wavelength interferometry. In *Optical Fiber Sensors*; Optical Society of America: Washington, DC, USA, 2006.
18. Soller, B.J.; Gifford, D.K.; Wolfe, M.S.; Froggatt, M.E. High resolution optical frequency domain reflectometry for characterization of components and assemblies. *Opt. Express* **2005**, *13*, 666–674. [CrossRef] [PubMed]
19. Luna Technology. Available online: https://lunainc.com/ (accessed on 6 January 2020).
20. Bao, T.; Swartz, R.A.; Vitton, S.; Sun, Y.; Zhang, C.; Liu, Z. Critical insights for advanced bridge scour detection using the natural frequency. *J. Sound Vib.* **2017**, *386*, 116–133. [CrossRef]
21. Ménard, L.; Bourdon, G.; Gambin, M. Méthode générale de calcul d'un rideau ou d'un pieu sollicité horizontalement en fontion des résultats pressiomètriques Sol. *Sol. Soils* **1969**, *VI*, 16–29.
22. Montero, A.; Saez de Ocariz, I.; Lopez, I.; Venegas, P.; Gomez, J.; Zubia, J. Fiber Bragg Gratings, IT techniques and strain gauge validation for strain calculation on aged metal specimens. *Sensors* **2011**, *11*, 1088–1104. [CrossRef] [PubMed]
23. Kong, X.; Cai, C.; Hou, S. Scour effect on a single pile and development of corresponding scour monitoring methods. *Smart Mater. Struct.* **2013**, *22*, 055011. [CrossRef]
24. Rayleigh, J. *The Theory of Sound*; Dover Books: Dover, UK, 1945.
25. Turhan, O. On the fundamental frequency of beams carrying a point mass: Rayleigh approximations versus exact solutions. *J. Sound Vib.* **2000**, *230*, 449–459. [CrossRef]
26. Boujia, N.; Schmidt, F.; Chevalier, C.; Siegert, D.; Van Bang, D.P. Effect of Scour on the Natural Frequency Responses of Bridge Piers: Development of a Scour Depth Sensor. *Infrastructures* **2019**, *4*, 21. [CrossRef]

Article

Detection and Measurement of Matrix Discontinuities in UHPFRC by Means of Distributed Fiber Optics Sensing

Bartłomiej Sawicki [1,*], Antoine Bassil [2,3], Eugen Brühwiler [1], Xavier Chapeleau [2] and Dominique Leduc [4]

1 Laboratory of Maintenance and Safety of Structures, Structural Engineering Institute, Swiss Federal Institute of Technology (EPFL), CH-1015 Lausanne, Switzerland; eugen.bruehwiler@epfl.ch
2 COSYS-SII, I4S Team (Inria), Univ Gustave Eiffel, IFSTTAR, F-44344 Bouguenais, France; antoine.bassil@quadric.arteliagroup.com (A.B.); xavier.chapeleau@univ-eiffel.fr (X.C.)
3 Quadric, Artelia Group, 14 Porte de Grand Lyon, F-01700 Neyron, France
4 GeM UMR 6183, University of Nantes, F-44322 Nantes, France; dominique.leduc@univ-nantes.fr
* Correspondence: bartek.sawicki@epfl.ch

Received: 13 June 2020; Accepted: 7 July 2020; Published: 12 July 2020

Abstract: Following the significant improvement in their properties during the last decade, Distributed Fiber Optics sensing (DFOs) techniques are nowadays implemented for industrial use in the context of Structural Health Monitoring (SHM). While these techniques have formed an undeniable asset for the health monitoring of concrete structures, their performance should be validated for novel structural materials including Ultra High Performance Fiber Reinforced Cementitious composites (UHPFRC). In this study, a full scale UHPFRC beam was instrumented with DFOs, Digital Image Correlation (DIC) and extensometers. The performances of these three measurement techniques in terms of strain measurement as well as crack detection and localization are compared. A method for the measurement of opening and closing of localized fictitious cracks in UHPFRC using the Optical Backscattering Reflectometry (OBR) technique is verified. Moreover, the use of correct combination of DFO sensors allows precise detection of microcracks as well as monitoring of fictitious cracks' opening. The recommendations regarding use of various SHM methods for UHPFRC structures are given.

Keywords: crack detection; crack opening; distributed fiber optic sensors; DIC; UHPFRC; testing; SHM; microcracking

1. Introduction

To tackle the challenges that are in front of civil engineering—such as reduction in carbon footprint with optimized design, proper allocation of scarce resources through the use of engineered structural materials or extension of service duration thanks to deeper understanding of performance of structures—up-to-date methods should be used.

From the construction material point of view, such a developing technology is the Ultra High Performance Fiber Reinforced Cementitious composite (UHPFRC). It allows for design of refined and more slender structures as well as reinforcing and upgrading existing ones. To fully master its performance on both micro- and macroscopic levels, new measurement techniques are needed.

Such a possibility is given through the development of Distributed Fiber Optics (DFO) sensing techniques. The DFOs, used mostly in the automotive and mechanical industry, have recently found place in the civil engineering field. During the last decade, Fiber Optics (FO) sensors became increasingly popular in Structural Health Monitoring (SHM), and now, they are the second most used sensing technology in this field [1]. These sensors are small-sized, lightweight and resistant to both chemical degradation and electromagnetic interference. In the past few years, it was demonstrated by numerous researchers that DFOs can be used for measurements of strain and crack opening in ordinary reinforced concrete [2–5] and fiber reinforced concrete [6]. Still, the verification of usefulness of this SHM for UHPFRC is missing, especially considering the unique behavior of this material under tensile action.

For UHPFRC structures, the photogrammetry with digital image correlation (DIC) is now a well-established technique to track and measure cracks [7–9]. However, there were not any attempts to monitor crack propagation and measure crack opening using the DFO sensing techniques in UHPFRC structural elements. Schramm and Fischer [10] tested a slab element and a prestressed beam. For the slab without rebars, the externally glued DFOs were able to detect the apparent strain peaks due to microcracking. However, no estimation of crack opening was done. In cases of beam elements under shear action, the pattern of fictitious cracks was observed using DIC. The DFOs were glued on the steel rebars and used to detect the stress peaks in the reinforcement (stirrups) near these fictitious cracks. Similar instrumentation, with DFOs on the rebars and DIC on the surface, was used for investigation of the stress transfer between UHPFRC and reinforcement by others [11]. The DFOs sensors were employed for SHM of a UHPFRC bridge [12] but without attempt to detect discontinuities.

This paper presents results of an experimental validation of the use of DFOs to detect and measure discontinuities in UHPFRC structures. The full-scale beam is instrumented with DFOs, DIC and extensometers in order to compare their crack monitoring features. The results are discussed from structural and material points of view, with an outlook for the possible use of DFOs for Structural Health Monitoring of UHPFRC structures.

2. Materials and Methods

2.1. Distributed Fiber Optics Sensing for Discontinuity Monitoring

Recently, DFO sensors were especially used where an urgent need of high number of sensing points appeared. The difference between these systems and traditional long gauge or point sensors is their ability to provide distributed measurements, and thus, simultaneously local and global information [13]. Measurement systems are composed of an interrogator and an optical fiber playing the role of a sensor. These sensors are either embedded into new concrete structures or bonded to the surface of existing ones. Different interrogation units available nowadays are based on the analysis of the Brillouin and Rayleigh backscattered light over the silica optical fiber. The Rayleigh-based systems can perform distributed strain measurements with higher spatial resolutions (<1 cm) than Brillouin-based systems which, on the other hand, can interrogate larger distances (>100 km).

The DFOs techniques can be used not only to measure strains but also to detect, localize and measure cracks of small openings. DFO sensors allow achieving of an accurate, reliable and quasi real-time crack detection and characterization in concrete structures. They contributed to the detection and localization of cracks in the massive structures, showing supremacy over the short and long gauge sensors (Figure 1a). Actually, any type of discontinuity in the host material, like a crack, can cause a strain localization propagating through the optical fiber layers up to the core of the optical fiber (Figure 1b).

There are numerous studies on the use of DOFs for detection of crack formation. The sensors, based on measuring losses with Optical Time Domain Reflectometry technique [3–5,14–16], were very limited in practical applications. Similarly, those based on Brillouin backscattering [6,17–20] were limited due to their low spatial resolution, affecting their strain sensing accuracy around a crack in the concrete material [21,22]. In fact, the complicated strain distribution and its rapid variation within the spatial resolution decreases the strain measurement accuracy [23]. Later on, Optical Backscattering Reflectometry (OBR), based on the Optical Frequency Domain Reflectometry (OFDR) technique, emerged. This technique, characterized by high spatial resolution, is proven to be capable of detecting and localizing tiny microcracks in reinforced concrete structures [24]. Different methods were also proposed to quantify crack openings from the strain profiles, either based on a combination with finite element models of the structure [25,26] or on the calculation of the optical fiber elongation by summing distributed strain gradients [27,28].

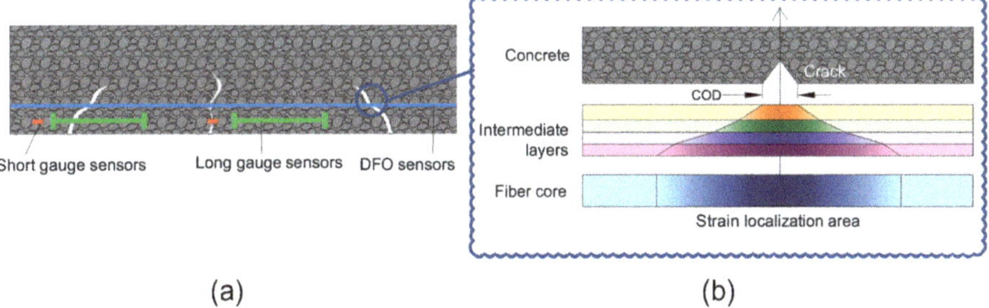

Figure 1. Crack detection in concrete using Distributed Fiber Optics sensing techniques; (**a**) comparison to traditional sensors, (**b**) strain transferring between layers.

2.2. Analytical Models Based on Strain Transfer Theories

A distributed optical fiber sensor is an optical fiber surrounded by various protective and adhesive layers, forming a multilayered strain transfer system. The existence of these intermediate layers leads to differences in the strain of host material and the strain measured by an optical fiber due to the shear lag effect in intermediate layers. The problem of strain transfer through an optical fiber sensor has been studied in the field of short dimensional sensors like Bragg grating or interferometric sensors [29–35]. Indeed, many research works focused on designing discrete sensors with improved strain transfer efficiency [36] and performing parametric studies of different mechanical and geometrical properties of multilayered sensors [37].

Since 2012, different analytical and numerical models were proposed [38] to describe the strain transfer from a discontinuous (cracked) host material. Imai et al. [6] introduced the effect of crack discontinuity in host material as a Gaussian distribution at the interface with protective coating. Later, it was assumed that the strain at the discontinuity location is equal to the crack opening over the spatial resolution of the measurement instrument [39]. Finally, the Crack Opening Displacement (COD) was introduced as an additional term provoked by the local discontinuity in the host material deformation field. Feng et al. [18] deduced a mechanical transfer equation, showing that the strain measured by the optical fiber $\varepsilon_f(z)$ consists of a crack-induced strain $\varepsilon_{crack}(z)$ part added to the strain in host material $\varepsilon_m(z)$. Recently, Bassil et al. [40,41] deduced a similar strain transfer equation for a multilayer system with imperfect bonding between layers:

$$\varepsilon_f(z) = \varepsilon_{crack}(z) + \varepsilon_m(z) = \lambda \frac{\text{COD}}{2} e^{-\lambda|z|} + \varepsilon_m(z) \tag{1}$$

$$\lambda^2 = \frac{2}{E_f r_f^2 \left[\frac{1}{G_1} ln\left(\frac{r_1}{r_f}\right) + \sum_{i=1}^{N} ln\left(\frac{r_i}{r_{i-1}}\right) + \sum_{i=1}^{N} \frac{1}{k_i r_i} \right]} \tag{2}$$

where λ is the strain lag parameter that includes mechanical (G and E, shear and Young's moduli, respectively) and geometrical properties (r) of the different i layers and z is the position of discontinuity along the optical cable. It also includes coefficients k depicting the level of interfacial adhesion between two consecutive layers.

The strain lag parameter λ is crucial for the sensing of cracks. As this value increases, the crack-induced strains $\varepsilon_{crack}(z)$ figure higher peaks at the crack location, and thus, the exponential part covers a narrower zone over the optical fiber length, as shown in Figure 2. Thanks to this, the capacity of detecting and localizing discontinuities increases.

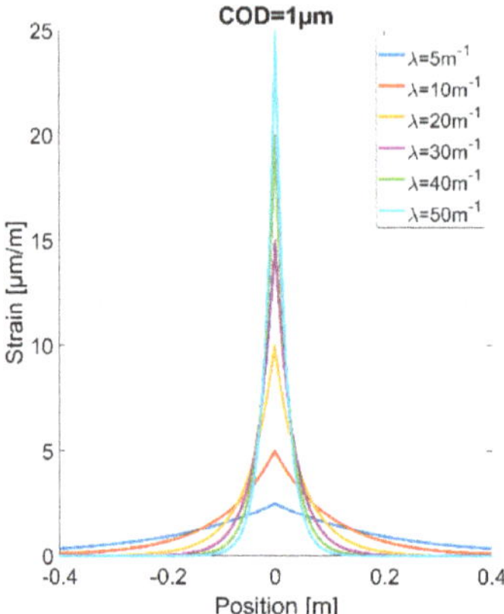

Figure 2. The spatial distribution form of the crack-induced strains $\varepsilon_{crack}(z)$ for different strain lag parameter λ values.

The authors also demonstrated the validity of the model for different types of optical cables through an experimental testing campaign on concrete specimens. The estimated CODs proved to be accurate, reaching relative errors of 1–10% for a dynamic range [COD_{min}, COD_{max}], with a strain repeatability of ±20 μm/m for the interrogator unit. In this range, the layers behave in an elastic manner and sufficient, stable bonding between them exists. COD_{max} varied widely from 80 to 1500 μm for different types of optical cable assemblies. On the other hand, the authors fixed the COD_{min} to 50 μm, below which other parameters prevail, i.e., the nature of cracking of concrete material (in the fracture process zone) or the strain accuracy and repeatability of interrogator.

In terms of crack detection, Bassil et al. [42] demonstrated that an OBR system with a strain repeatability of ± 2 μm/m and an optical cable with $\lambda = 20$ m^{-1} can detect concrete discontinuities of less than 1 μm.

2.3. Ultra High Performance Fiber Reinforced Cementitious Composite

UHPFRC is a composite fiber reinforced cementitious building material with a high content (>3% vol.) of short ($l_f < 20$ mm) and slender steel fibers. Its behavior under tensile stress comprises three stages, as shown in Figure 3.

The first stage is an elastic stage. The cementitious matrix is continuous and the behavior of UHPFRC is simply linear. The strain of the material can be directly measured.

After the elasticity limit (f_e, ε_e) is reached, discontinuities in the matrix start to appear and the material enters the strain-hardening phase. The openings of these fine, distributed microcracks (hairline cracks) are smaller than 50 μm and their spacing can vary from 2 to 30 mm [43–45]. They are not detrimental from a durability point of view [46,47] and are impossible to see with the naked eye. These microcracks can be, however, measured using appropriate instrumentation. From the macroscopic point of view, the material can be considered as continuous, with strain-hardening quasi-linear behavior and reduced stiffness [48]. However, after unloading, the residual strain remains in the material.

When the maximum tensile resistance f_u is reached, the material enters the softening phase. One or more neighboring microcracks start rapidly growing, eventually reaching openings above 50 μm. This localized discontinuity is bridged by multiple fibers carrying the tensile stress. It is called a fictitious crack, contrary to the real crack which cannot transfer the stress [49]. Since the stress transfer capability in this critical zone is reduced, the overall stress in the area decreases. The localized fictitious crack is growing, while the strain and stress around it decrease. The location of the fictitious crack depends on the distribution of fibers [44,45,50,51]. Since this fictitious crack leads eventually to the failure of structural elements, it is called the critical crack as well. With the gradual opening of the fictitious crack, the measured deformation increases, leading to fast growth of the apparent strain. Importantly, it is not the real strain of the material anymore due to the fictitious crack localized between the reference measurement points.

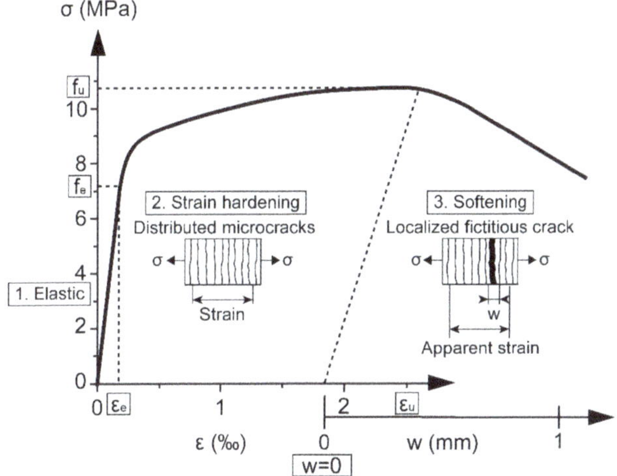

Figure 3. UHPFRC behavior under tension.

The fictitious crack grows until opening of half of the steel fiber length, in the present case $l_f/2 = 6.5$ mm [44] and the resistance of the material decreases. After the fibers are pulled out, no more stress transfer is possible and the real crack is formed.

In the case of a structural R-UHPFRC (Reinforced UHPFRC) element under bending action, the tensile behavior of UHPFRC has important influence on the overall response. Each of the three stages are present under the maximum bending moment in the critical section, where the fictitious crack forms. Under the assumption that the cross-sections remain plain, the distribution of stress along the height of the beam is nonlinear due to nonlinearity of the constitutive law of UHPFRC, as presented schematically in Figure 4. The proportion between parts in elastic, strain-hardening and softening regimes depends on geometry of the element [44].

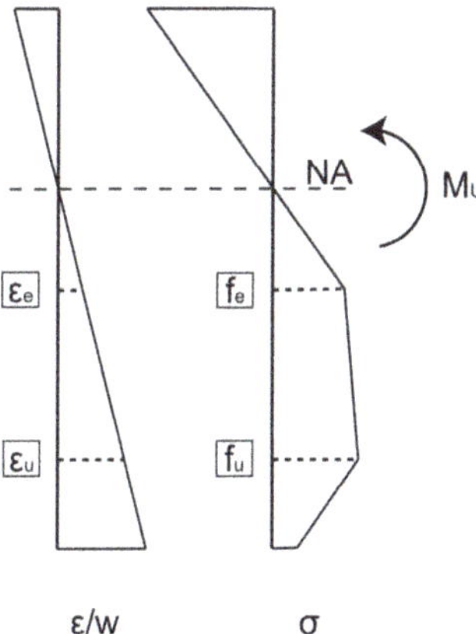

Figure 4. Schematic distribution of stresses and strains in UHPFRC at the critical section of R-UHPFRC beam under the ultimate bending moment.

3. Test Set-Up and Specimen

The tested beam has a T-shaped cross-section and dimensions according to Figure 5. This kind of design refers to the use of UHPFRC for waffle deck or unidirectional ribbed slab designs. An example of such a structure is the railway bridge in Switzerland described in reference [52].

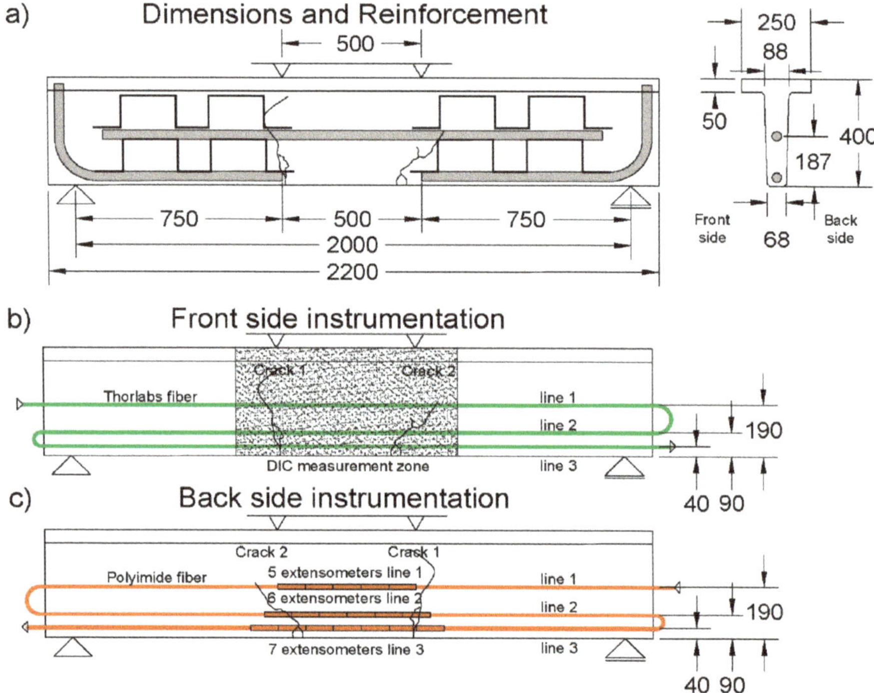

Figure 5. Scheme of test setup and instrumentation: (**a**) beam dimensions and reinforcement, (**b**) instrumentation on the front side and (**c**) instrumentation on the back side; dimensions in mm; the critical fictitious cracks 1 and 2 are marked.

Commercially available UHPFRC premix Holcim710® was used, with 3.8% vol. of 13 mm long straight steel fibers with an aspect ratio of 65. Its mechanical properties as obtained by material testing were compressive strength f_c = 149 MPa; elastic limit stress f_e = 6.3 MPa; tensile strength f_u = 12.0 MPa; strain at $f_u - \varepsilon_u = 3.5‰$; modulus of elasticity E = 41.9 GPa. Steel reinforcement bars B500B (with f_{sk} = 500 MPa) were used for both stirrups and longitudinal rebars.

To observe the behavior of UHPFRC at the three stages of its performance, the beam was designed with a reinforcement bar of 34 mm diameter at a height of 187 mm from the bottom, thus, the distance between the bottom face of the beam and bottom of the reinforcement was 170 mm at midspan (Figure 5), allowing for observation of unreinforced UHPFRC. To impose bending failure rather than shear failure under four-point bending, Ω shaped stirrups, Ø 6 mm, were placed outside the constant bending moment zone. Additionally, L-shaped Ø 34 mm reinforcement bars were used on the bottom of the beam, outside of the constant bending zone, to increase the lever arm of longitudinal reinforcement in shear.

The beam of 2 m span was tested in displacement-controlled four point bending. The displacement of the servo-hydraulic actuator was transmitted with the use of hinges and a steel beam. The application points were symmetrically positioned at ±0.25 m from the midspan of the R-UHPFRC beam. The course of the actuator was done with velocity of 0.01 mm/s during the first loading, and 0.02 mm/s in unloading and re-loading phases. Several unloadings were performed to obtain the residual deflection of beams at each loading stage.

The beam was instrumented with extensometers; photogrammetry DIC by means of two 20 MP cameras; DFO sensors (Figure 5). The fiber optics sensors for distributed strain sensing were installed in three lines at each face of the beam—40, 90 and 190 mm from the bottom of the beam. As shown in Figure 6, the DFO sensors were glued in a 2 × 2 mm groove on the UHPFRC surface using a bicomponent epoxy adhesive (Araldite 2014-2). On the front side of the beam, the SMF-28 Thorlabs® fiber was glued, with an external diameter of Ø 900 µm and elastomer tubing. On the back side of the beam, the Luna® High-Definition Polyimide fiber was used, with an external diameter of Ø 155 µm. The DIC measurement zone spanned 35 cm from the midspan symmetrically, and over the whole height of the beam on the front side. The extensometers of 100 mm measurement base were glued on the back side of the beam, at the level of each DFO measurement line. Additionally, three LVDTs with the common measurement base were vertically installed on the back side of the beam, at midspan and over the supports. The mean vertical displacement over the supports is subtracted from the vertical displacement at midspan to obtain the deflection of the beam. The resistance force was measured by the load cell of the actuator.

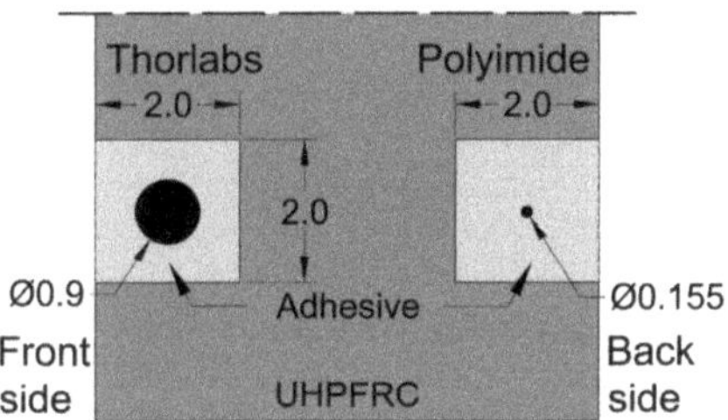

Figure 6. Scheme of installation method of DFO sensors in UHPFRC; dimensions in mm.

4. Test Results

4.1. Global Response of the Beam

The force–midspan deflection curve is presented in Figure 7. Several loading–unloading cycles were executed at different stages of the test. The goal was to visualize the influence of residual strain of the UHPFRC in the strain-hardening domain after unloading on the global response of the beam. Thanks to this, a gradual degradation of material can easily be observed. The load steps (LS) were chosen arbitrarily to discuss the state of material in detail.

The first linear part of the curve is very short. This is due to the material at the bottom of the beam entering the strain-hardening regime relatively soon. As the zone where UHPFRC is in the strain-hardening regime is growing, gradual reduction in material stiffness, and thus, beam rigidity occurs, effecting nonlinearity of the force–deflection curve. The residual deflection in the unloading cycle comes from the fact that this part of cross-section contains discontinuities (microcracks < 50 µm) or, in the further stages, the fictitious crack (> 50 µm) is present. Both types of discontinuities transmit the tensile stresses thanks to fibers, but do not close completely while unloaded. Finally, when the beam resistance is maximum with the force of 313 kN, gradual degradation with a rise in deflection continues as the localized fictitious cracks propagate and the longitudinal rebar is yielding.

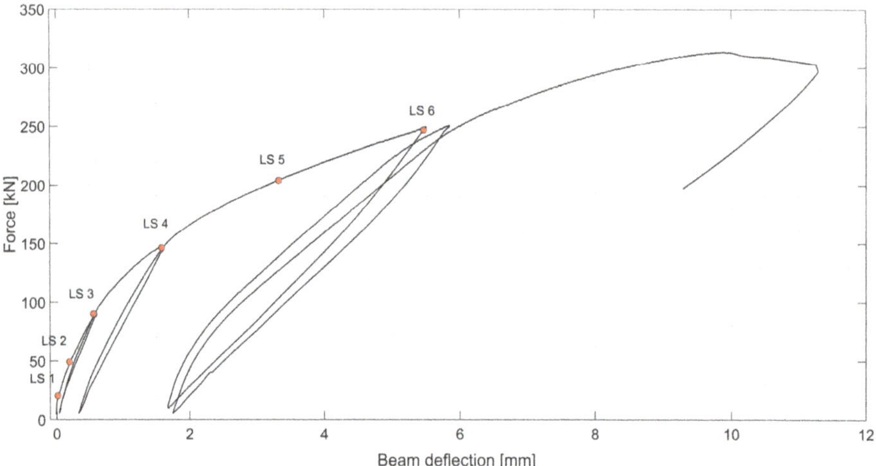

Figure 7. Load–deflection curve of the R-UHPFRC beam during quasi-static test, total jack force presented with consecutive load steps (LS) marked.

4.2. Detailed Examination: DIC

The unloaded state of the beam is presented in Figure 8. The measurement noise for DIC is mostly at the level of ±100 $\mu\varepsilon$, with local peaks of −300 $\mu\varepsilon$. The strain noise is not only due to the quality of the cameras and nonuniformity of light, but mostly from the variation of size and distribution of speckles.

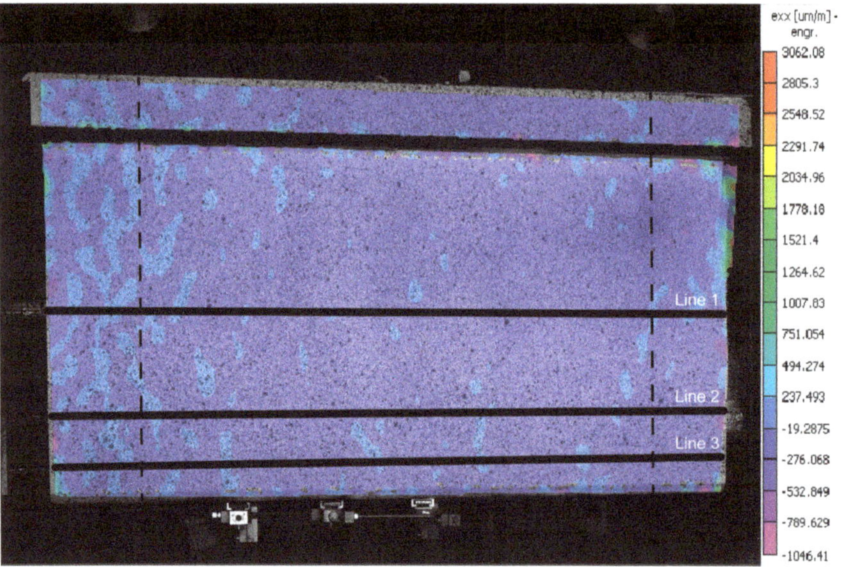

Figure 8. DIC measurement at unloaded state with virtual measurement lines noted; the constant bending moment zone is marked with dashed lines.

Figure 9 presents the horizontal (ε_{xx}) strain distribution measured with DIC at different load steps (LS), marked in Figure 7. The color scale is the same as presented in Figure 8. LS 1 is not shown since mostly the noise is registered. At load step 2, the uniform elastic strains are registered. At LS 3, strain peaks are observed due to the distributed microcracking of UHPFRC in strain-hardening. They are more pronounced in two weak spots at around −0.25 m and +0.25 m (Figure 9b). While the microcracks keep growing and propagating, two of them grow faster than others, leading to localization of fictitious cracks (Figure 9c). At LS 5, fictitious Crack 1 develops a second, left branch. This could be due to the first branch reaching a stronger area with higher concentration of steel fibers (Figure 9d). Simultaneously, fictitious Crack No. 2 keeps propagating on the right side. At LS 6, the fictitious cracks are clearly visible to the naked eye (Figure 9e). As the fibers bridge these macrocracks, the overall response of the beam remains in the hardening domain. When the beam reaches the peak resistance with a force load of 313 kN, the fictitious cracks reach the level of the reinforcement bar (see Figure 9f). In their bottom part, they transmit hardly any stress due to the large opening and advanced fiber pullout. This is why the bottom part of the beam between the fictitious cracks is almost unloaded. The highest strains are present at the level of the reinforcement bar. After this stage, due to transformation of the fictitious cracks into real cracks with no stress transfer and reinforcement yielding, the resistance of the beam started to decrease and the test was stopped.

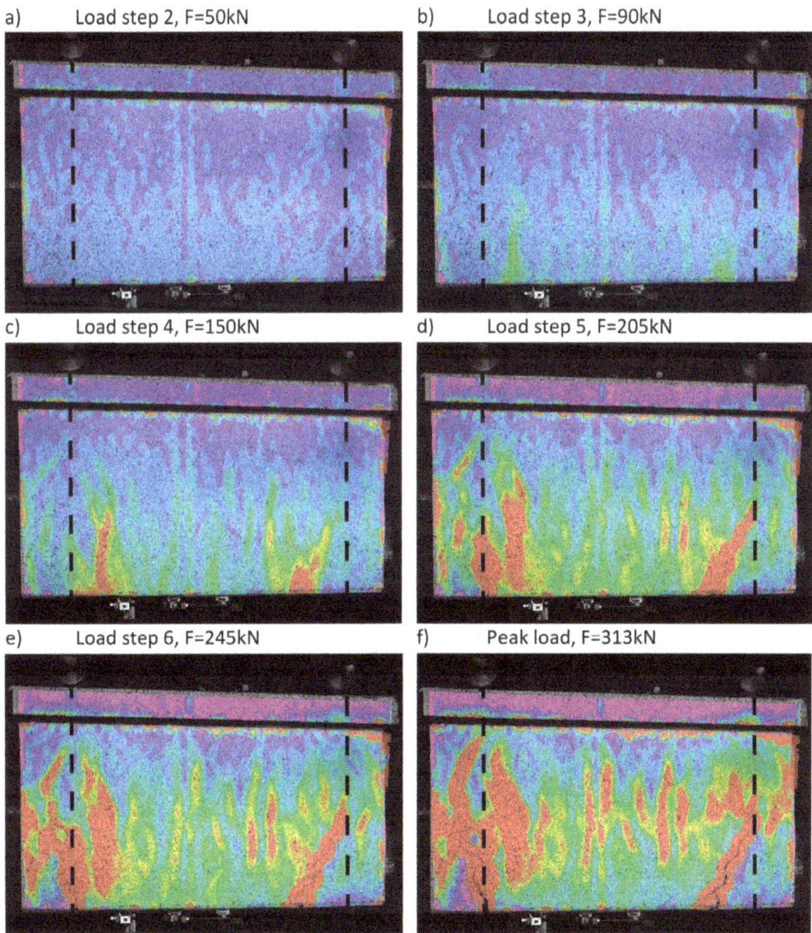

Figure 9. Strain distribution obtained with DIC for different load steps: (**a**) LS2, (**b**) LS3, (**c**) LS4, (**d**) LS5, (**e**) LS6, (**f**) Peak load; ε_{xx} with scale as in Figure 8; the constant bending moment zone is marked with dashed lines.

4.3. Detailed Examination: Strain Measurements

The strain is directly obtained from DFOs and extensometers. For DIC, the virtual measurement lines (Figure 8), positioned at the same height as the DFOs and extensometers (Figure 5), were prepared in the post-processing software VIC 3D®.

The measurements taken along each measurement Line 1, 2 and 3 are presented below from the top to the bottom, respectively, in coherence with their position over the height of the beam (Figure 5).

At load step 1 (LS 1), all systems, except DIC, show good agreement (Figure 10). The strain spatial distribution can be considered as uniform in the constant bending moment zone, between ±0.25 m from the midspan. The material remains elastic and the structural response is linear. In the bottom line (Line 3), local variations of modulus of elasticity can be visible with extensometers and Polyimide, but not with Thorlabs fiber. Possibly, initial microcracks appear. Two strain peaks are visible at Line 2 position −0.25 m. Apparently, there is a local defect of the UHPFRC material there, possibly due to the fabrication of the beam. It can be considered as a disturbance on the surface, since other measurement techniques do not record it.

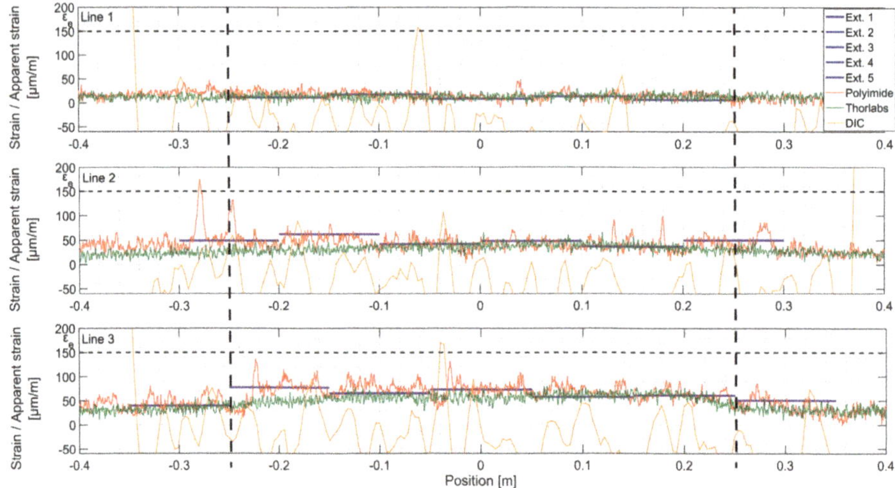

Figure 10. Strains measured with different techniques, load step 1, F = 20 kN; the constant bending moment zone is marked with dashed lines.

At LS 2 (Figure 11), as UHPFRC enters into the hardening stage at Lines 2 and 3, clear peaks from microcracks are visible over the Polyimide lines, while Thorlabs do not present any local strain variations. The microcracked zones can be identified with extensometers as well, which show higher strains than in the neighboring zones. It is important to mention that at this stage, propagating microcracks are identifiable by the naked eye when the surface is wetted with alcohol. The difference between the measurements of systems deployed on the back (Polyimide, extensometers) and the front (Thorlabs, DIC) sides of the beam may come from non-horizontal loading of the beam, despite the hinge between the actuator and redistribution beam, or to locally weaker material close to the surface. In perfect conditions, the total elongation obtained with extensometers and DFOs, thus, 'smeared' strain, are equal [39]. These incoherencies are not observed for Line 1, which remains in the elastic stage.

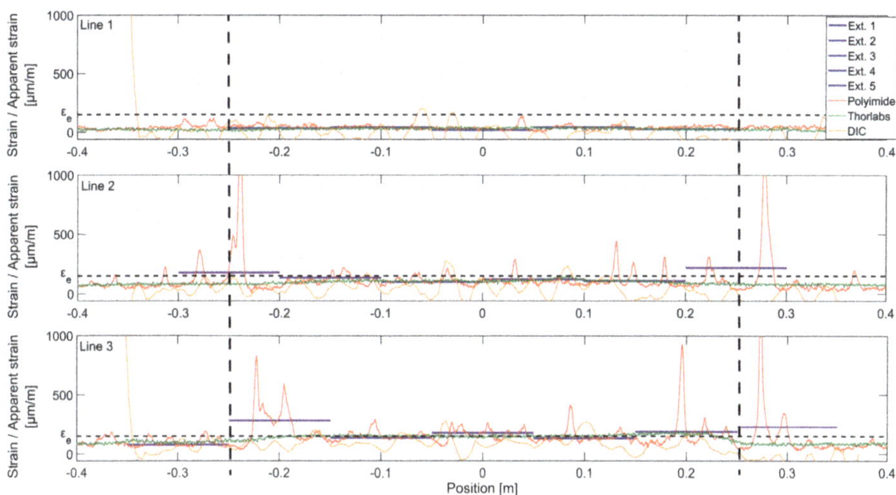

Figure 11. Strains measured with different techniques, load step 2, F = 50 kN; the constant bending moment zone is marked with dashed lines.

Under the force of 90 kN (LS 3, Figure 12), more microcracks are visible over Line 3. Most of these cracks are localized around positions −0.2 m and +0.25 m. Line 2 presents more uniform microcracking behavior. The origin of this nonuniformity in the bottom of the beam is discontinuity of the L-shaped rebars (see Figure 5) causing disturbance of fiber orientation and concentration of stresses. Zones where the fictitious crack will further develop are now clearly visible with DIC and Thorlabs fiber (Crack 1 and 2 line 3, Crack 1 line 2), as well as extensometers (both fictitious cracks, Lines 2 and 3). For zones where the fictitious cracks are developed, the apparent strain measured with extensometers cannot be considered as material strain anymore (see Figure 3). Clear microcracks start to appear at Line 1.

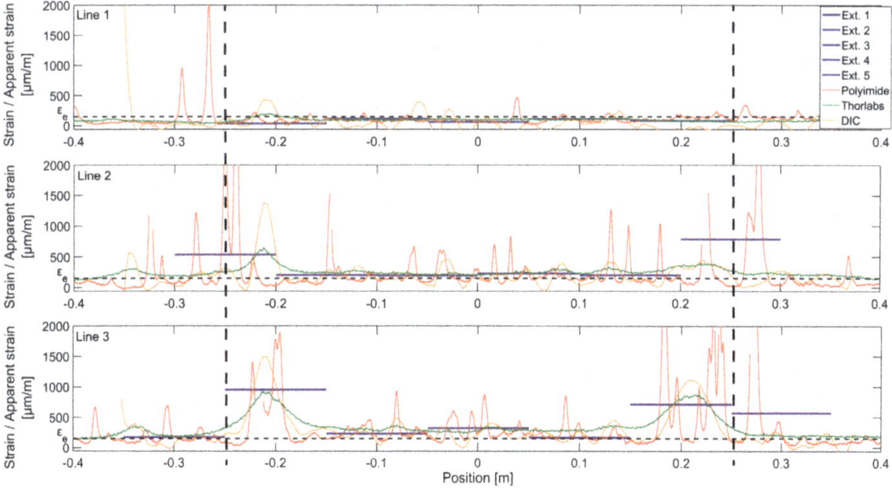

Figure 12. Strains measured with different techniques, load step 3, F = 90 kN; the constant bending moment zone is marked with dashed lines.

At LS4 (Figure 13), dropout points start to appear at crack locations in Lines 2 and 3 in the DFOs results. These points are dropped out by the spectral shift calculation algorithm due to low correlation with the reference spectrum. This phenomenon of miscalculated points increases due to rapid variation of strain over the spatial resolution length. For fictitious Crack 1, Line 3, DIC shows three fictitious crack fronts forming at positions: −0.25, −0.2 and −0.17 m. They all lay within the same extensometer measurement base, and thus, cannot be distinguished with this technique. Additionally, the Thorlabs fiber is arguably not sensitive enough to clearly separate these fronts due to low shear lag parameter λ (Equation (1)). For fictitious Crack 2, Line 3, the apparent strain reaches ε_u, exponential shape is being formed in Thorlabs, and UHPFRC enters softening stage. Two other fictitious crack fronts can be noticed with DIC but hardly with Thorlabs. Extensometer of location [0.05; 0.15 m] does not show fictitious crack formation, while it is visible in the same position with DIC and Thorlabs fiber. This comes from the nonorthogonality of the crack regarding the beam axis and is confirmed by Polyimide fiber recording only microcracks in the discussed location. On Line 2, localization of fictitious Crack 2 starts being detectable by Thorlabs fiber and DIC.

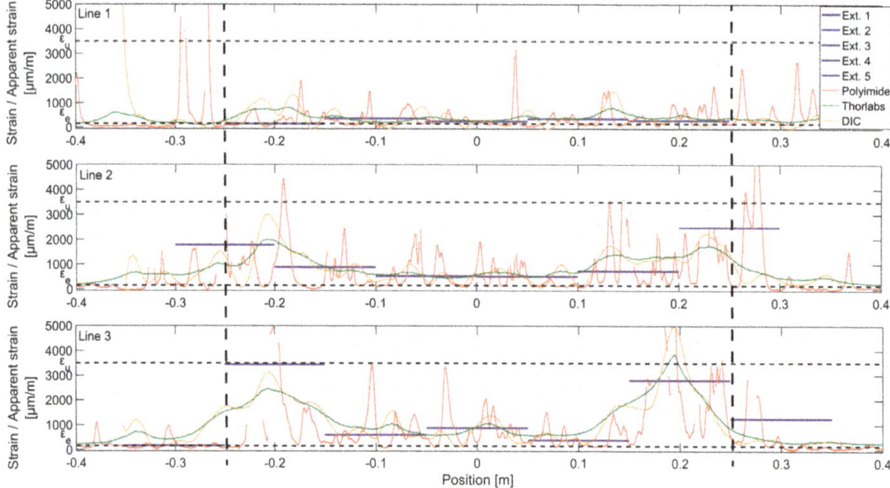

Figure 13. Strains measured with different techniques, load step 4, F = 150 kN; the constant bending moment zone is marked with dashed lines.

At LS 5 (Figure 14), both fictitious cracks are clearly formed in Line 3, and UHPFRC is in the softening stage. The DFOs do not work properly in their vicinity anymore. The transversal skewness of fictitious Crack No. 1 can be seen, since the peak of DIC is shifted with respect to the extensometers. Interestingly, it is positioned some 7 cm towards the left regarding the previously observed strain concentration. For both Lines 2 and 3, the clear exponential shapes can be noticed in Thorlabs fiber measurements, but with multiple dropouts. While comparing the measurement Line 1 at the current load step with Line 2 and Line 3 at LS 3, it can be concluded that microcracking is more uniformly spaced for the lines positioned higher on the beam. The reasons might be the nonuniformity of fiber dispersion and discontinuity of strains, both due to the rebar alignment. At this load step, the fictitious cracks are clearly visible to the naked eye, and UHPFRC is in the softening stage (see Figure 9d). The stress transferred by bridging fibers is lower than f_u (see Figure 4), and stress in the neighboring material decreases. Thus, the strain measured at midspan is similar for all the measurement lines.

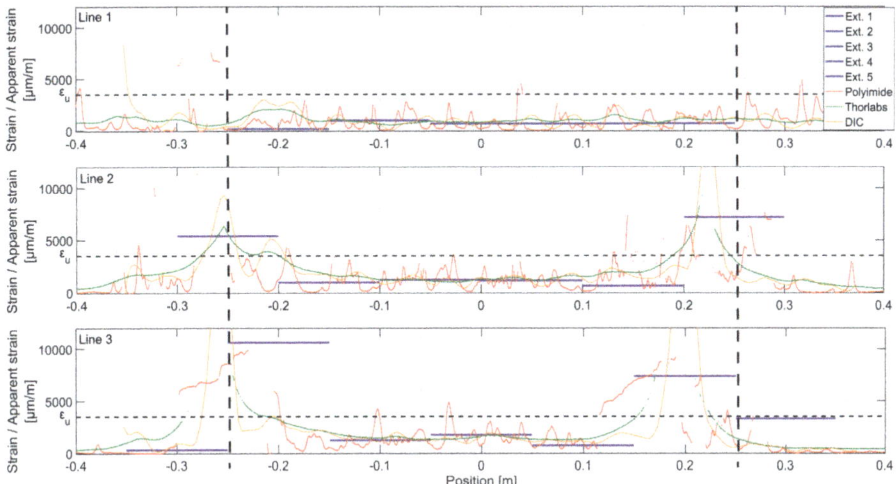

Figure 14. Strains measured with different techniques, load step 5, F = 205 kN; the constant bending moment zone is marked with dashed lines.

Due to multiple dropouts, DFOs are not useful anymore. The DIC measurements were presented before and, as mentioned above, the results obtained with extensometers crossed by fictitious cracks are not useful. Thus, the detailed analysis of strains ends here.

4.4. Monitoring of Fictitious Crack Opening

After examining the DFOs, discontinuity detection performance, it is interesting from both a structural and material point of view to follow the material discontinuities that evolve to discrete fictitious cracks in order to assess their implication on the safety of the UHPFRC structure. Thus, in this section, the strain transfer model is applied to Thorlabs fiber measurements. The Polyimide fiber was not examined because of its limited dynamic range that does not exceeded 80µm in ordinary concrete [41], preventing fictitious COD monitoring.

The notation of COD is continued here in view of previously discussed state of the art for crack measurement in concrete. As explained before, UHPFRC has more complex response under tensile action. Conveniently, the term COD refers to opening of the matrix discontinuity, be it a microcrack in strain-hardening stage, a fictitious crack bridged by fibers or a real crack with no stress transfer.

The mechanical strain transfer equation for the multiple cracks case is fitted to the strain profiles using the robust least square method:

$$\varepsilon_f(z) = \sum_{i=1}^{20} \frac{\text{COD}_i}{2} \lambda e^{-\lambda|z-z_i|} + \varepsilon_m(z) \tag{3}$$

where COD_i is the opening displacement of each discontinuity i, and λ is the strain lag parameter. Each COD_i and λ are selected as variable parameters. Similar to [42], a trapezoidal approximation of material strain ε_m (z) is adapted based on the measurements outside the constant bending moment zone; z_i corresponds to the position of the 20 most important strain peaks in the strain profiles.

Figures 15 and 16 present fitted strain profiles to those measured over Line 2 and Line 3 respectively, together with the corresponding residuals for different load levels. A discontinuity propagates in the UHPFRC material through searching the lowest energy path depending on the local fiber content and orientation [51]. Despite the host material's complex microcracking nature, the proposed mechanical model fits clearly the distributed strain profiles measured by the DFOs system at different levels. Low residual levels are randomly scattered around zero all over the length of FO Line 2 and 3.

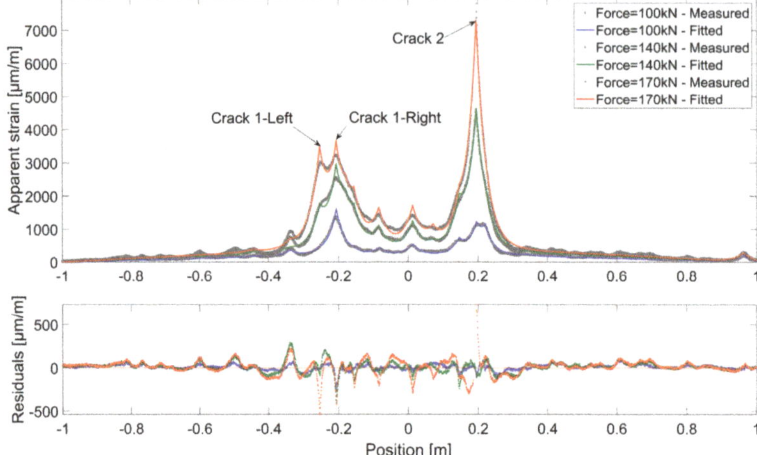

Figure 15. Comparison between the measured and fitted strain profiles and the corresponding residuals over Line 3 of Thorlabs cable.

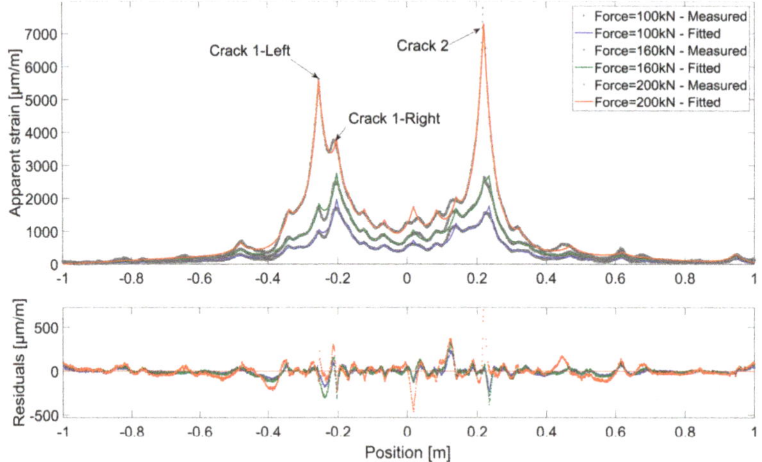

Figure 16. Comparison between the measured and fitted strain profiles and the corresponding residuals over Line 2 of Thorlabs cable.

On the left beam part, two microcracks are developing to form fictitious cracks. Unlike in concrete, there is no immediate unloading around these discontinuities. Thus, when the fictitious crack localizes and the stress transferred by bridging fibers reaches the value f_u, another fictitious crack can appear nearby. This phenomenon is observed with fictitious Crack 1, where the propagation of one branch stops (Crack 1-Right) and a second one develops (Crack 1-Left). On the other hand, fictitious Crack 2 goes through a more localized propagation. When the force reaches 170 kN for Line 3 and 200 kN for Line 2, an increase in strain residuals is observed around Crack 2. As discussed in reference [41], this could be attributed to the optical cable/host material mechanical system entering a post-elastic phase.

Figure 17 shows the estimated strain lag parameter λ as well as the discontinuity openings COD_i of fictitious Cracks 1 and 2, under loading and unloading cycles. For both Line 2 and 3, the estimated strain lag parameter λ varies around 35 m^{-1} in a ± 10% interval (Figure 17a,c). Higher λ values can be observed at early stages of the tests. Similar to previous findings on concrete structures [42,53], this variation can be associated with the first stages of UHPFRC cracking behavior, where discontinuities in the cementitious matrix are starting to develop in the so-called fracture process zone, and end up leading to the creation of a microcrack. When most of the matrix discontinuities exceed an estimated opening COD_{min} of 50 µm, the strain lag estimations become stable and consistent around $\lambda \approx 35$ m^{-1}. This confirms the assumption of one global strain lag parameter characterizing the Thorlabs fiber/epoxy glue/UHPFRC mechanical response in the presence of a fictitious crack. Lower λ (compared to concrete's surface-mounted fibers (50 m^{-1})) can be attributed to a lower stiffness level at the Epoxy/UHPFRC interface, possibly due to much smaller porosity.

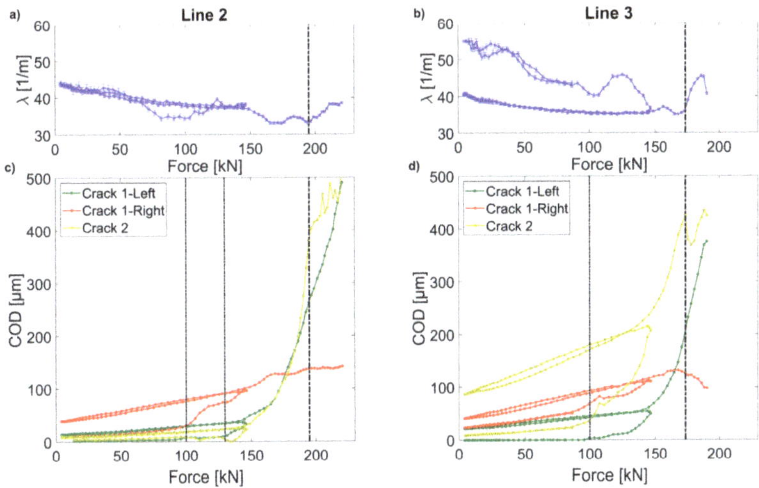

Figure 17. Variation of the estimated COD ((**a**,**b**)) and λ ((**c**,**d**)) during the test; fictitious crack initiation marked with thin dashed lines; loss of precision marked with thick dashed lines.

The estimated CODs for discontinuities Cracks 1 and 2 are shown in Figure 17 c and d. At the level of Line 3, the discontinuities Crack 1-right and Crack 2 are formed as microcracks (<50 µm) and propagate steadily until a force of around 80kN, where they grow rapidly to form fictitious cracks. At F = 120 kN, another microcrack grows rapidly to form the fictitious Crack 1-left. This growth leads to a decrease in the growth rate of fictitious Crack 1-right.

Similar development of COD for the three discontinuities can be observed for Line 2, with a delay regarding Line 3 due to its closer position to the neutral beam axis. Akin two-phased growth of COD is observed: stable during microcracking and fast once the fictitious crack is formed in the softening phase.

The growth of COD of fictitious Crack 2 is faster than Crack 1, where the damage development is shared by the two branches. Once it reaches a COD_{max} of 400 µm, unstable growth in estimated COD is observed in both measurement lines. This threshold marks the validity limit of the strain transfer model, where all the layers behave in an elastic manner with no progressive debonding occurring at successive layer interfaces. This phenomenon, equally observed in concrete [41], is pronounced by a change in the exponential form of the strain profiles initiated near the strain peak, and thus, leading to an increase in strain residuals (Figures 15 and 16). Consequently, this leads to a change in the tendency of λ and COD variations due to increased estimation errors.

Importantly, the COD and λ estimations show proper agreement for the loading–unloading cycles. In other words, the UHPFRC as well as the optical fibers attached to it deform in the same manner, even under multiple crack opening and closing over an important area of the beam. It also shows the great potential of the DFOS techniques to monitor residual and periodical openings of discontinuities, which is an important feature for long-term structural health monitoring and studying of the fatigue of the structural elements.

In this experiment, the large noise level of DIC measurement prevented accurate microcrack and early fictitious crack opening measurements. The COD values obtained using DIC were outside of applicability of the DFOs measurement method. Thus, the results cannot be validated using both methods.

5. Discussion

The detailed analysis of results revealing the differences in performance of discussed measurement techniques is presented and summed up in Table 1.

Table 1. Comparison of used measurement methods regarding their application range.

Measurement Method		Extensometers	DIC	DFOs	
				Polyimide	Thorlabs
Strain measurement	Elastic	+	+/−	+	+
	Strain-hardening	+	+/−	−	+
Distributed microcracks	Detection	+	+/−	+	−
	Localization	−	+/−	+	−
	Measurement	−	+/−	+	−
Localized cracks / fictitious cracks	Detection	+/−	+	+	+
	Localization	−	+	−	+
	Measurement	−	+	−	+
Comments		Limited area covered; simplest in application and analysis	Highly dependent on noise and area of interest	Measurement of microcracks theoretically possible	Crack measurement range limited to 400 μm for UHPFRC

+ - yes / − - no

The extensometers allow for measurement of strains in the elastic and strain-hardening phases, which is important from a practical 'smeared' approach point of view. They allow for early detection of microcrack propagation with faster rise of strains in the given area in the strain-hardening phase of the UHPFRC response. However, it is impossible to distinguish between accumulation of distributed microcracks and the onset of the fictitious crack formation. Thus, the determination whether the material is in the hardening or softening phase cannot be directly achieved. Additionally, the strain resolution and the localization of discontinuities is limited to the measurement base length of the extensometer. Furthermore, they do not allow for measurements of the fictitious crack opening. Still, they remain the measurement technique that is the easiest in installation and provide results that can be analyzed straightforwardly.

Due to the large measurement noise, the DIC technique did not allow in this experiment for observation of strain variations during the elastic stage of the structural response. However, it allows for tracking the localized fictitious cracks, particularly their length and their opening, at the macro-level. The large measurement noise is due to the relatively large measurement field (0.7 × 0.4 m) and nonuniformity of speckles. It was proven that this technique allows for tracking of microcracks for smaller observation fields [45]. This method remains highly complex and sensitive in practical application for Structural Health Monitoring.

The results obtained using the DFOs technique depend on the sensitivity of the used optical cable or fiber, with regard to discontinuities in the host material. The fiber with Polyimide coating features high sensitivity, allowing early and accurate detection and localization of microcracks. Through computation of the total elongation of segment of fiber, the strain of UHPFRC in the strain-hardening domain can be obtained [28]. Therefore, both the practical 'smeared' as well as 'discrete' approach to distributed microcracking can be used. This is relevant for Structural Health Monitoring, as structural UHPFRC remains in the elastic or strain-hardening state during normal service life.

On the other hand, the Thorlabs fiber with Acrylate and Hytrel double coatings features lower crack sensitivity than Polyimide fiber. This allows for strain measurement during elastic and strain-hardening stages. The detection, localization and measurement of microcracks is limited due to its sensitivity. It is however capable of detecting and localizing fictitious cracks, as well measuring their opening since their formation and up to 400 μm. More importantly, in this range, the optical fiber sensors maintain their elastic behavior allowing accurate estimation of cracks widths during closing–opening cycles. From a practical point of view, formation of fictitious cracks can indicate problems in the UHPFRC structure, for example, due to overloading. Thus, this kind of DFOs can play an important role in SHM and verification of structural safety.

In order to take full advantage of the DFOs technique, both types of optical fibers with their different crack sensitivity could be used to monitor the behavior of UHPFRC in the elastic, strain-hardening and softening domains.

In recent years, rapid development in the field of DFOs interrogation units enabled accurate, continuous, dynamic and simultaneous strain sensing along different types of optical fibers. With a better understanding of the sensor properties (like crack and temperature sensitivity) and durability (long-term fatigue and aging), DFOs technique can perform global and local strain measurements to provide information on the overall UHPFRC behavior in a holistic manner. Thus, DFOs can form an undeniable asset for long-term continuous health monitoring of this type of new structures.

6. Conclusions

In this work, the DFOs technique based on the Rayleigh backscattering phenomenon is used to follow the behavior of the R-UHPFRC beam tested under four-point bending. The capacity to measure strains and monitor matrix discontinuities with two types of optical fiber sensors was evaluated. The comparison with DIC and extensometers revealed application ranges of each method.

The usefulness of extensometers is limited to the elastic and strain-hardening phases. They can measure strains in the UHPFRC and detect microcrack accumulation. It is impossible to distinguish between advanced microcracking and nucleation of fictitious cracks.

The DIC is highly dependent on size of the measurement field. In this research, it allowed for detection and tracking of fictitious cracks. The complexity regarding the measurement area preparation and data processing makes this technique too complex to be used in situ for now.

The DFOs technique is able to precisely monitor the elastic, strain-hardening and softening stages of UHPFRC. While using a high spatial resolution OBR measurement technique, its performance depends on the type of fiber used for sensing. While strain measurement in the elastic phase or detection and localization of microcracks is of interest, Polyimide coated optical fiber should be used. If the strain measurement in both elastic and strain-hardening phases or fictitious crack detection and localization is to be observed, the Thorlabs fiber with thicker coating prevails.

It was confirmed that the COD of fictitious cracks can be successfully estimated using the proposed analytical model with proper choice of sensing optical fiber. Importantly, the coherent estimation of opening–closing fictitious crack width shows the potential of this method for SHM under repeated loading and real-time SHM of UHPFRC structures. However, testing of the optical fiber sensors under high numbers of crack closing/opening cycles should be considered.

Author Contributions: Conceptualization, B.S. and A.B.; methodology, A.B., X.C. and B.S.; software, A.B. and B.S.; validation, B.S. and A.B. formal analysis, A.B. and B.S.; investigation, B.S. and A.B.; resources, B.S. and A.B.; data curation, A.B. and B.S.; writing—original draft preparation, B.S. and A.B.; writing—review and editing, B.S.,

A.B., E.B., X.C. and D.L.; visualization, B.S. and A.B.; supervision, E.B. and X.C.; project administration, B.S.; funding acquisition, E.B. and X.C. All authors have read and agreed to the published version of the manuscript.

Funding: From the research and the APC were funded by the European Union's Horizon 2020 research and innovation program under the Marie Skłodowska-Curie grant agreement No 676139 – INFRASTAR.

Acknowledgments: Both of the first authors contributed equally in test preparation and execution as well as data processing and interpretation.

Conflicts of Interest: The authors declare no conflict of interest.

References

1. Ferdinand, P. The Evolution of Optical Fiber Sensors Technologies during the 35 Last Years and Their Applications in Structure Health Monitoring. In Proceedings of the EWSHM−7th European Workshop on Structural Health Monitoring, Nantes, France, 27 June 2014.
2. Berrocal, C.G.; Fernandez, I.; Rempling, R. Crack Monitoring in Reinforced Concrete Beams by Distributed Optical Fiber Sensors. *Struct. Infrastruct. Eng.* **2020**, 1–16. [CrossRef]
3. Bao, T.; Wang, J.; Yao, Y. A fiber optic sensor for detecting and monitoring cracks in concrete structures. *Sci. China Technol. Sci.* **2010**, *53*, 3045–3050. [CrossRef]
4. Voss, K.F.; Wanser, K.H. Fiber sensors for monitoring structural strain and cracks. In Proceedings of the Second European Conference on Smart Structures and Materials, Glasgow, UK, 12−14 October 1994; pp. 144–147. [CrossRef]
5. Wan, K.T.; Leung, C.K.Y. Applications of a distributed fiber optic crack sensor for concrete structures. *Sens. Actuator A Phys.* **2007**, *135*, 458–464. [CrossRef]
6. Imai, M.; Nakano, R.; Kono, T.; Ichinomiya, T. Crack Detection Application for Fiber Reinforced Concrete Using BOCDA-Based Optical Fiber Strain Sensor. *J. Struct. Eng.* **2010**, *136*, 1001–1008. [CrossRef]
7. Mészöly, T.; Randl, N. An advanced approach to derive the constitutive law of UHPFRC. *ACEE* **2019**, *11*, 89–96. [CrossRef]
8. McDonach, A.; Gardiner, P.T.; McEwen, R.S. *International Digital Image Correlation Society*; Jones, E.M.S., Iadicola, M.A., Eds.; A Good Practices Guide Digital Image Correlation; Springer: Berlin/Heidelberg, Germany, 2018. [CrossRef]
9. Shen, X.; Brühwiler, E. Characterization of the tensile behavior in UHPFRC thin slab using NDT method and DIC system. In Proceedings of the 2nd International Conference on UHPC Materials and Structures UHPC, Fujian, China, 7–10 November 2018.
10. Schramm, N.; Fischer, O. Precast bridge construction with UHPFRC—Shear tests and railway bridge pilot application. *Civil Eng. Des.* **2019**, *1*, 41–53. [CrossRef]
11. Toutlemonde, F.; Simon, A.; Rivillon, P.; Marchand, P.; Baby, F.; Quiertant, M.; Khadour, A.; Cordier, J.; Battesti, T. Recent experimental investigations on reinforced UHPFRC for applications in earthquake engineering and retrofitting. In Proceedings of the UHPFRC 2013 - International Symposium on Ultra-High Performance Fibre-Reinforced Concrete, Marseille, France, 2–4 October 2017; pp. 597–606.
12. Friedl, H.; Río, O.; Freytag, B.; Rodriguez, Á. Advanced bridge monitoring application for analysing actual structural performance of UHPFRC constructions -Wild bridge. In Proceedings of the 6th Central European Congress on Concrete Engineering, Marianske Lazne, Czechia, 30 September−1 October 2010.
13. Hartog, A. Distributed Fiber-Optic Sensors: Principles and Applications. In *Optical Fiber Sensor Technology: Advanced Applications—Bragg Gratings and Distributed Sensors*; Grattan, K.T.V., Meggitt, B.T., Eds.; Springer: Boston, MA, USA, 2000; pp. 241–301. [CrossRef]
14. Leung, C.K.Y.; Wang, X.; Olson, N. Debonding and Calibration Shift of Optical Fiber Sensors in Concrete. *J. Eng. Mech.* **2000**, *126*, 300–307. [CrossRef]
15. Liu, H.; Chen, J.; Sun, M.; Ding, R. Theoretical analysis and experiment of micromechanics and mechanics-optics coupling of distributed optic-fiber crack sensing. *Sci. China Technol. Sci.* **2011**, *54*, 185–191. [CrossRef]
16. Olson, N.; Leung, C.K.Y.; Meng, A. Crack sensing with a multimode fiber: Experimental and theoretical studies. *Sens. Actuator A Phys.* **2005**, *118*, 268–277. [CrossRef]
17. Bao, X.; Chen, L. Recent Progress in Brillouin Scattering Based Fiber Sensors. *Sensors* **2011**, *11*, 4152–4187. [CrossRef]

18. Feng, X.; Zhou, J.; Sun, C.; Zhang, X.; Ansari, F. Theoretical and Experimental Investigations into Crack Detection with BOTDR-Distributed Fiber Optic Sensors. *J. Eng. Mech.* **2013**, *139*, 1797–1807. [CrossRef]
19. Imai, M.; Feng, M. Sensing optical fiber installation study for crack identification using a stimulated Brillouin-based strain sensor. *Struct. Health Monit.* **2012**, *11*, 501–509. [CrossRef]
20. Klar, A.; Linker, R. Feasibility study of automated detection of tunnel excavation by Brillouin optical time domain reflectometry. *Tunn. Undergr. Space Technol.* **2010**, *25*, 575–586. [CrossRef]
21. Bastianini, F.; Rizzo, A.; Galati, N.; Deza, U.; Nanni, A. Discontinuous Brillouin strain monitoring of small concrete bridges: Comparison between near-to-surface and smart FRP fiber installation techniques. *Sensors* **2005**, *5765*, 612–623.
22. Minardo, A.; Bernini, R.; Amato, L.; Zeni, L. Bridge Monitoring Using Brillouin Fiber-Optic Sensors. *IEEE Sens. J.* **2012**, *12*, 145–150. [CrossRef]
23. Zhang, H.; Wu, Z. Performance Evaluation of BOTDR-based Distributed Fiber Optic Sensors for Crack Monitoring. *Struct. Health Monit.* **2008**, *7*, 143–156. [CrossRef]
24. Henault, J.M.; Quiertant, M.; Delepine-Lesoille, S.; Salin, J.; Moreau, G.; Taillade, F.; Benzarti, K. Quantitative strain measurement and crack detection in RC structures using a truly distributed fiber optic sensing system. *Constr. Build. Mater.* **2012**, *37*, 916–923. [CrossRef]
25. Rodríguez, G.; Casas, J.R.; Villaba, S. Cracking assessment in concrete structures by distributed optical fiber. *Smart Mater. Struct.* **2015**, *24*, 035005. [CrossRef]
26. Rodríguez, G.; Casas, J.R.; Villalba, S. Assessing Cracking Characteristics of Concrete Structures by Distributed Optical Fiber and Non-Linear Finite Element Modelling. In Proceedings of the EWSHM−7th European Workshop on Structural Health Monitoring, Nantes, France, 8–11 July 2014; p. 8.
27. Brault, A.; Hoult, N.A.; Greenough, T.; Trudeau, I. Monitoring of Beams in an RC Building during a Load Test Using Distributed Sensors. *J. Perform. Constr. Facil.* **2019**, *33*, 04018096. [CrossRef]
28. Sieńko, R.; Zych, M.; Bednarski, Ł.; Howiacki, T. Strain and crack analysis within concrete members using distributed fibre optic sensors. *Struct. Health Monit.* **2019**, *18*, 1510–1526. [CrossRef]
29. Sun, L.; Hao, H.; Zhang, B.; Ren, X.; Li, J. Strain Transfer Analysis of Embedded Fiber Bragg Grating Strain Sensor. *J. Test. Eval.* **2016**, *44*, 20140388. [CrossRef]
30. Zhou, Z.; Wang, Z.; Shao, L. Fiber-Reinforced Polymer-Packaged Optical Fiber Bragg Grating Strain Sensors for Infrastructures under Harsh Environment. *J. Sens.* **2016**, 3953750. [CrossRef]
31. Her, S.-C.; Huang, C.-Y. Effect of Coating on the Strain Transfer of Optical Fiber Sensors. *Sensors* **2011**, *11*, 6926–6941. [CrossRef] [PubMed]
32. Zhao, H.; Wang, Q.; Qiu, Y.; Chen, J.; Wang, Y.; Fan, Z. Strain Transfer of Surface-Bonded Fiber Bragg Grating Sensors for Airship Envelope Structural Health Monitoring. *J. Zhejiang Univ. Sci. A* **2012**, *13*, 538–545. [CrossRef]
33. Duck, G.; Renaud, G.; Measures, R. The Mechanical Load Transfer into a Distributed Optical Fiber Sensor Due to a Linear Strain Gradient: Embedded and Surface Bonded Cases. *Smart Mater. Struct.* **1999**, *8*, 175–181. [CrossRef]
34. Yuan, L.; Zhou, L.; Wu, J. Investigation of a Coated Optical Fiber Strain Sensor Embedded in a Linear Strain Matrix Material. *Opt. Laser Eng.* **2001**, *35*, 251–260. [CrossRef]
35. Ansari, F.; Libo, Y. Mechanics of Bond and Interface Shear Transfer in Optical Fiber Sensors. *J. Eng. Mech.* **1998**, *124*, 385–394. [CrossRef]
36. Li, J.; Zhou, Z.; Ou, J. Interface Transferring Mechanism and Error Modification of Embedded FBG Strain Sensor Based on Creep: Part I. Linear Viscoelasticity. In *Smart Structures and Materials 2005: Sensors and Smart Structures Technologies for Civil, Mechanical, and Aerospace Systems*; International Society for Optics and Photonics: Bellingham, WA, USA, 2005; Volume 5765, pp. 1061–1072. [CrossRef]
37. Zhou, G.; Li, H.; Ren, L.; Li, D. Influencing Parameters Analysis of Strain Transfer in Optic Fiber Bragg Grating Sensors. In Proceedings of the Advanced Sensor Technologies for Nondestructive Evaluation and Structural Health Monitoring; International Society for Optics and Photonics, Xi'an, China, 22 April 2006. [CrossRef]
38. Wang, H.; Zhou, Z. Advances of strain transfer analysis of optical fibre sensors. *Pac. Sci.* **2014**, *16*, 8–18. [CrossRef]

39. Wang, C.; Olson, M.; Doijkhand, N.; Singh, S. A novel DdTS technology based on fiber optics for early leak detection in pipelines. In Proceedings of the 2016 IEEE International Carnahan Conference on Security Technology (ICCST), Orlando, FL, USA, 24–27 October 2016; pp. 1–8. [CrossRef]
40. Bassil, A.; Chapeleau, X.; Leduc, D.; Abraham, O. Concrete Crack Monitoring Using a Novel Strain Transfer Model for Distributed Fiber Optics Sensors. *Sensors* **2020**, *20*, 2220. [CrossRef]
41. Bassil, A. Distributed Fiber Optics Sensing for Crack Monitoring of Concrete Structures. Ph.D. Thesis, Université de Nantes, Nantes, France, 2019.
42. Bassil, A.; Wang, X.; Chapeleau, X.; Niederleithinger, E.; Abraham, O.; Leduc, D. Distributed Fiber Optics Sensing and Coda Wave Interferometry Techniques for Damage Monitoring in Concrete Structures. *Sensors* **2019**, *19*, 356. [CrossRef]
43. Charron, J.P.; Denarié, E.; Brühwiler, E. Permeability of ultra high performance fiber reinforced concretes (UHPFRC) under high stresses. *Mater. Struct.* **2007**, *40*, 269–277. [CrossRef]
44. Naaman, A.E. *Fiber Reinforced Cement and Concrete Composites*, 1st ed.; Techno Press 3000: Sarasota, FL, USA, 2018; ISBN 978-0967493930.
45. Shen, X.; Brühwiler, E. Influence of local fiber distribution on tensile behavior of strain-hardening UHPFRC using NDT and DIC. *Cem. Concr. Res.* **2020**, *132*, 106042. [CrossRef]
46. Charron, J.P.; Denarié, E.; Brühwiler, E. Transport properties of water and glycol in an ultra high performance fiber reinforced concrete (UHPFRC) under high tensile deformation. *Cem. Concr. Res.* **2008**, *38*, 689–698. [CrossRef]
47. Wang, K.; Jansen, D.C.; Shah, S.P.; Karr, A.F. Permeability study of cracked concrete. *Cem. Concr. Res.* **1997**, *27*, 381–393. [CrossRef]
48. SIA 2052. *Technical Leaflet SIA 2052, UHPFRC—Materials, Design and Construction*; Swiss Society of Engineers and Architects: Zurich, Switzerland, 2016.
49. Hillerborg, A.; Modéer, M.; Petersson, P.E. Analysis of crack formation and crack growth in concrete by means of fracture mechanics and finite elements. *Cem. Concr. Res.* **1976**, *6*, 773–781. [CrossRef]
50. Oesterlee, C.; Denarié, E.; Brühwiler, E. Strength and deformability distribution in UHPFRC panels. *Process. Sequence Prod. Eng. Cem. Compos.* **2009**, *1+2*, 390–397.
51. Sawicki, B.; Brühwiler, E. Static behavior of reinforced UHPFRC beams with minimal cover thickness. In Proceedings of the 2IIS-UHPC—The 2nd International Interactive Symposium on Ultra-High Performance Concrete, Albany, NY, USA, 2–5 June 2019.
52. Brühwiler, E.; Friedl, H.; Rupp, C.; Escher, H. Bau einer Bahnbrücke aus bewehrtem UHFB. *Beton-Stahlbetonbau* **2019**, *114*, 337–345. [CrossRef]
53. Bassil, A.; Niederleithinger, E.; Wang, X.; Kadoke, D.; Chapeleau, X.; Leduc, D.; Abraham, O.; Breithaupt, M.; Potschke, S. Distributed Fiber Optic Sensors for Multiple Crack Monitoring in Reinforced Concrete Structures. In Proceedings of the Structural Health Monitoring 2019 Enabling Intelligent Life-Cycle Health Management for Industry Internet of Things, Shanghai, China, 15 January 2019. [CrossRef]

Article

Embedded Fiber Sensors to Monitor Temperature and Strain of Polymeric Parts Fabricated by Additive Manufacturing and Reinforced with NiTi Wires

Micael Nascimento [1,*], Patrick Inácio [2], Tiago Paixão [1], Edgar Camacho [3], Susana Novais [1], Telmo G. Santos [2], Francisco Manuel Braz Fernandes [3] and João L. Pinto [1]

[1] Department of Physics and i3N, University of Aveiro, Campus Universitário de Santiago, 3810-193 Aveiro, Portugal; tiagopaixao@ua.pt (T.P.); susana.novais@inesctec.pt (S.N.); jlp@ua.pt (J.L.P.)

[2] UNIDEMI, Department of Mechanical and Industrial Engineering, Faculty of Science and Technology, Universidade NOVA de Lisboa, 2829-516 Caparica, Portugal; p.inacio@campus.fct.unl.pt (P.I.); telmo.santos@fct.unl.pt (T.G.S.)

[3] CENIMAT/i3N, Department of Materials Science, Faculty of Science and Technology, Universidade NOVA de Lisboa, 2829-516 Caparica, Portugal; e.camacho@campus.fct.unl.pt (E.C.); fbf@fct.unl.pt (F.M.B.F.)

* Correspondence: micaelnascimento@ua.pt; Tel.: +351-234-370-356

Received: 23 January 2020; Accepted: 17 February 2020; Published: 19 February 2020

Abstract: This paper focuses on three main issues regarding Material Extrusion (MEX) Additive Manufacturing (AM) of thermoplastic composites reinforced by pre-functionalized continuous Nickel–Titanium (NiTi) wires: (i) Evaluation of the effect of the MEX process on the properties of the pre-functionalized NiTi, (ii) evaluation of the mechanical and thermal behavior of the composite material during usage, (iii) the inspection of the parts by Non-Destructive Testing (NDT). For this purpose, an optical fiber sensing network, based on fiber Bragg grating and a cascaded optical fiber sensor, was successfully embedded during the 3D printing of a polylactic acid (PLA) matrix reinforced by NiTi wires. Thermal and mechanical perturbations were successfully registered as a consequence of thermal and mechanical stimuli. During a heating/cooling cycle, a maximum contraction of ≈100 μm was detected by the cascaded sensor in the PLA material at the end of the heating step (induced by Joule effect) of NiTi wires and a thermal perturbation associated with the structural transformation of austenite to R-phase was observed during the natural cooling step, near 33.0 °C. Regarding tensile cycling tests, higher increases in temperature arose when the applied force ranged between 0.7 and 1.1 kN, reaching a maximum temperature variation of 9.5 ± 0.1 °C. During the unload step, a slope change in the temperature behavior was detected, which is associated with the material transformation of the NiTi wire (martensite to austenite). The embedded optical sensing methodology presented here proved to be an effective and precise tool to identify structural transformations regarding the specific application as a Non-Destructive Testing for AM.

Keywords: optical fiber sensors; material extrusion; hybrid processes; temperature and strain monitoring

---

1. Introduction

Shape memory alloys (SMA) composites show their exceptional performance in adapting some physical parameters, such as shape, vibration, and impact resistance through a centralized control system [1,2]. However, a deformation of composite structures, resulting in the complex redistribution of stress states between matrix and reinforcement, is considered to be an important research topic that has been reported in the literature [2]. Thus, critical states reached by extreme deformation can be detected by using embedded sensors, which provide reliable information to the system.

In the last years, important issues related with some constraints identified in SMA alloys, such as nickel–titanium (NiTi), have been reported: (1) The importance of indirect identification of the present phases and their transformations in a composite system incorporating SMA elements [3], (2) the incorporation of SMA allows to monitor the deformation/stress state of structural components but, if shape memory alloying elements are to be used to act as actuators, composite monitoring must be done by third sensory elements added. It is, therefore, important that these sensors can identify not only temperature variations and mechanical stress, but also, indirectly, the structural constituents present and the structural changes they undergo, especially given the nonlinearities of response that may exist in certain contexts [4].

The use of optical fiber sensors (OFS) for structural monitoring and detection of defects and temperature fluctuations in different materials has been suggested by several researchers [5–7]. There are many advantages associated with OFS technology, such as its reduced dimensions, immunity to electromagnetic interference, passivity, chemical inertness, multiplexing capability, nearly punctual sensing, and the capability to measure different parameters within one single optical fiber [8]. Regarding the specific application of OFS as a Non-Destructive Testing (NDT) for Additive Manufacturing (AM), different solutions can be applied as a complementary technique [9].

There are multiple OFS configurations, depending on the application, the required resolution, and/or sensitivity [8,10]. From all these possible configurations, fiber Bragg gratings (FBGs) are the most favorable solutions to NDT for AM based composite products. However, these sensors suffer from large cross-sensitivity, mainly strain, and temperature. To overcome this, a method based on using two different FBGs in two different fibers has shown to be the most reliable and easy technique to simultaneously monitor and discriminate these parameters [11]. Nevertheless, when it is intended to discriminate the parameters in embedded materials, such as polylactic acid (PLA) samples, this method can be a challenge, due to the need of having strain-free FBGs introduced inside other protective materials (for instance, a capillary tube), increasing the invasiveness on the sample [12].

To solve the need for internal discrimination of temperature and strain when monitoring their simultaneous variations, hybrid sensors comprising FBG and interferometric FP sensors can be fabricated, forming a cascaded optical sensor in which each element has different sensitivities. This way, the invasiveness inside the host material decreases, once only a single optical fiber is used to monitor the same point [13]. However, when this type of sensor is embedded in materials, an internal calibration for temperature and strain is needed, due to the mechanical stresses induced by the surrounding material [14,15]. Therefore, the use of OFS can be an important tool to assess and detect characteristic parameters in different types of materials, such as polymeric and/or SMA, depending on the sensing configuration used, behaving as an NDT.

This work is about the use of AM (MEX) technology for the production of polymer matrix composite materials reinforced with previously functionalized NiTi wires. Three essential aspects of the application of this technology for the production of these composite materials were addressed: (1) The evaluation of the effect of the process (AM-MEX) on the properties of NiTi wires and their heat treatment (by monitoring the time and temperature to which it is subjected during production), (2) the evaluation of the mechanical and thermal behavior of the material in service (with the measurement of the stresses during tensile tests, evaluation of the adhesion of the matrix wires), (3) non-destructive inspection/material quality, using thermography and Joule effect on embedded NiTi wires to confirm disposition and whether heat-treatment (functionally graded materials) has been maintained. For that, an optical fiber sensing network based on FBGs and cascaded OFS was embedded in a 3D printed PLA matrix reinforced by NiTi wires to real-time monitor, temperature and strain shifts, in the PLA matrix, and temperature variations, which are associated to structural transformations in NiTi wires, during Joule heating of the NiTi wires and tensile cyclic load/unload.

2. Materials and Methods

2.1. Fiber Optic Sensors

Usually, an FBG sensor consists of a short segment of a single-mode optical fiber (SMF) with a photoinduced periodic modulation of the fiber core refractive index. When this device is illuminated by a broadband optical source, the reflected power spectrum shows a sharp peak, which is caused by interference of light with the planes of the grating and can be defined through Equation (1) [10]:

$$\lambda_B = 2n_{eff}\Lambda, \quad (1)$$

where λ_B is the so-called Bragg wavelength, n_{eff} is the effective refractive index of the core mode, and Λ is the grating period. When the optical fiber is exposed to external parameters, such as temperature and strain, both n_{eff} and Λ can be modified, resulting in a shift of the Bragg wavelength.

The sensor sensitivity towards a given parameter is obtained by monitoring the Bragg wavelength behavior while exposing the sensor to pre-determined and controlled conditions. In the case of a linear response, the sensitivity is provided by the slope of the obtained linear fit. The effects of temperature are accounted for in the Bragg wavelength shift by differentiating Equation (1):

$$\Delta\lambda = \lambda_B\left(\frac{1}{n_{eff}}\frac{\partial n_{eff}}{\partial T} + \frac{1}{\Lambda}\frac{\partial \Lambda}{\partial T}\right)\Delta T = \lambda_B(\alpha + \xi)\Delta T = k_T\Delta T, \quad (2)$$

where α and ξ are the thermal expansion and thermo-optic coefficient of the optical fiber material, respectively [15]. By inscribing FBGs with different Bragg wavelengths, by changing the grating period, it is possible to get multiple temperature sensors within one single fiber. Thus, inspection and mapping of the sample temperature can be done by simultaneously monitoring the spectral variations of all sensors. This technique can be combined with other conventional methods, such as thermography analysis to produce a complete thermal analysis of a given sample material [16].

The sensing configuration that was employed to monitor the internal temperature and strain shifts on the PLA matrix consisted of cascaded optical fiber sensors, whose configuration scheme is shown in Figure 1. The simultaneous strain and temperature discrimination inside the PLA sample can be performed, combining the signals of the FBG sensor with the FP cavity interferometer, forming a cascaded optical fiber sensor. The FP cavity was fabricated by producing an air microbubble between a single-mode fiber (SMF 28e) and a multimode fiber (MMF, GIF625) [17]. To achieve point-of-care monitoring, the FBG was inscribed as close as possible to the FP interferometer.

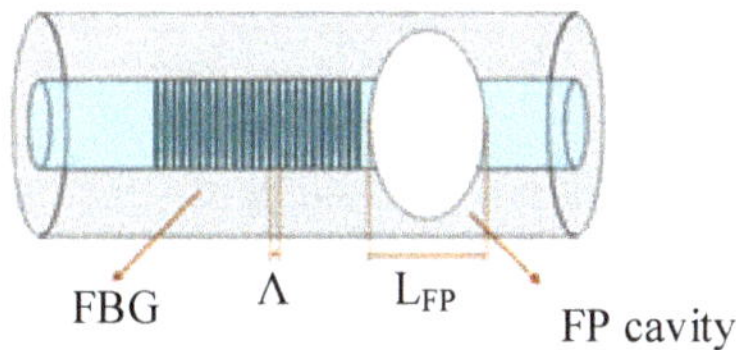

Figure 1. Diagram of the cascaded optical fiber sensor. L_{FP} represents the cavity length.

Assuming a linear response of the FBG to strain and temperature, the strain and temperature shifts ($\Delta\varepsilon$ and ΔT, respectively) are provided by:

$$\Delta\lambda_{FBG} = k_{FBG_\varepsilon}\Delta\varepsilon + k_{FBG_T}\Delta T, \quad (3)$$

where $k_{FBG\varepsilon}$, and k_{FBGT} are the strain and temperature sensitivities of the FBG, respectively, which were determined in the calibration procedure.

The FP interferometer can also work as a strain and temperature sensor, where the wavelength shift, $\Delta\lambda_{FP}$, is given by:

$$\Delta\lambda_{FP} = k_{FP_\varepsilon}\Delta\varepsilon + k_{FP_T}\Delta T, \tag{4}$$

where $k_{FP\varepsilon}$ and k_{FPT} are the strain and temperature sensitivities, respectively. Thus, the temperature and strain variations can be discriminated through the matrixial method, using Equations (3) and (4). If the sensitivity values are known, a sensitivity matrix for simultaneous measurement of strain and temperature can be obtained as:

$$\begin{bmatrix} \Delta\varepsilon \\ \Delta T \end{bmatrix} = \frac{1}{M}\begin{bmatrix} k_{FP_T} & -k_{FBG_T} \\ -k_{FP_\varepsilon} & k_{FBG_\varepsilon} \end{bmatrix}\begin{bmatrix} \Delta\lambda_{FBG} \\ \Delta\lambda_{FP} \end{bmatrix}, \tag{5}$$

where $M = k_{FPT} \times k_{FBG\varepsilon} - k_{FP\varepsilon} \times k_{FBGT}$ is the determinant of the coefficient matrix, which must be non-zero for simultaneous measurement. Thus, internal discrimination of strain and temperature in the PLA matrix can be improved by combining the reflection spectra of this cascaded optical sensor.

The main advantages of this process are different strain and temperature sensitivities between the two sensing elements, together with the use of a single fiber to monitor the same point, decreasing the invasiveness inside the PLA matrix composite. No extra-material integration is needed with this method. Moreover, the strain values obtained can be converted to displacement variations (ΔL), by multiplying the detected strain values by the sample length.

2.2. PLA Matrix and NiTi Wires

PLA is a thermoplastic material that has been widely used in components produced by AM, especially in Material Extrusion (MEX), due to the low melting point, good tensile stiffness and final surface quality. These properties potentiate the use of PLA as a matrix for the production of composite by MEX.

NiTi ribbons (cross-section 3×1 mm^2), were previously processed by localized heat-treatment at 400 °C during 10 min along a 20 mm segment, using Joule effect (21 A current). Previously, the NiTi ribbon was coated with black ink in order to establish an emissivity of around 0.95. All temperature measurements were performed with the infrared camera Fluke Ti400. After the local heat-treatment, these ribbons and the sensors were impregnated in the PLA matrix during the AM process as described in the next section.

2.3. Experimental Setup

In order to assess the temperature and strain variations in samples of PLA matrix and verify the microstructural heterogeneity along NiTi wires, composed by heat-treated regions (marked in red) and the transition zones to the non-heat-treated zone, the experimental setup illustrated in Figure 2 with OFS was used.

PLA samples were produced with the commercial BQ Prusa i3 3D printer, having a design comprising a cavity at the half-thickness in order to incorporate the NiTi wire and the optical fiber. After the fabrication, the model was sliced in the open-source software CURA. The feedstock was PLA and the print core had a nozzle with 1.2 mm of diameter. The layer height was set to have 0.5 mm, infill to 100%, and the print speed was 7 mm/s. At the half-thickness, the print was paused and the NiTi wire with the optical fiber 1 was incorporated in the PLA matrix. This process was repeated to embed the fiber 2 in the sample.

Two sets of samples of PLA + NiTi ribbon + sensors were prepared: One for the experiments on thermal cycling (no external mechanical load applied) and the other for the tensile tests (mechanical loading/unloading without any external thermal excitation).

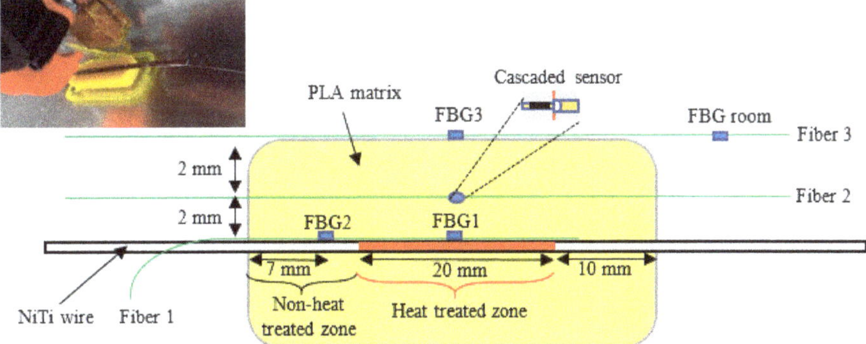

Figure 2. Experimental setup, sample cross-section view. Fiber 1 was embedded in the sample together with NiTi wire, while the 3D printing process was stopped for a few seconds. The same procedure was adopted to embed in the PLA matrix the cascaded sensor, recorded on fiber 2.

The FBGs (length of ~3.0 mm each) were recorded in a photosensitive SMF (GF1, Thorlabs Inc., Newton, MA, USA), by the phase mask method. The inscribing system consists of focusing UV (266 nm) laser pulses originated by a pulsed Q-switched Nd:YAG laser system (LOTIS TII LS-2137U Laser, Minsk, Belarus) onto the SMF core by a plano-convex cylindrical lens (working length of 320 mm), passing through a phase mask.

During the inscription of the FBGs, as well as throughout the experiments, the Bragg wavelengths were monitored by a single channel optical interrogator (sm125-500, Micron Optics Inc., Atlanta, GA, USA), operating at 1 Hz and wavelength accuracy of 1.0 pm. To read all the optical data at the same time, a 3 × 1 coupler was used.

Fiber 1 with the two FBGs were maintained strain-free, so they could detect temperature shifts of both heat-treated (FBG1) and non-heat-treated regions (FBG2). The cascaded sensor on fiber 2 simultaneously detected strain and temperature shifts on the PLA matrix. Externally, fiber 3, which has 2 FBGs (FBG3 and FBG room), was also placed in direct contact with the PLA sample surface to monitor external temperature shifts. The FBG room sensor was used out of the sample to monitor the room temperature variations, eliminating any possible external fluctuations.

The tensile tests were performed in a Shimadzu NG50KN, using a 50 kN load cell, a crosshead speed of 5 and 10 mm/min, and the maximum stroke of 6 % of the gauge length.

2.4. Optical Fiber Sensing Calibrations

FBG1, FBG2, and cascaded optical sensors integration on the NiTi wire and in the PLA matrix, respectively, were done during the 3D manufacturing process. In this case, the fibers were fully embedded in the sample, thus presenting a more accurate response towards strain and temperature, when compared to external sensing devices. Before being embedded in the material, the fiber coatings were removed to minimize the intrusiveness of the sensing structures, presenting a total thickness of only 125 µm.

Previous calibration of each sensing head towards each parameter was performed. Figure 3a,b shows the spectral response of the cascaded sensor and the FBGs after and before being embedded on the polymeric sample, and under two different temperatures, respectively. It is possible to observe the induced strain on the fiber sensors by the surrounding materials and a higher spectral change in the cascaded sensors, comparatively with the FBGs. Similar to other FBGs, the cascaded sensor (embedded in the PLA matrix) suffer higher induced strains.

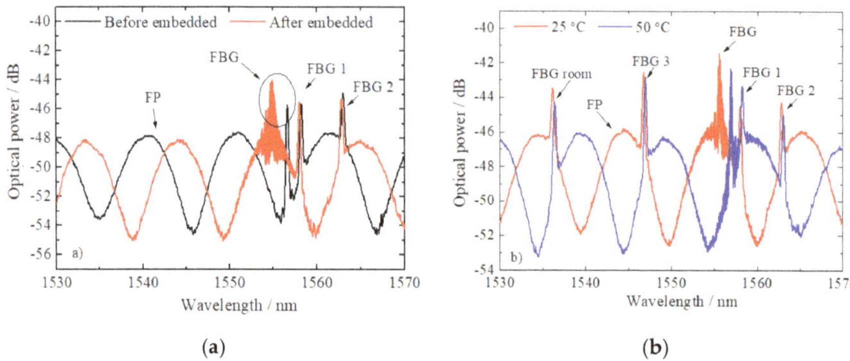

(a) (b)

Figure 3. (**a**) Spectral response of the OFS after and before embedded in the polymeric sample. (**b**) Response of the cascaded optical sensor and FBGs after embedded on the sample at two different temperatures (25.0 °C and 50.0 °C).

Table 1 presents the strain and temperature sensitivities of the sensors before and after embedding in the PLA matrix and NiTi wire. From the internal strain and temperature calibrations, and according to the matrixial method (Equation (5)), a determinant value of 8.23 was obtained for the cascaded optical sensor embedded in the PLA matrix.

Table 1. Temperature and strain sensitivities of the cascaded optical sensor and FBGs obtained before and after embedding in the PLA matrix and NiTi wire, respectively.

Cascaded Optical Sensor in PLA Matrix				
Type of sensor	($k_T \pm 0.1$) pm/°C		($k_\varepsilon \pm 0.1$) pm/με	
	Before embedding	*After embedding*	*Before embedding*	*After embedding*
FP (L = 115.4 ± 0.1 μm)	0.1	154.1	2.1	0.1
FBG	9.0	71.8	1.2	0.1
FBGs in NiTi Wire				
	($k_T \pm 0.1$) pm/°C		($k_\varepsilon \pm 0.1$) pm/με	
	Before embedding	*After embedding*	*Before embedding*	*After embedding*
FBG1	9.5	9.5	1.1	-
FBG2	8.9	8.8	1.2	-

The thermal calibrations after and before the sensors' integration in the sample were performed in a thermal chamber (Model 340, Challenge Angelantoni Industrie, Massa Martana, Italy), between 15.0 °C and 60.0 °C, in steps of 5.0 °C. The strain characterization was performed using a micrometric translation stage between 0 με and 1000 με, in steps of 50 με.

3. Results and Discussion

3.1. Joule Heating of the NiTi Wire Tests

After the additive fabrication of the samples, they were cooled down to room temperature and then, a controlled intensity current was injected on the NiTi ribbon to heat it by Joule heating effect in the temperature range of 40 to 55 °C.

The temperature variation on the heat-treated and non-heat-treated zones of the inserted NiTi wire was measured on the external surface by thermography (as can be seen in Figure 4), and internally by FBG1 and FBG2. The temperature and strain variations on the PLA matrix were monitored by the cascaded sensor.

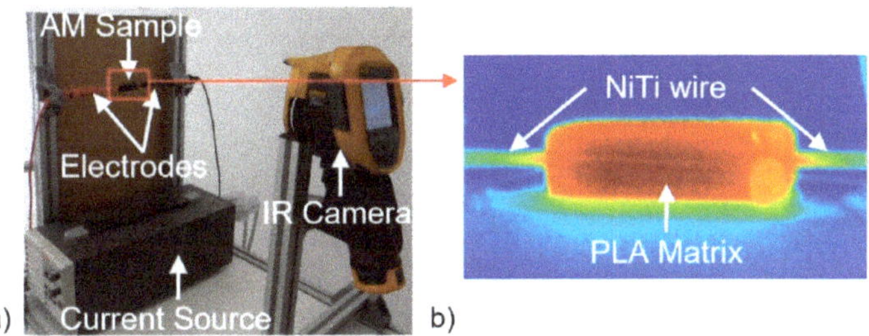

Figure 4. (**a**) Experimental setup used to perform the cycling Joule heating of the NiTi wire tests. (**b**) Inset of the external surface temperature measured by thermography in the sample.

In total, three cycling tests were performed in the sample. During the first cycle (test 1), currents of 2.12, 2.81, and 3.1 A were applied, followed by natural cooling to stabilize the sample temperature. For the tests 2 and 3, after applying the same three currents, a 4.0 A current was also used.

Figure 5 shows the results of the temperature variations detected by all the sensing elements used (FBGs, cascaded sensor, and thermography) at all the locations (surface, PLA matrix, and NiTi wire), where the cascaded sensor registers both strain and temperature variation in the PLA matrix.

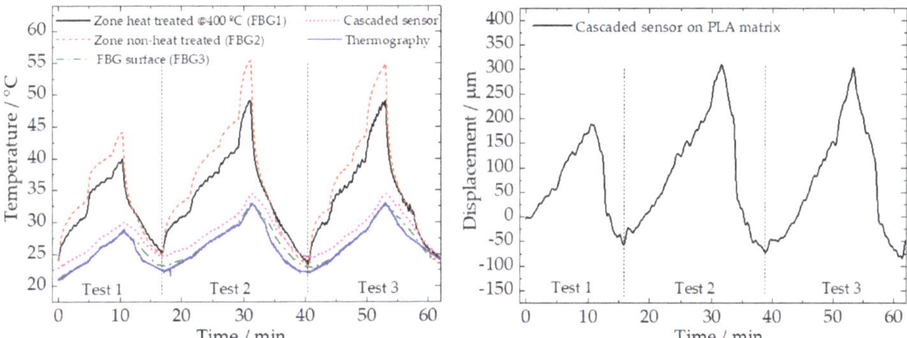

Figure 5. Temperature detected by all the sensing elements (**left**), and displacement sensed by the cascaded sensor in the PLA matrix (**right**), during the cyclic tests of heating by Joule effect, followed by natural cooling.

The results show that a consistent deviation between the temperature of the external face of the PLA matrix (measured by thermography) and the temperature at the face of the NiTi wire can be observed. Moreover, a consistent deviation on the temperature measured at two different points of the NiTi wire is observed, with a significantly lower value in the heat-treated zone (more notorious when the higher current was applied), due the higher electrical conductivity of the heat-treated region. This effect could be assigned to the local reduction of structural defects (mostly dislocations) induced by the recrystallization associated to the localized heat-treatment.

A small temperature difference (~1.5 °C) can be observed between the surface (as measured by thermography) and inside the PLA matrix, 2 mm below the surface (registered by the cascaded sensor). However, maximum displacement shifts of ~350 µm were detected, during tests 2 and 3. It is also observed that a successive contraction of the material after the heating/cooling cycles occurs. At the end of test 3, the contraction is ~100 µm, which may be due to the accommodation of the PLA material, indicating a good adhesion of the cascaded sensor to the surrounding material.

The temperature recorded by all the sensing elements is highlighted in Figure 6. A thermal perturbation associated with a material phase transformation can be observed in the curves represented by the heat-treated and non-heat-treated zones (close to 33.0 ± 0.1 °C), during the natural cooling process. This perturbation may be assigned to a structural transformation (R-phase to austenite) taking place in the NiTi wire during cooling.

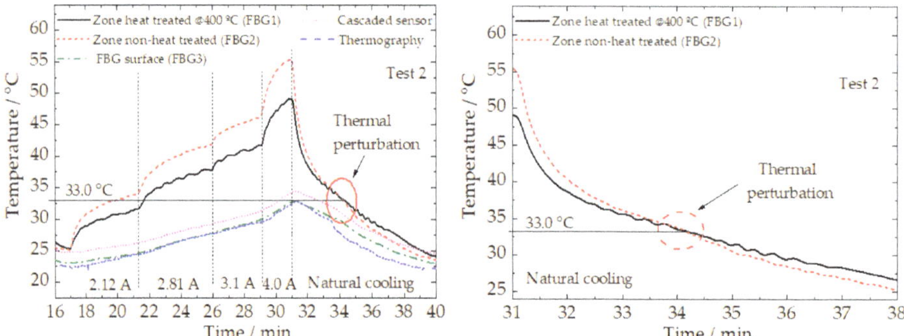

Figure 6. Temperature recorded by all the sensing elements during test 2 (**left**), and an inset during the natural cooling step (**right**), highlighting the thermal perturbation near the 33.0 °C.

The temperature shift for the two different regions of the NiTi wire (mostly remarkable during heating) may be assigned to different fractions of R-phase versus austenite (electrical resistivity of the R-phase is higher than that of the austenite). It is apparent that the optical sensors (FBG1 and FBG2) clearly identify the moment of phase transition in the two zones under study, for both cooling steps.

According to the sub-surface temperature variations detected by the surface FBG, there is a very good relationship with thermography values, although, and as expected, the temperature variations recorded internally by cascaded sensors in the PLA matrix are significantly higher (~2.0 °C difference).

3.2. Tensile Tests

Tensile cycling tests were performed on the NiTi wire with embedded sensors to study their thermal behavior regarding longitudinal deformation of the heat-treated and non-heat-treated regions. The sample was clamped on the extremities of the NiTi ribbon, by tensile test machine grips, typically used in tensile tests. Figure 7 shows the temperature results monitored by the fiber sensors placed in the heat-treated zone, non-heat-treated zone, and in the PLA matrix, while the tensile cycles were applied.

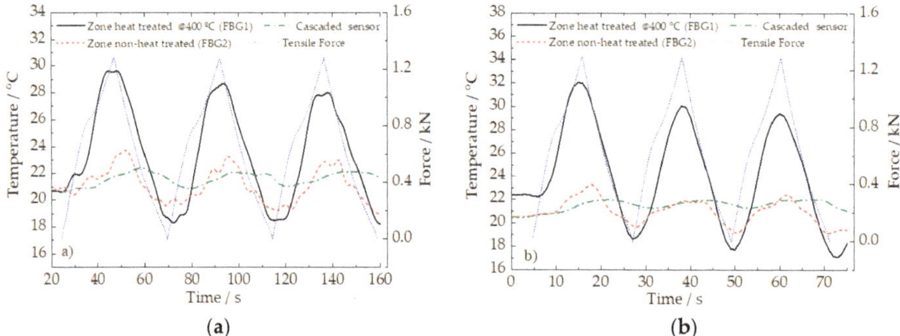

Figure 7. *Cont.*

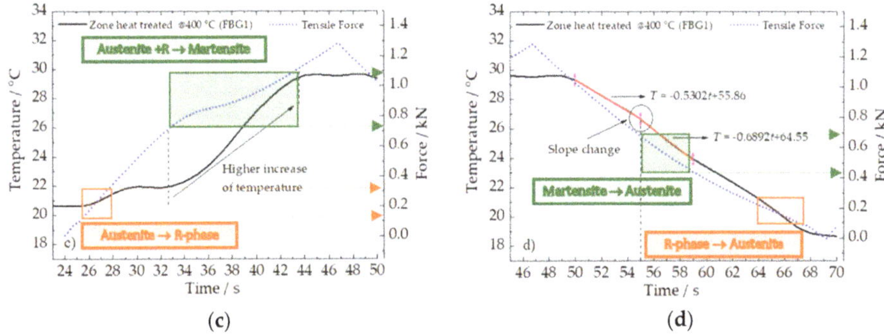

(c) (d)

Figure 7. Temperature detected by the fiber sensors during the tensile cycling tests. (**a**) Tensile test at 5 mm/min. (**b**) Tensile test at 10 mm/min. (**c**) and (**d**) highlight the load and unload steps during the first cycle, at 5 mm/min, respectively.

In total, six cycles were applied to the sample: Three of them at 5 mm/min rate (Figure 7a) and the other three at 10 mm/min rate (Figure 7b). Basically, when the wire was submitted to the longitudinal deformation, a consequent exothermic and endothermic process could be observed during loading and unloading, respectively. At the end of each load/unload cycle, the temperature reached by the sample is lower than the temperature at the beginning of the corresponding cycle.

Regarding the heat-treated and non-heat-treated zones on the NiTi wire, significative mean differences of 6.1 ± 0.1 °C, were detected. Notice that, on the non-heat-treated zone (FBG2) and the PLA matrix (hybrid sensor), the temperature peaks were reached by conduction, 4 and 10 s, respectively, after the heat-treated zone.

As can be highlighted in Figure 7c, during loading, there are two load ranges where a steeper increase of temperature is observed: First, from 0.1 to 0.3 kN related to the stress-induced austenite R-phase transformation, and a second one, from 0.7 to 1.1 kN associated with the stress-induced austenite to martensite (B19′) transformation, both transformations are exothermal. The final step of the loading (above ~1.1 kN) corresponds to the elastic deformation of the stress-induced martensite, which, for this deformation rate, does not produce a significant amount of heat.

Analyzing the three cycles applied for each deformation rate, a successive decrease of maximum temperature is recorded by the optical sensors, being a consequence of the cooling associated to the reverse transformation (martensite to austenite) that takes place during the last step (downloading). The next cycle will then start from a lower temperature, so that the heating associated with the direct transformation (austenite to martensite) will not cause such an accentuated temperature increase as the one for the previous cycle. Additionally, due to this lower increase of temperature, the next downloading step will then go to a slightly lower temperature. This effect (decrease of the maximum temperature at the end of the loading step and of the minimum temperature at the end of the downloading) will be attenuated from one cycle to the next one.

Figure 7d, shows a zoom-in of the unload step during the first cycle at 5 mm/min. Near 55 s, a slope change may be observed in the temperature behavior. That is probably associated with the reverse stress-induced transformation (martensite to austenite) which is endothermal. At the final step of the unloading (below 0.3 kN, between 62 and 67 s), a new slope variation occurs, which is related to the R-phase→austenite transition (also endothermal).

For this particular application, the embedded sensing network proves to be a very effective tool to perform NDT, especially to identify and detect very localized temperature and strain variations, during operating processes in composite services. Comparing to other techniques (for instance thermography), this solution could detect a wide range of different parameters, such as structural transformations in SMA, and has very good reliability.

4. Conclusions

An OFS network was successfully developed and embedded during the 3D printing by AM of a PLA matrix reinforced by NiTi wire. Real-time monitoring of characteristic parameters, such as internal temperature and displacement shifts in the matrix and temperature variations in different treated zones of the wire allowed to use the OFS network as an NDT.

Joule heating experiments of NiTi wires were performed to assess changes to the sample temperature. The moments in which different currents were injected on the sample can be clearly proved and measured by all integrated fiber sensors. During the natural cooling, a thermal perturbation (structural transformation of R-phase to austenite) can be observed near 33.0 °C, and at the end of the cycling tests, a sample contraction of ~100 µm was detected on the PLA sample.

Regarding the tensile tests, the higher increase of temperature (exothermic behavior) arises when the applied force is between the 0.7 and 1.1 kN, on the heat-treated zone. During the unload step, a slope variation in the temperature behavior associated with the thermal-induced transformation in the heat-treated region (R-phase to austenite) was detected.

The embedded optical sensing methodology presented proved to be an effective, minimally invasive, and precise tool to identify materials' structural transformations, revealing to be a suitable solution to be applied as an NDT for Additive Manufacturing.

Author Contributions: Conceptualization and methodology, M.N., P.I., T.P., E.C., S.N., T.G.S. and F.M.B.F.; validation, M.N., P.I., T.G.S., F.M.B.F. and J.L.P.; data curation, M.N., P.I., T.P., E.C., S.N.; sensors fabrication, M.N., T.P. and S.N. Polymeric and NiTi wire fabrication, P.I., E.C., T.G.S. and F.M.B.F.; Software, M.N., P.I. and E.C.; Supervision, T.G.S., F.M.B.F. and J.L.P.; data analysis, M.N., S.N., P.I., T.G.S., F.M.B.F.; writing—original draft preparation, M.N.; writing—review and editing, all authors; funding acquisition, T.G.S., F.M.B.F. and J.L.P. All authors have read and agreed to the published version of the manuscript.

Funding: Authors gratefully acknowledge the funding of Project POCI-01-0145-FEDER-016414 (FIBR3D), cofinanced by Programa Operacional Competitividade e Internacionalização and Programa Operacional Regional de Lisboa, through Fundo Europeu de Desenvolvimento Regional (FEDER) and by National Funds through FCT—Fundação para a Ciência e a Tecnologia, Grant numbers BI/UI96/6642/2018, BI/UI96/6643/2018, PD/BD/128265/2016 (DAEPHYS), respectively. E. Camacho and F.M. Braz Fernandes acknowledge the funding of CENIMAT/I3N by National Funds through FCT, references UID/CTM/50025/2019 and FCT/MCTES. T. G. Santos and P. Inácio acknowledge FCT—MCTES for its financial support via the project UIDB/00667/2020 (UNIDEMI). P. Inácio acknowledge FCT—MCTES for its financial support via the PhD scholarship FCT-SFRH/BD/146885/2019. This work was also developed within the scope of the project i3N, UIDB/50025/2020 & UIDP/50025/2020, financed by national funds through the FCT/MEC.

Conflicts of Interest: The authors declare no conflict of interest.

References

1. Patrick, L.I.; Camacho, E.; Rodrigues, P.F.; Miranda, R.M.; Velhinho, A.; Braz Fernandes, F.M.; Santos, T.G. Evaluation of functionally graded Ni-Ti wires for additive manufacturing of composites. In Proceedings of the 12th European Conference on Non-Destructive Testing (ECNDT2018), Gothenburg, Sweden, 11–15 June 2018; pp. 1–3.
2. Ho, M.P.; Lau, K.T.; Au, H.Y.; Dong, Y.; Tam, H.Y. Structural health monitoring of an asymmetrical SMA reinforced composite using embedded FBG sensors. *Smart Mater. Struct.* **2013**, *22*, 125015. [CrossRef]
3. Lester, B.; Baxevanis, T.; Chemisky, Y.; Lagoudas, D. Review and Perspectives: Shape Memory Alloy Composite Systems. *Acta Mech.* **2015**, *226*, 3907–3960. [CrossRef]
4. Pinto, F.; Ciampa, F.; Meo, M.; Polimeno, U. Multifunctional SMArt composite material for in situ NDT/SHM and de-icing. *Smart Mater. Struct.* **2012**, *21*, 105010. [CrossRef]
5. Antunes, P.; Rodrigues, H.; Travanca, R.; Ferreira, L.; Varum, H.; André, P. Structural health monitoring of different geometry structures with optical fiber sensors. *Photonic Sens.* **2012**, *2*, 357–365. [CrossRef]
6. Manzo, N.R.; Callado, G.T.; Cordeiro, C.M.B.; Vieira, L.C.M., Jr. Embedding optical fiber Bragg grating (FBG) sensors in 3D printed casings. *Opt. Fiber Technol.* **2019**, *53*, 102015. [CrossRef]
7. Santos, J.L.; Farahi, F. *Handbook of Optical Sensors*; CRC Press: Boca Raton, FL, USA, 2015.
8. Wu, Q.; Okabe, Y. High-sensitivity ultrasonic phase-shifted fiber Bragg grating balanced sensing system. *Opt. Express* **2012**, *20*, 28353–28362. [CrossRef] [PubMed]

9. Grattan, K.T.; Meggitt, B.T. *Optical fiber sensor technology (Volume 2: Optoelectronics, Imaging and Sensing Series)*; Kluwer Academic Publishers: Berlin, Germany, 1999.
10. Nascimento, M.; Ferreira, M.S.; Pinto, J.L. Simultaneous Sensing of Temperature and Bi-Directional Strain in a Prismatic Li-Ion Battery. *Batteries* **2018**, *4*, 23. [CrossRef]
11. Homa, D.; Hill, C.; Floyd, A.; Pickrell, G. Fiber Bragg gratings embedded in 3D printed prototypes. *Sci. Adv. Today* **2016**, *2*, 25242.
12. Nascimento, M.; Novais, S.; Ding, M.; Ferreira, M.S.; Koch, S.; Passerini, S.; Pinto, J.L. Internal strain and temperature discrimination with optical fiber hybrid sensors in Li-ion batteries. *J. Power Sources* **2019**, *410*, 1–9. [CrossRef]
13. Singh, A.K.; Berggren, S.; Zhu, Y.; Han, M.; Huang, H. Simultaneous strain and temperature measurement using a single fiber Bragg grating embedded in a composite laminate. *Smart Mater. Struct.* **2017**, *26*, 115025. [CrossRef]
14. Zubel, M.G.; Sugden, K.; Webb, D.J.; Saez-Rodriguez, D.; Nielsen, K.; Bang, O. Embedding silica and polymer fibre Bragg gratings (FBG) in plastic 3D-printed sensing patches. In Proceedings of the SPIE 9886, Micro-Structured and Specialty Optical Fibres IV, Brussels, Belgium, 27 April 2016; p. 98860N. [CrossRef]
15. Rao, Y.J. In-fibre Bragg grating sensors. *Meas. Sci. Technol.* **1997**, *8*, 355. [CrossRef]
16. Carvalho, M.; Martins, A.; Santos, T.G. Simulation and validation of thermography inspection for components produced by additive manufacturing. *Appl. Therm. Eng.* **2019**, *159*, 113872. [CrossRef]
17. Duan, D.W.; Rao, Y.J.; Hou, Y.S.; Zhu, T. Microbubble based fiberoptic Fabry–Perot interferometer formed by fusion splicing single-mode fibers for strain measurement. *Appl. Opt.* **2012**, *51*, 1033–1036. [CrossRef]

Article

Improved Visual Inspection through 3D Image Reconstruction of Defects Based on the Photometric Stereo Technique

Sanao Huang [1,2], Ke Xu [1,*], Ming Li [2] and Mingren Wu [1]

[1] University of Science and Technology Beijing, Collaborative Innovation Center of Steel Technology, Beijing 100083, China; huangsanao@cgnpc.com.cn (S.H.); mingrw_exc@163.com (M.W.)

[2] In-Service Testing Center, CGN Inspection Technology Co., Ltd, Suzhou 215000, China; liming@cgnpc.com.cn

* Correspondence: xuke@ustb.edu.cn

Received: 10 October 2019; Accepted: 12 November 2019; Published: 14 November 2019

Abstract: Visual inspections of nuclear power plant (NPP) reactors are important for understanding current NPP conditions. Unfortunately, the existing visual inspection methods only provide limited two-dimensional (2D) information due to a loss of depth information, which can lead to errors identifying defects. However, the high cost of developing new equipment can be avoided by using advanced data processing technology with existing equipment. In this study, a three-dimensional (3D) photometric stereo (PS) reconstruction technique is introduced to recover the lost depth information in NPP images. The system uses conventional inspection equipment, equipped with a camera and four light-emitting diodes (LEDs). The 3D data of the object surface are obtained by capturing images under multiple light sources oriented in different directions. The proposed method estimates the light directions and intensities for various image pixels in order to reduce the limitation of light calibration, which results in improved performance. This novel technique is employed to test specimens with various defects under laboratory conditions, revealing promising results. This study provides a new visual inspection method for NPP reactors.

Keywords: nuclear power plant; visual inspection; photometric stereo; 3D reconstruction

---

1. Introduction

The reactor pressure vessel (RPV) of a nuclear power plant (NPP) requires periodic inspection to ascertain current conditions. Any defects on the internal surfaces may undermine the safe operation of the NPP. Moreover, exposure to irradiation and corrosive coolants, or damage caused by manufacturing and outage activities, could accelerate the growth of these defects [1]. Thus, inspection systems and implementation practices must be capable of detecting small flaws, to prevent them from growing to a size that could compromise the leak tightness of the pressure boundary.

Visual inspection is the main method for detecting defects, structural integrity issues, or leakage traces on the surface of key components in an NPP. Owing to its advantages, the demand for more advanced visual inspection techniques is increasing. The U.S. Nuclear Regulatory Commission (NRC) has approved the use of high-resolution cameras for inspecting specific areas of key NPP components instead of ultrasonic examination [2]. In addition, machine vision technology has been applied to measure fuel assembly deformation [3].

Visual inspection systems capture images of the surfaces of objects by using an image sensor, a charge-coupled device (CCD), or a complementary metal-oxide semiconductor (CMOS), with appropriate optical tools and lighting conditions. The visual module is typically composed of light sources and image sensing units, and completes the inspection with the aid of automated tools. Companies such as AREVA and CYBERIA in France, DEKRA in Germany, Ahlberg in Sweden,

DIAKNOT in Russia, and Westinghouse in the United States have been actively developing such devices. The majority of inspection systems are equipped with high-definition cameras, include a choice of light sources, from halogen to light-emitting diode (LED) lights, and boast anti-irradiation and waterproof properties. The types of automated tools are diverse. Some products have data processing functions, which are becoming increasingly popular. For example, AREVA's latest RPV device, SUSI 420 HD, is equipped with a high-definition camera and four adjustable high-power LEDs, but the sizing of indications is limited to the length measurement [4].

Existing NPP visual inspection methods still use two-dimensional (2D) images to identify defects. Occasionally, the lack of three-dimensional (3D) observations makes it difficult to evaluate certain observations—specifically, the potential size of the defect. Some inspection tasks can only be performed using 3D analysis methods, because surface defects may only appear with changes in the shape of the surface [5]. Therefore, visual inspections can be improved by detecting changes in the 3D surface. Three-dimensional shape reconstruction methods based on visible light include structured light and stereo vision technology. For example, the laser 3D scanner of the Newton Laboratory can map NPP fuels and check the size of defects [6]. Karl Storz laser technology, called MULTIPOINT, is a 3D laser system with 49 laser points that enables cooperation between the camera and the software to detect the surface structure of the subject [7]. However, few of these devices can operate individually without additional overheads. Therefore, to save time and money, it is preferable to use conventional devices to extract and analyze 3D data for defects. One such method achieves 3D reconstruction of the inner surfaces of boreholes or cavities using conventional endoscopy equipment [5]. However, this method does not provide system calibration results or evaluation criteria of the results; thus, further improvement is required. Furthermore, the 3D visualization function obtained through the shape from motion (SFM) method can be used to inspect the advanced, air-cooled core in an NPP, but a lack of surface features limits the application of this method [8].

The photometric stereo (PS) technique recovers 3D shapes from multiple images of the same object, taken under different illumination conditions. As a result of the pioneering work of Woodham, it has been widely applied to 3D surface reconstruction [9]. This technique features two advantages: low hardware costs and low computation costs. In the field of industrial inspection, PS has improved the detection of very small surface defects [10,11]. The Lambertian model assumption is commonly used where albedo is assumed to be constant. Although this does not necessarily correspond to the actual conditions, there are approaches available to realize the normal calculations [12–14]. However, for models with non-Lambertian reflection properties, highlight and shadow processing requires additional images [13].

The assumption of the light source and the demand for extensive calibration procedures in conventional PS limit its applicability [15]. Some previous studies have established illumination models that conform to actual conditions, such as near-field light models [16–19]. Light calibration, which aims to estimate the light direction and intensity, often requires a specific equipment or a dedicated process [20]. However, equipment such as precise calibration spheres or positioning devices are unlikely to be available in actual applications. Some studies have proposed fully uncalibrated or semi-calibrated PS methods. A fully uncalibrated, near-light PS method achieves the calculation of the light positions, light intensities, normal, depth, and albedo without making any assumptions about the geometry, lights, or reflectance of the scene [21]. Regarding semi-calibrated PS, various approaches achieve light intensity calibration [22]. However, additional information is required to solve the high-dimensional ambiguity, and more importantly, at least 10 images are typically required [15]. To apply PS to an NPP environment, a fully automatic calibration method should be designed.

This study employs the PS method with a conventional NPP visual inspection device to reconstruct the 3D shapes of defects from visual inspection images. Additional contributions include the development of an auto-calibrated, near-field light calibration method that can easily and accurately calibrate the light source to meet the demands of practical applications. Moreover, depth information can be extracted from the captured images, which enables better and more reliable visualization of surface defects.

The structure of this paper is as follows: The PS formulation is briefly introduced in Section 2. The algorithm details for extracting 3D information from captured images are described in Section 3, as well as the method for estimating light direction and intensity. The experimental setup and results are discussed in Section 4, and the conclusions are presented in Section 5.

2. Photometric Stereo Technique

PS techniques use multiple images taken from the same viewpoint, but under illumination from different directions, in order to recover the surface orientation from a known combination of reflectance and lighting values. The depth and shape of the surface can be obtained via the reconstruction algorithms. The objects in the scene are Lambertian, and the illumination is a distant point light; the measured image intensity at a point $P(x,y,z)$ can be written as:

$$I = \rho\langle l, n\rangle E \tag{1}$$

where ρ is the albedo at P; $\boldsymbol{l}$ is the light direction; $\boldsymbol{n}$ is the surface normal; and E is the light irradiance. The image intensity I can be measured per-pixel.

At least three independent light sources are required.

Suppose we have $M \geq 3$ images under varying light directions, which we denote as direction vectors $\boldsymbol{l_1}, \ldots \ldots, \boldsymbol{l_M} \in \mathbb{R}^3$. By assuming equal light irradiance; i.e., $\rho E = \rho E_1 = \cdots\cdots = \rho E_M$, we can estimate the normal vector $\boldsymbol{n}$ on a surface point P by solving the pseudo-inverse matrix of $\boldsymbol{L}$:

$$n = \frac{\left(L^T L\right)^{-1} L^T I}{\left\|\left(L^T L\right)^{-1} L^T I\right\|} \tag{2}$$

where $L = \begin{bmatrix} l_1 \\ \vdots \\ l_M \end{bmatrix}, I = [I_1, \cdots, I_M]^{\mathrm{T}}$.

The surface normal can be computed using Equations (1) and (2). Then, the recovery of the surface from the computed normal can be achieved using algorithms, including optimization iterative methods and pyramid reconstruction algorithms. The entire procedure of PS-based 3D shape reconstruction includes: calibration, surface normal computation, and shape reconstruction from the normal.

3. Three-Dimensional Shape Reconstruction of Defects

3.1. Existing Two-Dimensional Image Capture and Data Analysis Method

The RPV generally consists of a cylindrical part, a spherical part, and nozzles. Although its size varies, the diameter of the cylinder is always approximately 4000 mm. An examination of the entire internal surface of the reactors is required to detect surface disorders, deformation, or other important defects; i.e., (1) mechanical defects, such as scratches or impact damages caused by foreign bodies; (2) metallurgical defects, such as cracks or arc strikes; and (3) corrosion pitting or deposits. Moreover, the defects may need to be evaluated both qualitatively and quantitatively.

During the inspection process, tools are used to position the cameras close to the different areas requiring inspection. Scanning is performed using tools to record videos of the internal surface, using cameras facing the wall within a field limited to the zone under examination. Technicians observe the video and images throughout the entire process to identify defects. If an abnormality or suspicious observation is recorded, the movement of the tool can be paused to allow more detailed information to be observed manually.

Because of its beneficial features, a pan/tilt/zoom (PTZ) setup, which normally consists of a camera, a number of surrounding light sources, lasers (optional), and a built-in pan/tilt unit, has been widely used as the visual module. The pan and tilt functions allow adjustment of the camera position to capture better images. Data analysis is performed by inspectors; therefore, suitable frames are required for observation and evaluation, along with other information recorded during the inspection, to make a qualified evaluation [23]. As the existing analysis method is based on 2D images, it is often difficult to evaluate defects due to a lack of depth information; thus, some inspection tasks cannot be solved using 2D analysis methods.

3.2. Photometric Stereo System

PTZ is one type of module used for NPP visual inspection. In our configuration, it is equipped with four 30 W LED lights placed around a CMOS camera with 1920 × 1080 picture elements, attached with an F1.6~F3.0 lens, as shown in Figure 1. The LEDs exhibit approximately the same performance in terms of the radiant flux value and emitting angle. In the PTZ setup, all LED lamps are fixed to make their optical axes parallel to the viewing angle of the camera. Each LED can be dimmed in increments from 0% to 100%. In addition, two laser generators display reference points for length measurement; the distance between them is calibrated prior to the experiment. Practical inspections must be conducted at a viewing angle as perpendicular to the target surface as possible. As shown in Figure 2, X_c, Y_c, and Z_c are the three axes of the camera coordinate system, which also represent the global coordinate system in our method.

Figure 1. Pan/tilt/zoom (PTZ) image capture setup and light-emitting diode (LED) light layout dimensions (mm).

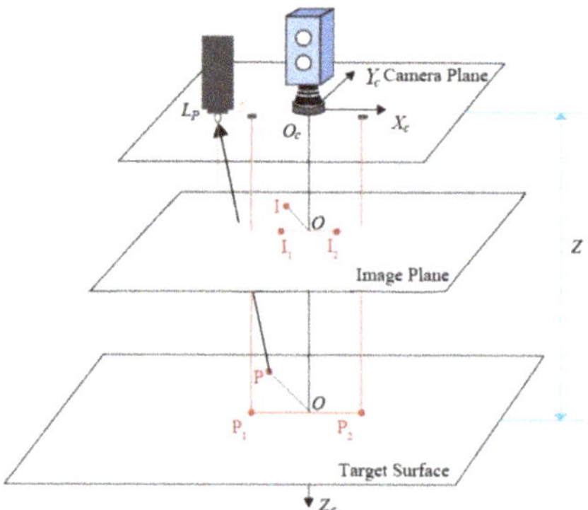

Figure 2. Diagram of the photometric stereo system.

The first stage of the shape reconstruction algorithm—i.e., the calibration of camera parameters—begins by calibrating the camera intrinsic parameters using Toolbox in Matlab. Then, images are captured using the camera with the laser points turned on. Given these images, the following steps are taken:

(1) Estimate the distance between the PTZ (camera plane) and the target surface by using laser points and prior knowledge;
(2) Estimate l and E for each pixel by determining the relationship between each image pixel and its corresponding point on the target surface, using the near-field light model;
(3) Compute the normal by resolving the irradiance equations;
(4) Compute the final 3D surface shape from the normal field via an optimization algorithm.

The PS formulation for our configuration is briefly introduced in Section 2, and details of the algorithm are provided in Sections 3.3 and 3.4.

3.3. Estimation of Light Direction and Intensity

Previous studies typically assume that the light sources are infinitely far from the surface, and generally adopt the parallel ray model. However, our system uses LEDs, which are very close to the target surface; thus, the lighting direction and intensity differ among various image areas. Moreover, the camera and lights are not fixed during an actual inspection. As previous light calibration methods are neither practical nor accessible for our application, it is necessary to design a fully automatic calibration method to estimate the light direction and intensity at each point.

In our configuration, the viewing angle is as perpendicular to the target surface as possible. In addition, the target surface is approximated as a planar surface that is assumed to be parallel to the image plane. Furthermore, the LED chip center and camera lens lie on a single plane, which is also parallel to the image plane. Accordingly, there are three parallel planes in the configuration, as shown in Figure 2.

The lighting direction for each point is decided by the position of the light L_p and the point P on the target surface. Supposing that the camera plane is the horizontal plane in the coordinate system, L_p can be determined by the PTZ structure. To determine the coordinate of point, the mapping relationship between image pixels and surface points must be determined, as well as the distance between the camera plane and the target plane. Thus, a two-stage process is designed:

(1) The distance z between the camera plane and the target plane is determined via using lasers;
(2) The orthographic projection-based method is used to determine how a point on the target surface is related to a pixel on the image plane.

Figure 2 shows that the viewing angle is aligned with the negative z-axis of the coordinate system, which simplifies the geometry calculation. The laser generators are fixed on PTZ, and their relative positions are known. Thus, $P_1(x_1, y_1, z)$ and $P_2(x_2, y_2, z)$, the intersections of the laser optical axis and the target plane, are also fixed. Their corresponding image pixels are $I_1\left(x'_1, y'_1, z'\right)$ and $I_2\left(x'_2, y'_2, z'\right)$, respectively, as shown in Figure 2.

The assumption of orthographic projection has typically been used in the conventional PS, although the perspective projection has been demonstrated to be more realistic [24]. However, when the change in scene depth is small relative to the distance from the scene to the camera, an orthographic projection can be used instead of a perspective projection [25]. In our method, the viewing angle is kept as perpendicular to the target surface as possible, and the distance between the target surface and the camera is considerably greater than the depth of the defects. Therefore, it is reasonable to apply an orthographic projection model without resulting in a large deviation.

The image magnification can be expressed as follows:

$$m = \frac{\| P_1 - P_2 \|}{\| I_1 - I_2 \|} = \frac{f}{z} \tag{3}$$

where m is the imaging magnification, f is the camera focal length, $\|\bullet\|$ denotes the length of a vector, and z is the distance between the target surface and the lens aperture, which is approximately equal to the distance between the surface plane and the camera plane. We denote z_0 as the calibrated distance of reference $I_1(x_1, y_1, z)$ and $I_2(x_2, y_2, z)$; z can then be calculated using

$$z = \frac{\left[(x_2 - x_1)^2 + (y_2 - y_1)^2\right]^{1/2}}{\left[(x'_2 - x'_1)^2 + (y'_2 - y'_1)^2\right]^{1/2}} z_0. \tag{4}$$

Laser points with different distances between the camera plane and the target plane are shown in Figure 3. Thus, the next step is to define how a point on the target surface is related to a pixel on the image plane.

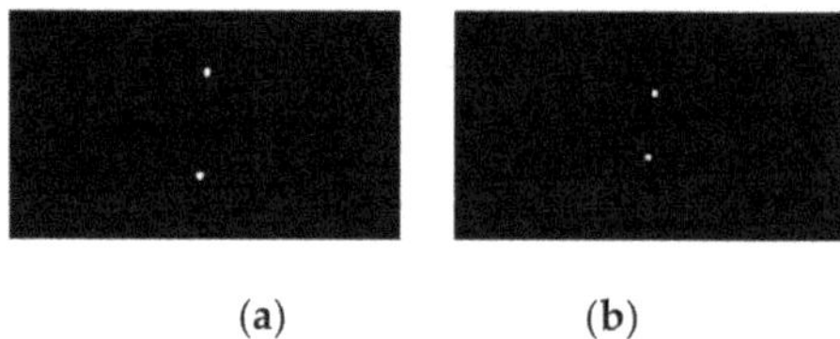

(a) **(b)**

Figure 3. Laser points with different distances between the camera plane and the target plane: (**a**) 200 mm and (**b**) 350 mm.

As illustrated in Figure 2, there is a clear relationship between image pixels and their corresponding points. For any image pixel $I(x'_i, y'_i, z')$, the position of its corresponding point in the target surface can be defined as $P(kx'_i, ky'_i, z)$, where $k = 1/m$. Therefore, the coordinate of P is decided by k, which can be determined by the laser points and z. The position of the LED, $L_P(x_{LED}, y_{LED}, 0)$, can also be estimated from prior knowledge. Therefore, the light direction $\boldsymbol{l}$, or $\overrightarrow{LP}$, can be expressed as $(kx'_i - x_{LED}, ky'_i - y_{LED}, z)$.

For LED lights, the irradiance (E) on the target surface can be expressed as follows:

$$E = \frac{I_{LED}(\theta)\cos(\theta)}{r^2} \tag{5}$$

where θ is the emitting angle of the LED, I_{LED} denotes the radiant intensity of the LED, and r is the distance from the light source to the target point [17]. By transforming the parameters, E can also be calculated from Equation (5). Then, $\boldsymbol{l}$ and E can be determined for various image pixels $I(x'_i, y'_i)$.

The goal of this step is to determine the lighting direction and intensity for each image pixel, which will improve the accuracy of the surface normal calculation, as well as the final 3D data quality.

3.4. Three-Dimensional Reconstruction using Photometric Stereo Technique

The PS procedure, assuming a Lambertian reflectance model, is applied for 3D shape reconstruction. The normal n is determined by using at least three images with various lighting conditions. In this study, they are calculated using the illustrated lighting direction and intensity estimation methods. Then, the image intensity follows inverse squared law, as

$$I = \rho\frac{(n \times l)}{r^2} = \rho\frac{n \times (P - L_p)}{\| P - L_p \|^{3/2}} \tag{6}$$

where r is the distance between the light source and the surface point [17].

The normal, n, can be calculated using at least three equations; once it is solved, the surface shape can be reconstructed via the height estimation by a global iterative method [26,27].

Suppose each image has W rows that are indexed by j, and has H columns that are indexed by i. The pixel is therefore denoted as (j, i), assuming its depth is $z(j, i)$, and the gradients in the x and y directions can be expressed as

$$p = \partial z(j,i)/\partial i,\ q = \partial z(j,i)/\partial j \tag{7}$$

Then,

$$n/\|n\| = (p, q, -1)^{\mathrm{T}}. \tag{8}$$

The image size is $W \times H$, so that the discrimination function in the iterative method is

$$E = \frac{1}{W \times H} \iint \left(\frac{\partial z(j,i)}{\partial i} - p(j,i) \right)^2 + \left(\frac{\partial z(j,i)}{\partial j} - q(j,i) \right)^2 \mathrm{d}i\mathrm{d}j. \tag{9}$$

4. Experimental Results and Discussion

Existing NPP visual inspection methods are not highly reliable for identifying small defects. This could be improved by using a camera with a higher resolution, which would produce a higher contrast between the defect and the metal surface [1]. However, it can also be difficult to discriminate between true and false defects; therefore, we tested specimens exhibiting small, hard to discern defects.

For the experiments, we used PTZ, introduced in Section 3.2. The layout dimensions of the lens and lights and the image capture setup is shown in Figure 1. With a beam fixing the PTZ, the camera looks downward, and the sample images are captured from the bottom. The four light sources, lasers, and lens are controlled via a controller. In our experiments, the defects were placed immediately below the camera.

Firstly, to show the efficacy of the proposed method for estimating z, the estimated distances were compared with the set values, as listed in Table 1, using a distance of 253 mm as the calibration point. The method provides accurate results, ranging from 133–493 mm. The errors were predominantly due to the orthographic image formation model, which does not always precisely reflect the actual conditions. These deviations are also related to the accuracy of laser point extraction.

Table 1. Estimation results of the distance between the camera plane and the target plane (mm).

Real Distance	Estimated Value	Error
133	138	5
193	196	3
253	-	-
313	318	5
375	376	1
433	441	8
493	510	17

The remaining experiments were conducted with two metal plates, as shown in Figure 4. One contains two small dents, labelled as defect 1 and defect 2 respectively, and the other one contains a pit, labelled as defect 3.

The 3D data generated for each defect are shown in Figures 5–7. The transformation from image pixels to real-world coordinates was achieved as described in Section 3.4. The depth information was then extracted five times along the Y direction for each defect, and the maximum depth contour was provided. The 3D data were then combined with altimetric readings of the maximum depth contour for further analysis.

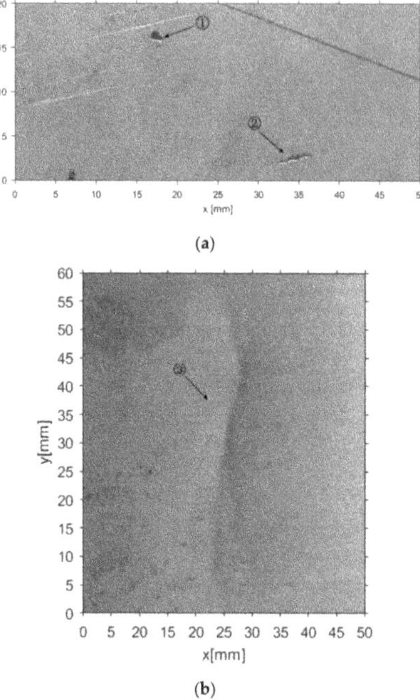

Figure 4. Images of test specimens: (**a**) 50 × 20 mm containing two dents (defects 1 and 2); (**b**) size 50 × 60 mm, containing one pit (defect 3).

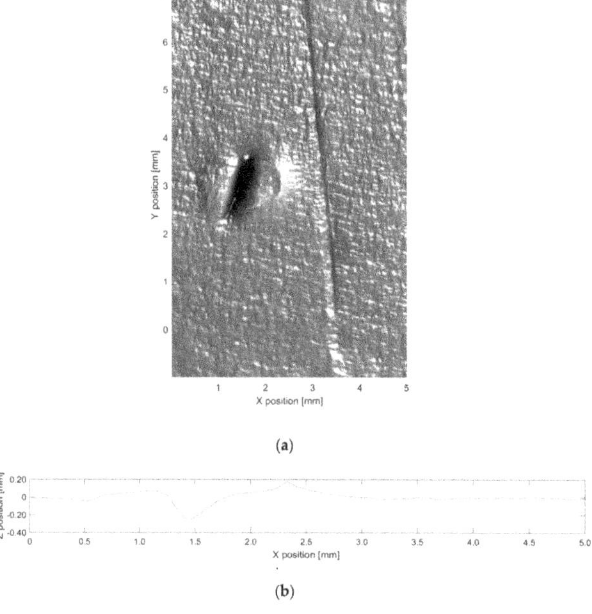

Figure 5. (**a**) Three-dimensional (3D) reconstruction results and (**b**) altimetric readings for the maximum depth in the 3D result for defect 1.

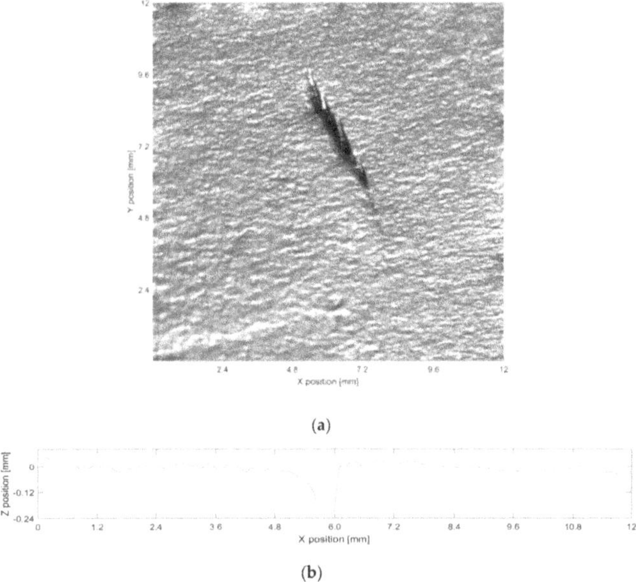

Figure 6. (**a**) 3D reconstruction results and (**b**) altimetric readings for the maximum depth in the 3D result for defect 2.

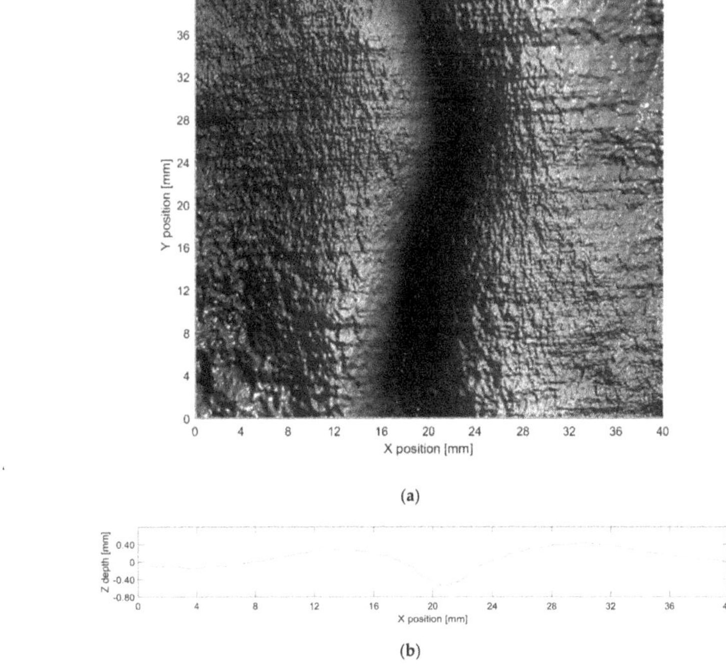

Figure 7. (**a**) 3D reconstruction results and (**b**) altimetric readings for the maximum depth in the 3D result for defect 3.

It is clear that the proposed method produces significantly more information than can be observed in the original acquisition images, enabling an accurate characterization of the dimensions and depth of defects. This facilitates image analysis for a range of inspection tasks that could not be solved by 2D analysis of the original acquisition images.

The initial results of the 3D reconstruction show that the proposed method is very useful for identifying defects, and can contribute to more reliable visual inspections with currently available devices. Thus, exploiting available devices will contribute to significant improvements and time savings for NPP visual inspections.

Table 2 shows the evaluation of defect depth. We compare our method with the baseline derived from MarSurf LD 120, which can conduct measurements using the Mahr metrology products. Z is the average maximum depth of the five contour lines extracted from the 3D results. The baseline is the average maximum depth of the five contour lines obtained by LD 120, which uses the non-defect area of the metal plate as the reference. LD 120 measures by contact and has a resolution of 0.001 mm in the depth direction, whereas the capacity for our image sensor and lens is approximately 0.030 mm under the proposed setup.

Table 2. Comparison of defect depth calculated from the 3D results and obtained by LD 120 [mm].

Defect Number	Z	Baseline
1	0.29	0.23
2	0.22	0.20
3	0.53	0.49

As shown in Table 2, the depth estimated by our method is close to the baseline. Errors in the experimental setup come from the camera, light sources, object reflectance, etc. The PS technique can deal with these errors to ensure more precise results; however, all PS techniques rely on radiance measurements. The characteristics of the defects, including defect opening displacement and geometry, will affect the validity of the results. In future research, we will determine the parameters and their impacts on the efficacy of depth measurement, which will require further experiments and verification.

Additionally, there are a few limitations of this study. First, it is assumed that the surface reflectance of the object follows the Lambertian model. Figure 6 reveals no highlights or shadows; therefore, there is no distortion in the results. Conversely, Figures 5 and 7 reveal the distortion in specific areas because of specular reflection. The processing of highlights and shadows requires more images, yet the small number of PTZ lights limits highlight and shadow processing. A methodology to handle shadows and highlights with four lights may be useful to improve the performance [28]. It is also important to note that the experimental defects analyzed in this study are more easily processed than the defects observed under actual inspection conditions. That is, the reflectance of the objects does not deviate substantially from the Lambertian model, and the experiment does not include any underwater images. Thus, the next step is to perform extensive experiments under a range of different conditions.

The limitations discussed here do not limit the use of this technique in NPP visual inspections. Moreover, we suggest that this research provides an important basis for developing a method that can readily identify and quantify defects through NPP visual inspection.

5. Conclusions

This study presents a 3D shape reconstruction method for the visual inspection of defects in NPP reactors. The method is based on the photometric stereo techniques and does not necessitate new inspection devices. The proposed approach, which involves estimating the light source directions and intensities, has reduced the limitation of light calibration and exhibits good practical applicability. The developed methodology can obtain the 3D shape and depth information of defects, thereby improving NPP visual inspection.

The demands for 3D image reconstruction will allow the visual inspection sector to perform more complex and accurate tasks. However, this is only possible if both the software and hardware are improved. The new market applications are expected to continue to emerge as the benefits of a 3D generating function are revealed. This research may also be relevant for designing inspection devices for future generations of NPP reactors.

Author Contributions: Conceptualization, K.X.; Data curation, M.W.; Formal analysis, M.W.; Funding acquisition, K.X.; Investigation, S.H.; Methodology, S.H.; Resources, M.L.; Supervision, K.X.; Validation, M.L. and M.W.; Writing—original draft, S.H.; Writing—review & editing, K.X.

Funding: This work was funded by National Key R&D Program of China (No. 2018YFB0704304) and the National Natural Science Foundation of China (No. 51674031 and No. 51874022).

Acknowledgments: This work was supported by National Key R&D Program of China (No. 2018YFB0704304) and the National Natural Science Foundation of China (No. 51674031 and No. 51874022).

Conflicts of Interest: The authors declare no conflicts of interest.

References

1. Cumblidge, S.E.; Doctor, S.R.; Anderson, M.T. The Capabilities and Limitations of Remote Visual Methods to Detect Service-Induced Cracks in Reactor Components. In Proceedings of the ASME Pressure Vessels and Piping Conference, Volume 5: High Pressure Technology, Nondestructive Evaluation, Pipeline Systems, Vancoucer, BC, Canada, 23–27 July 2006; pp. 183–193.
2. U.S. Nuclear Regulatory Commission. *Regulatory Guide 1.147: Inservice Inspection Code Acceptability, ASME Section, X.I*; Division 1; NRC: Washington, DC, USA, 2004.
3. Arias, J.; Hummel, W. Fuel Services Progress in Visual Examination and Measurements on Fuel Assemblies and Associated Core Components Track 1. In Proceedings of the Top Fuel 2009, Paris, France, 6–10 September 2009.
4. Heinsius, J.; Tsvetkov, E. ENIQ-Qualified Visual Examinations by Means of a Remote Controlled Submarine. In Proceedings of the 19th World Conference on Non-Destructive Testing, Munich, Germany, 13–17 June 2016.
5. Spinnler, K.; Bergen, T.; Sandvoss, J.; Wittenberg, T.; Fraunhofer-Institut für Integrierte Schaltungen IIS Röntgentechnik. Digital Image Processing for the Automation of NDT by Means of Endoscopy. In Proceedings of the 19th World Conference on Non-Destructive Testing, Munich, Germany, 13–17 June 2016.
6. Hebel, C.; DiSabatino, R.; Gresh, K. Machine Vision Technology. *Nucl. Plant J.* **2012**, *30*, 28–29.
7. Williams, T.; Hammel, K. Time is Money and Image is Everything. In Proceedings of the 19th World Conference on Non-Destructive Testing, Munich, Germany, 13–17 June 2016.
8. West, G.; Murray, P.; Marshall, S. Improved Visual Inspection of Advanced Gas-Cooled Reactor Fuel Channels. *Int. J. Progn. Health Manag.* **2015**, *6*, 1–11.
9. Woodham, R.J. Photometric Method for Determining Surface Orientation from Multiple Images. *Opt. Eng.* **1980**, *19*, 139–144. [CrossRef]
10. Kang, D.Y.; Jang, Y.J.; Won, S.C. Development of an Inspection System for Planar Steel Surface using Multispectral Photometric Stereo. *Opt. Eng.* **2013**, *52*, 039701. [CrossRef]
11. Eva, W.; Sebastian, Z.; Matthias, S.; Eitzinger, C. Photometric Stereo Sensor for Robot-assisted Industrial Quality Inspection of Coated Composite Material Surfaces. In Proceedings of the SPIE 12th International Conference on Quality Control by Artificial Vision, Le Creusot, France, 3–5 June 2015.
12. Han, T.Q.; Shen, H.L. Photometric Stereo for General BRDFs via Reflection Sparsity Modeling. *IEEE Trans. Image Process.* **2015**, *24*, 4888–4903. [CrossRef] [PubMed]
13. Shi, B.X.; Mo, Z.P.; Wu, Z.; Duan, D.; Yeung, S.K.; Tan, P. A Benchmark Dataset and Evaluation for Non-Lambertian and Uncalibrated Photometric Stereo. *IEEE Trans. Pattern Anal. Mach. Intell.* **2019**, *41*, 271–284. [CrossRef] [PubMed]
14. Ackermann, J.; Goesele, M. A Survey of Photometric Stereo Techniques. *Found. Trends Comput. Graph. Vis.* **2015**, *9*, 149–254. [CrossRef]
15. Mo, Z.P.; Shi, B.X.; Lu, F.; Yeung, S.K.; Matsushita, Y. Uncalibrated Photometric Stereo under Natural Illumination. In Proceedings of the CVPR 2018, Salt Lake City, UT, USA, 18–22 June 2018.
16. Xie, W.Y.; Dai, C.K.; Wang, C. Photometric Stereo with Near Point Lighting: A solution by mesh deformation. In Proceedings of the CVPR 2015, Boston, MA, USA, 7–12 June 2015.

17. Xie, L.M.; Song, Z.H.; Jiao, G.H.; Huang, X.; Jia, K. A Practical Means for Calibrating an LED-based Photometric Stereo System. *Opt. Lasers Eng.* **2015**, *64*, 42–50. [CrossRef]
18. Mecca, R.; Wetzler, A.; Bruckstein, A.M. Near Field Photometric Stereo with Point Light Source. *SIAM J. Imaging Sci.* **2014**, *7*, 2732–2770. [CrossRef]
19. Queau, Y.; Durix, B.; Wu, T.; Cremers, D.; Lauze, F.; Durou, J.D. LED-based Photometric Stereo: Modeling, Calibration and Numerical Solution. *J. Math. Imaging Vis.* **2018**, *60*, 1–26. [CrossRef]
20. Logothetis, F.; Mecca, R.; Cipolla, R. Semi-calibrated Near Field Photometric Stereo. In Proceedings of the CVPR 2017, Honolulu, HI, USA, 22–25 June 2017.
21. Papadhimitri, T.; Favaro, P. Uncalibrated Near-Light Photometric Stereo. In Proceedings of the BMVC 2014, Nottingham, UK, 1–5 September 2014.
22. Queau, Y.; Wu, T.; Cremers, D. Semi-calibrated Near-Light Photometric Stereo. In Proceedings of the SSVM 2017, Kolding, Denmark, 4–8 June 2017; pp. 656–668.
23. Murray, P.; West, G.; Marshall, S.; McArthur, S. Automated In-Core Image Generation from Video to Aid Visual Inspection of Nuclear Power Plant Cores. *Nucl. Eng. Des.* **2016**, *300*, 57–66. [CrossRef]
24. Papadhimitri, T.; Favaro, P. A New Perspective Uncalibrated Photometric Stereo. In Proceedings of the CVPR 2013, Portland, OR, USA, 23–28 June 2013.
25. Horn, B.K. *Robot Vision*; The MIT Press: Cambridge, MA, USA, 1986.
26. Smith, D.J.G.; Bors, A.G. Height Estimation from Vector Fields of Surface Normal. In Proceedings of the International Conference on Digital Signal Processing, Hellas, Greece, 1–3 June 2002.
27. Barsky, S.; Petrou, M. The 4-Source Photometric Stereo Technique for 3-Dimensional Surfaces in the Presence of Highlights and Shadows. *IEEE Trans. Pattern Anal. Mach. Intell.* **2003**, *25*, 1239–1252. [CrossRef]
28. Frankot, R.T.; Chellappa, R. A Method for Enforcing Integrability in Shape from Shading Algorithms. *IEEE Trans. Pattern Anal. Mach. Intell.* **1988**, *10*, 439–451. [CrossRef]

Article

Colorimetric Paper-Based Device for Hazardous Compounds Detection in Air and Water: A Proof of Concept

Valeria De Matteis [1,*], Mariafrancesca Cascione [1], Gabriele Fella [1], Laura Mazzotta [2] and Rosaria Rinaldi [1]

[1] Department of Mathematics and Physics "Ennio De Giorgi", University of Salento, Via Arnesano, 73100 Lecce, Italy; mariafrancesca.cascione@unisalento.it (M.C.); gabriele93_fella@libero.it (G.F.); ross.rinaldi@unisalento.it (R.R.)

[2] Studio Effemme-Chimica Applicata, Via Paolo VI, 73018 Squinzano (LE), Italy; laura.mazzotta@studioeffemme.com

* Correspondence: valeria.dematteis@unisalento.it

Received: 7 August 2020; Accepted: 23 September 2020; Published: 25 September 2020

Abstract: In the last decades, the increase in global industrialization and the consequent technological progress have damaged the quality of the environment. As a consequence, the high levels of hazardous compounds such as metals and gases released in the atmosphere and water, have raised several concerns about the health of living organisms. Today, many analytical techniques are available with the aim to detect pollutant chemical species. However, a lot of them are not affordable due to the expensive instrumentations, time-consuming processes and high reagents volumes. Last but not least, their use is exclusive to trained operators. Contrarily, colorimetric sensing devices, including paper-based devices, are easy to use, providing results in a short time, without particular specializations to interpret the results. In addition, the colorimetric response is suitable for fast detection, especially in resource-limited environments or underdeveloped countries. Among different chemical species, transition and heavy metals such as iron Fe(II) and copper Cu(II) as well as volatile compounds, such as ammonia (NH_3) and acetaldehyde (C_2H_4O) are widespread mainly in industrialized geographical areas. In this work, we developed a colorimetric paper-based analytical device (PAD) to detect different contaminants, including Fe^{2+} and Cu^{2+} ions in water, and NH_3 and C_2H_4O in air at low concentrations. This study is a "proof of concept" of a new paper sensor in which the intensity of the colorimetric response is proportional to the concentration of a detected pollutant species. The sensor model could be further implemented in other technologies, such as drones, individual protection devices or wearable apparatus to monitor the exposure to toxic species in both indoor and outdoor environments.

Keywords: PAD; environmental monitoring; colorimetric detection; water; atmosphere

---

1. Introduction

In the last decades, due to the increase of industrialization activities, the release of hazardous materials in the atmosphere, water and soil has raised many concerns about their impact on living organisms [1]. Metals and heavy metals together with gaseous organic compounds are the most widespread toxic elements due to their ability to enter the living organism by different routes, such as inhalation and ingestion [2,3]. Then, they can enter the food chain, integrating into enzymatic processes with the consequence to boost various diseases and inflammation processes onset [4].

Cu(II) and Fe(II) are transition metals having a key role in several physiological pathways, such as fetal growth, brain development, cholesterol metabolism and immune function [5–7]. Cu(II) represents one of the main components of the PM 2.5 produced by the road dust emissions, allowing its easy penetration into the organisms' body [8].

In addition, the ecological risk deriving from Cu(II) exposure is a problem in European saltwater environments [9,10]. Cu(II) can be toxic to aquatic life at concentrations approximately 10 to 50 times higher than the tolerated range [11]. In addition, humans can adsorb a great amount of Cu(II) from drinking water, food, air and supplements, reaching a daily absorption of 1.85 mg [12]. In order to understand the collateral effects of the Cu(II), the US National Toxicology Program (NTP) exposed B6C3F1 mice to the five concentrations of Cu(II) (76, 254, 762, 2543, 7629 mg Cu/L) to [13] for 13 weeks observing the organs weight loss and animals death, at the higher concentrations tested. These results were consistent with another study in which the same toxicity was observed in female and male mice using 762 mg/L of Cu(II). Additionally, Fe(II) triggered adverse effects in vivo by acute toxicity induction [14]. In aquatic environments, Fe(II) boosted the growth default of aquatic organisms at a concentration of 1 mg/L [15]. In addition, in some European countries such as Lithuania, people were exposed to high levels of Fe(II) due to the contamination of groundwater that overcome the permissible limit established by the European Union Directive 98/83/EC, related to the quality of drinking water [16]. Regarding the volatile compounds pollution, NH_3 is one of the major manufactured industrialized soluble alkaline gases on Earth [17]. NH_3 originates from both natural and anthropogenic sources, in particular from the agricultural industry, high-density intensive farming practices as well as fertilizer applications [18]. According to the Agency for Toxic Substances and Disease Registry, the concentrations of NH_3 in the environment are very variable due to its continuous recycling and its internalization in biosphere. Therefore, it is possible to find different natural NH_3 levels in the soil (1–5 ppm), in air (1–5 ppb) and in water (approximately 6 ppm) [19]. The NH_3 smell can be identified by humans at concentrations greater than 5 ppm; at 30 ppm and with an exposure time of up to 2 h, human volunteers underwent slight irritation, whereas strong effects were recorded up to 500 ppm [20]. However, NH_3 lethality requires higher concentrations [21]. In addition to NH_3, also some kinds of carbonyls which constitute the motor vehicle exhaust, such as C_2H_4O are toxic air contaminants, particularly dangerous for living organisms [22,23]. Woutersen et al. [24] used Wistar rats to study the toxic effect of C_2H_4O administered in air (6 h/day) at three concentrations (750, 1.500, 3.000 ppm) for more than a year. All the concentrations tested induced the increase of nasal tumors incidence with remarkable impact especially at higher concentrations. Other evidences suggested that the C_2H_4O administration (1.650–2.500 ppm) for more than two years (7 h/day) induced tracheal, but not nasal, tumors in Syrian golden hamsters [25]. Then, the study of these compounds in polluted areas is a key factor to control the exposure rate.

Today, several analytical techniques are available for the detection of toxic analytes. However, many of them are not affordable due to the expensive instrumentations and high reagent volumes required. On the contrary, point-of-care and easy-to-use analysis provide results in a short time, preventing the production of an elevated amount of waste [26]. In addition, they can be employed in resource-limited environments and developing countries where pollution is uncontrollable and not regulated with specific rules.

In particular, paper is the best choice to develop sustainable devices [27]; it is considered a valid alternative to traditional materials due to its ease of fabrication, satisfactory levels of sensitivity, specificity, low cost, lightweight, versatility, being easily portable and low reagent consumption requiring [28,29]. The paper-based analytical devices (PAD) can work following the principle of color change in the presence of specific target analytes [30]. The sensitivity and specificity of the assay are dependent on an interaction between the target analyte and the surface of the PAD due to the functionalization of cellulose fibers [31]. The paper surface can be functionalized by different molecules, such as chemoresponsive dyes, nanoparticles (NPs) and biomolecules (antibodies, aptamers, nucleic acids) [32–35]. Xi et al. [36] prepared a paper device based on Pb(II) metal-organic nanotubes characterized by a large {Pb14} metallamacrocycle, to detect H_2S based on the fluorescence "turn-off" response. However, the fabrication of nanotubes and the general technique required specific scientific competences and elevated costs; moreover the toxicity of nanotubes, is not negligible [37]. Maity et al. [38] used perovskite halide ($CH_3NH_3PbI_3$) to achieve a thin-film sensor fabricated on a paper by a growth process able to detect NH_3 gas by a color change from black to yellow. Despite the effectiveness of this device, the $H_3NH_3PbI_3$ is chemically unstable and toxic for living organisms. [39]. Then, the disposal of the device could present a serious problem. In a recent work [40], a microporous cellulose-based smart xerogel bromocresol purple was used into cross-linked carboxymethyl cellulose to detect NH_3 by a colorimetric response. The authors performed a freeze-drying process to obtain the xerogel with a low limit of detection.

In these PADs, the colorimetric shift can be evaluated by colorimetric assay, as a result of the interaction with the ligand. In general, the PADs sensing areas are fabricated by the printing method using a wax printer [41]. The results obtained can be directly interpreted by the naked eye together with the spectrophotometer analysis. In the last years, the use of smartphones to detect color change has been developed [42–44]. Therefore, its use showed some limitations regarding the low lighting conditions that prevent the smartphone camera exploitation [44].

In this work, we developed an effective PAD suitable to detect different contaminants, namely Fe(II) and Cu(II) cations (Fe^{2+} and Cu^{2+}) in water and NH_3 and C_2H_4O vapor in air. The design and fabrication of the sensor did not require specific instrumentations. In particular, for metals detection, only a wax pen able to design the specific areas of chemical interaction was required, without the use of a wax printer. We functionalized the paper (Whatman filter paper) using different analytes capable of reacting with metallic ions and gaseous substances, allowing a specific response; the aim of this process was to develop calibration curves to correlate the obtained color to the concentrations of toxic compounds. The results were easily interpreted using a digital scanner and ImageJ. The tests achieved using intermediate concentrations suggested the sensitivity and reproducibility of the PAD, making it a powerful tool to detect hazardous materials in different mediums without the use of sophisticated technologies.

2. Materials and Methods

2.1. Ammonia Detection

2.1.1. Reagents

Whatman filter paper n.1 (thickness 180 μm), ammonium hydroxide (NH_4OH, 28%), hydrochloric acid (HCl), Aniline ($C_6H_5NH_2$) and ammonium persulfate $(NH_4)_2S_2O_8$ were purchased from Merck.

2.1.2. Functionalization of Whatman Paper for Reversible Ammonia Vapor Detection

The reversible colorimetric detection of gaseous NH_3 was realized by coating Whatman filter paper with polyaniline (PANI) film, achieved by $C_6H_5NH_2$ polymerization (2.5 g/L) in the presence of HCl (1 M) and $(NH_4)_2S_2O_8$ (0.125 g/L) at room temperature [45]. Briefly, $(NH_4)_2S_2O_8$ solution was added dropwise into the $C_6H_5NH_2$ solution under stirring (1000 rpm). The two compounds were in a volume ratio of 1:1. After 3 min, half of the colorless reaction mixture was immediately added into a silicon funnel, where a piece of round filter paper (c.a 2 cm) was placed and fixed. Then, the remaining solution was slowly suction-filtered through the filter paper, and the unused volume was left in the dark for approximately 1 h. During this time, the solution slowly turned light blue. After this step, the solution was again filtered and then, the paper was carefully washed with Milli-Q water. Finally, it was left to air dry until the emerald green filter paper was achieved. The functionalized paper was exposed to different concentrations of NH_3. The schematic representation of this procedure is represented in Figure 1a.

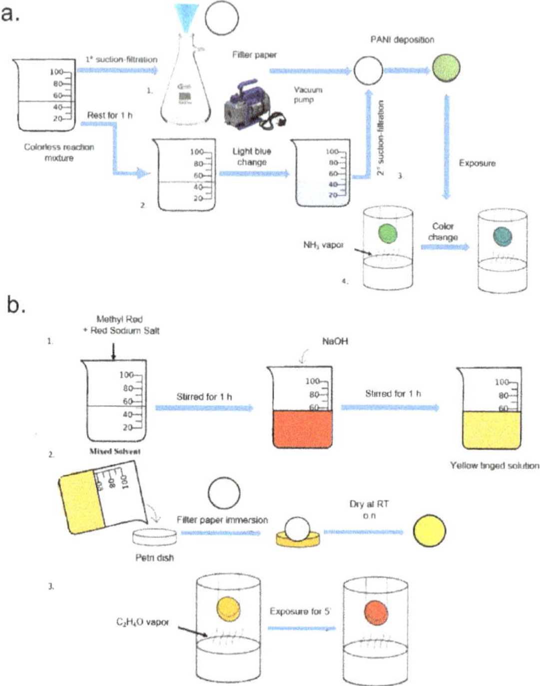

Figure 1. Schematic NH_3 (a) and C_2H_4O (b) paper sensor fabrication procedure: (**a**) **1.** Half of the colorless reaction mixture was immediately suction-filtered into a silicon funnel where a filter paper (*white circle*) was placed. **2.** The remaining part of the solution was left to stand for 1 h. During this time, the solution turned light blue. **3.** The solution was filtered (II suction-filtration) through the filter paper, in order to induce the polyaniline (PANI) deposition. After several washes and air flow drying, the formation of emerald green filter paper (*green circle*) was completed. **4.** The emeraldine green filter paper turned into a blue emeraldine base (*blue circle*) as a result of NH_3 vapor exposure. (**b**) **1.** The methyl red and methyl red sodium Salt were added to the mixture. The color solution turned into red-orange and was stirred for 1 h. After this time, NaOH was added, resulting in a color change to yellow. The solution was left to stand for 1 h. **2.** The solution was transferred in a petri dish and the filter paper was immersed in it for 1 h. The filter paper was dried overnight in the dark at room temperature. The formation of methyl red filter paper (*yellow circle*) was completed. **3.** The methyl red filter paper turned into red (*red circle*) as a result of C_2H_4O vapor exposure.

2.1.3. Construction of Calibration Curve by Colorimetric Response to Ammonia Vapor

Glass vials were used to detect NH_3 vapor exposure. In each vial, 10 mL of NH_3 solution was added at different concentrations (100, 300 500 and 1000 ppm) to achieve a standard curve. Small PANI-deposited filter paper pieces were fixed on the necks of the vials in order to expose them to the vapor generated from the corresponding NH_3 aqueous solution for a few seconds. The control was represented by pure NH_3. After this time, the paper was immediately removed and analyzed by a scanner (Samsung SCX-3400 series (USB002)) acquiring the color change after NH_3 vapor interaction.

2.2. *Acetaldehyde Detection*

2.2.1. Reagents

Whatman Filter paper n.1 (thickness 180 μm), methyl red ($C_{15}H_{15}N_3O_2$), methyl red sodium salt ($C_{15}H_{14}N_3NaO_2$), methanol (CH_3OH), Glycerol ($C_3H_8O_3$) and sodium hydroxide (NaOH) were purchased from Merck.

2.2.2. Functionalization of Whatman Paper for Acetaldehyde Vapor Detection

The colorimetric detection of gaseous C_2H_4O was obtained by coating a Whatman filter paper with thin methyl red film. methyl red and methyl red sodium salt was dissolved in a solvent constituted by CH_3OH, water and $C_3H_8O_3$ (1 mM). The red-orange solution was stirred for approximately 1 h. NaOH (8 mM) was added to the solution and stirred at room temperature for 1 h. The yellow-colored solution obtained was translocated in a petri dish where a piece of Whatman filter paper was immersed for 1 h. After this time, the paper was dried at room temperature overnight. The filter paper sheet was then cut into small round disks (diameter of approximately 2 cm) and successively exposed to different concentrations of C_2H_4O. After 5 min, the color appeared on the paper. The schematic representation of this procedure is represented in Figure 1b.

2.2.3. Construction of Calibration Curve by Colorimetric Response to Acetaldehyde Vapor

The C_2H_4O vapor detection was performed using different glass vials in which 10 mL of C_2H_4O was added at different concentrations in CH_3OH solvent: 100, 300, 500 and 1000 ppm, respectively, on the vial's neck. The deposited filter paper pieces were fixed in order to expose them to the vapor evaporated from each C_2H_4O/CH_3OH solution for 5 min. After this time, the paper was immediately removed and analyzed by a scanner (Samsung SCX-3400 series (USB002)) in order to acquire the color changes after C_2H_4O interaction.

2.3. *Fabrication of Paper-Based Colorimetric Device for Fe^{2+} and Cu^{2+}*

2.3.1. Reagents

Iron chloride tetrahydrate ($FeCl_2{\cdot}4H_2O$), HCl, copper sulfate pentahydrate ($CuSO_4{\cdot}5H_2O$), potassium ferricyanide ($K_3[Fe(CN)_6]$), and potassium iodide (KI) were purchased from Merck.

2.3.2. Iron and Copper Calibration Curve Standard Solutions Preparation

$FeCl_2{\cdot}4H_2O$ was dissolved in HCl (0.5 M) in order to achieve 1000 μg/mL of Fe^{2+} standard stock solution whereas $CuSO_4{\cdot}5H_2O$ was used to prepare 1000 μg/mL Cu^{2+} standard stock solution in Milli-Q water. The series of four standard solutions (25, 50, 100 and 200 μg/mL) of Fe^{2+} and Cu^{2+} were prepared by diluting the standard stock solutions with different volumes of Milli-Q water. After these steps, $K_3[Fe(CN)_6]$ (5 mM) and KI (0.4 M) solutions were prepared for Fe^{2+} and Cu^{2+} detection, respectively.

2.3.3. Fabrication of the Paper Analytical Device (PAD)

The fabrication of PAD was developed as follows:

1. The waxy channels on a piece of Whatman filter paper were obtained by using a wax pen. The shape of each channel was circular with a diameter of approximately 0.5 cm. Four spots were drawn on the filter paper, one for each standard.
2. The PAD was heated on a hot plate at ~60 °C for 1 h to melt the wax. The liquid wax penetrated into the cellulose pores to achieve hydrophobic barriers.
3. The PAD was dried at room temperature for approximately 30 min.

2.3.4. Assay Procedure

A small volume (5 μL) of Fe^{2+} and Cu^{2+} assay reagents (($K_3[Fe(CN)_6]$) and KI) was spotted by drop-casting on paper circular dots using a micropipette and allowed to dry at the room temperature for 3 h. Five microliters of each standard solution was added to the corresponding labeled spots of the PAD. The Fe^{2+} of the standard solutions reacted with the $K_3[Fe(CN)_6]$ generating blue colored complex in the detection zones. Instead, the Cu^{2+}, reacting with KI, produced a red-brown compound. The intensity of the color was proportional to the standard solution concentration. A schematic representation of the described process was represented in Figure 2.

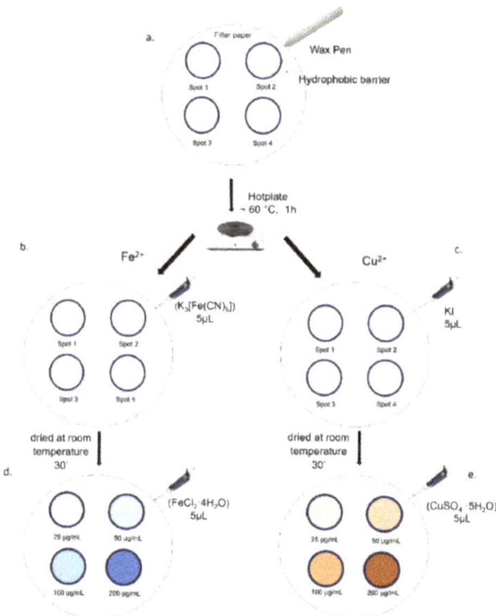

Figure 2. Schematic Fe^{2+} and Cu^{2+} colorimetric assay procedure: (**a**) The four spots were achieved by wax pen in order to create hydrophobic barriers after heating using a hot plate. (**b**,**c**) Five microliters of each standard solution were added by drop-casting to the corresponding labeled spot. (**d**) On the Fe(II) paper-based analytical device (PAD), a blue complex was formed after the reaction between the Fe^{2+} and (($K_3[Fe(CN)_6]$); the blue color intensity directly correlated with the Fe^{2+} concentration (**e**) On the Cu(II) PAD a red-brown compound was developed, generating by Cu^{2+} and KI reaction, whose color intensity was dependent on Cu^{2+} concentration.

2.4. *Quantitative Image Processing by ImageJ 1.47 Software*

Once the color changes were achieved due to the chemical interaction with the different hazardous compounds, the corresponding PADs images were captured using scanner Samsung SCX-3400 with a resolution of 300 dpi. Then, the images were stored in JPEG format and analyzed in RGB format with the open-source software, ImageJ [46]. An adjustment of the color threshold was applied to each image to filter out all colors that were not correlated to the colored complex to be detected during the analysis. For instance, the Fe^{2+} color adjustment was applied to delete all colors which was not in the blue range from the analysis spectrum. The color adjustment was set as follows:

1. The "Color Threshold" window was accessed through the ImageJ menu by selecting "Image"→ "Adjust"→"Color Threshold."
2. At the bottom of this window HSB was selected, which allowed the adjustment of hue, saturation, and brightness.

3. The hue was adjusted by moving the sliders directly below the "Hue" spectrum until only the color of interest was visible. The hue threshold ranges set for each metal were fixed as follows: NH_3 (244–255), C_2H_4O (38–240), Fe^{2+} (171–197), Cu^{2+} (37–255).

The images were then converted to an 8-bit grayscale ("Image" → "Type" → "8-bit") and inverted ("Edit" → "Invert"). The intensity measurements yielded a positive slope when plotted versus metal amounts. Mean Gray Value (MGV) was measured for each RGB channel (red, blue and green, "Image" → "Color" → "Merge Channel") by first selecting "mean gray value" and "limit to threshold" in the "Set measurements window," found from the ImageJ menu by selecting "Analyze" → "Set measurements". Each area was selected using the wand tool, which automatically found the edge of an object and traced its shape. The gray intensity of the outlined area was measured by selecting "Analyze" → "Measure." Then, the RGB channel was selected with the highest sensitivity for the metal detection according to Yu et al. [47]. The blue channel was selected for both metal cations, the red channel for NH_3 and green channel for C_2H_4O were selected. Data were then imported into Microsoft Excel 2019 in order to obtain the different calibration curves for the NH_3, C_2H_4O, Fe^{2+}, Cu^{2+} concentrations.

The colorimetric detection limits of NH_3, C_2H_4O, Fe^{2+} and Cu^{2+} were estimated based on 3SB/S according to IUPAC rules, where SB and S are standard deviation and slope, respectively [48,49].

2.5. Interference Studies

The selectivity of PAD to Cu^{2+} and Fe^{2+} was evaluated by interferences assessment exposing the functionalized PAD to several metal ions solutions containing Na^+, K^+, Mg^{2+}, Ca^{2+}, Al^{3+}, Mn^{2+}, Fe^{3+}, Co^{2+}, Ni^{2+}, Zn^{2+}, Cd^{2+} and Pb^{2+} at a concentration of 100 μg/mL. The same procedure was used to assess the specificity of PAD to NH_3 and C_2H_4O using methylamine, ethylamine, triethylamine, benzene, toluene, ethyl benzene, formaldehyde and ethanol at a concentration of 500 ppm.

3. Results and Discussion

In recent years, the environmental pollution has been at the center of many debates, due to the progressive and intense industrialization; the scientific community has thus focused its attention on the potentially toxic effects of certain substances on the living organisms [50]. Several people are exposed to different kinds of substances owing to the contamination of several environments in particular water, atmosphere and soil [51]. Among these, the most widespread are certainly the transition metals, heavy metals and gaseous substances, that are produced by intense processing activities especially in the agrifood sector [52]. These chemicals are generally released into the atmosphere and they can reach the groundwater as well as lakes and sea reaching living organisms with subsequent collateral effects [3,53]. In this scenario, environmental monitoring is a fundamental objective to prevent and to know at what doses an organism was exposed. The conventional analytical techniques (gas chromatography–mass spectrometry, high-performance liquid chromatography–mass spectrometry, atomic absorption spectroscopy) are sophisticated systems that require high energy consumption and expensive laboratory systems. Paradoxically, in fact, the environment analysis by the use of these instruments induces in turn pollution (energy, consumables, toxic reagents). Starting from these assumptions, we have developed a PAD that can be used without the need for trained operators to monitor some hazardous materials such as NH_3, C_2H_4O, Fe^{2+} and Cu^{2+}. For gaseous substances, namely NH_3 and C_2H_4O, we performed two different techniques to functionalize the filter paper. In particular, for NH_3 detection, we used a PANI film functionalization following the polymerization of aniline directly on paper substrate. The PANI film obtained was in the form of green emeraldine salt due to the protonation of the backbone induced by HCl. We selected four doses of NH_3 on the basis of toxicological results obtained in literature, as reported in the Introduction section (Section 1). When NH_3 molecules reached the functionalized paper, the deprotonation of PANI chains and, consequently, the transformation of them into a blue emeraldine base occurred. In addition, this dye shows peculiar chemical properties consisting of the reversible doping/dedoping nature. The dye reacted with the

NH_3 determining the color change; when the analyte was removed, it can be reverted to its initial chemical state. Due to the reversible nature of the process, the functionalized PAD can be reused many times (ca. 30 times) before its discard. After the exposure to different concentrations of NH_3 vapor (100, 300, 500, 1000 ppm) the color appeared in a few minutes. Immediately, a digital scanner was used to freeze the specific color. The scanner acquired the image in JPEG format, allowing the next analysis by ImageJ software. As shown in Figure 3, the paper assumed a specific coloration that can be visualized with the naked eye. The color switch from light green to blue at the higher concentration tested. By The JPEG images were analyzed after setting the specific parameters (Hue adjustment section of the Threshold Color window) described in detail in the Materials and Methods section (Section 2). The assay reproducibility was evaluated for three identical test zones.

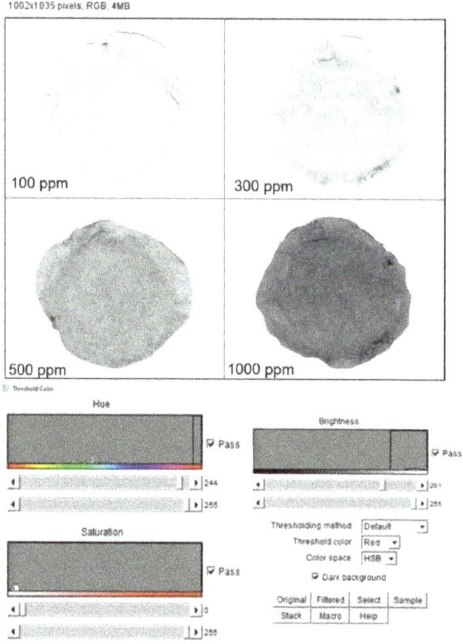

Figure 3. Hue adjustment section of the Threshold Color window in ImageJ analysis software of the NH_3 PAD.

The functionalization of PAD for the detection of C_2H_4O was achieved by the use of methyl red. The latter is determined by the concentration of acidic (red) and basic (yellow) forms. The colorimetric sensor was designed to show a selective response based on a chemical reaction, such as the nucleophile addition. Using an excess of hydroxide ions, the C_2H_4O underwent the nucleophile addition reaction, resulting in the sensor alkalinity changes and consequently in a color change, from yellow to red. The color change was almost instantaneous and it was stable for several days after drying. After the color response, the scanner was used to acquire the image and color intensity. The latter was analyzed for the second step of the experimental session using ImageJ analysis by the Hue adjustment section of the Threshold Color (Figure 4). The reproducibility was evaluated for three identical test zones.

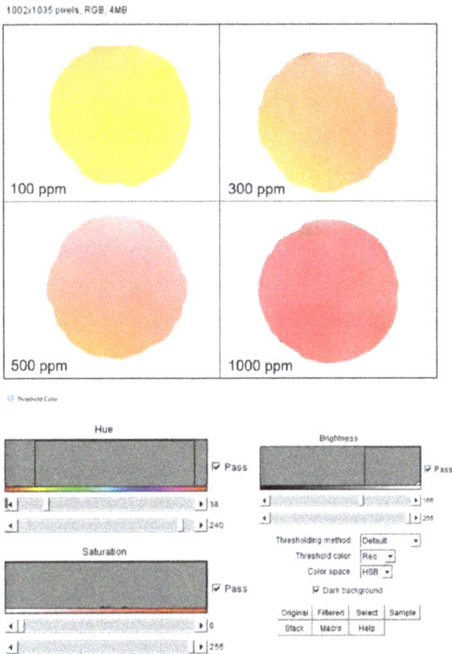

Figure 4. Hue adjustment section of the Threshold Color window in ImageJ of C_2H_4O PAD.

The test zones were used to create the calibration curve. Figure 5 shows the calibration curve for NH_3 detection using the color change after the exposure to the four concentrations. In detail, in Figure 5a we reported the pieces of devices related to the functionalized and unexposed PAD (top circle) and the PAD exposed to pure NH_3 (28%, bottom circle) with the relative MGV values extracted from the ImageJ analysis that were 10.3 ± 1.5 and 75 ± 4.5, respectively. In Figure 5b, the pieces of PAD after exposure to 100, 300, 500 and 1000 ppm of NH_3 were represented. Observing the pictures, it was possible to visualize a color trend with the naked eye, from the lightest to the darkest as the concentration increased. The successive ImageJ analysis performed on the scanner acquisitions correlated with the concentration with a specific MGV obtaining a calibration curve with R^2 = 0.99. The limit of detection (LOD) value was 7.64 ppm. The values were obtained by repeating the experiment three times. In order to understand if the device actually worked even with intermediate concentrations, we exposed the PAD to average concentrations calculated between the first and second (200 ppm) and third and fourth doses (750 ppm). Additionally, in this case, PANI film was able to efficiently induce the color response; it was possible to find the concentration simply by interpolating the MGV data on the straight line as shown in Figure 5c.

The same procedure was applied to the paper functionalized with methyl red, capable of detecting the C_2H_4O vapor. (Figure 6). Figure 6a shows the as-prepared paper device (yellow) and after exposure to pure C_2H_4O (≥99.5%, dark red) with the corresponding MGV values that were 20.3 ± 2.7 and 155 ± 7.5, respectively. In Figure 6b, the progression from yellow to red was observed when the tested concentrations increased. We used 100, 300, 500 and 1000 ppm concentrations and we built the calibration curve obtaining an R^2 value of 0.98 (Figure 6c). The LOD value was 11.09 ppm. The effectiveness of this colorimetric response was verified using two average concentrations: 200 and 750 ppm. Additionally, in this case, the MGV values were interpolated with the curve that exactly corresponded to the tested concentrations.

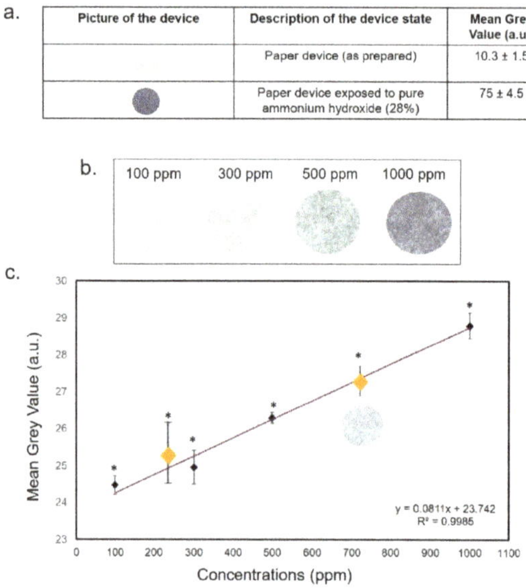

a.

Picture of the device	Description of the device state	Mean Grey Value (a.u.)
	Paper device (as prepared)	10.3 ± 1.5
	Paper device exposed to pure ammonium hydroxide (28%)	75 ± 4.5

b. 100 ppm 300 ppm 500 ppm 1000 ppm

c. Mean Grey Value (a.u.) vs Concentrations (ppm); y = 0.0811x + 23.742, R² = 0.9985

Figure 5. (**a**) Mean Gray Value (MGV) values of PAD as prepared and after the exposure to pure NH_3 (28%). (**b**) color change after NH_3 exposure. (**c**) Interpolation of NH_3 intermediate values (200 and 750 ppm). Data reported were the average of three independent experiments ± SD. The difference between the as-prepared paper and colored papers was considered statistically significant performing a Student's *t*-test with $p < 0.05$ (<0.05 *).

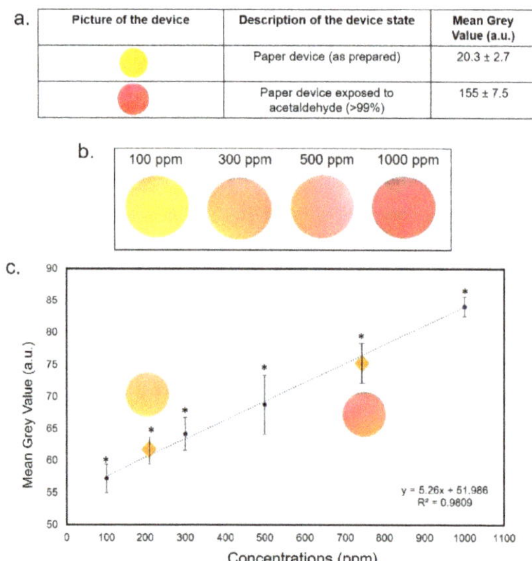

Picture of the device	Description of the device state	Mean Grey Value (a.u.)
	Paper device (as prepared)	20.3 ± 2.7
	Paper device exposed to acetaldehyde (>99%)	155 ± 7.5

Figure 6. (**a**) MGV values of PAD as-prepared and after the exposure to pure C_2H_4O (>99%). (**b**) color change after C_2H_4O exposure. (**c**) Interpolation of C_2H_4O intermediate values (200 and 750 ppm). Data reported were the average of three independent experiments ± SD. The difference between the as-prepared paper and colored papers was considered statistically significant performing a Student's *t*-test with $p < 0.05$ (<0.05 *).

After the analysis of gaseous molecules, we used the PAD to detect Cu^{2+} and Fe^{2+}, which are the most common metals released in the environment [1]. Then, we moved to the detection of these metals in water at low concentrations. Firstly, we designed four circle spots using a wax pen in order to achieve hydrophobic barriers without the use of wax printing, inkjet printing and screen-printing technologies. Once the heat produced by the hot plate allowed the penetration of the wax into the cellulose porous, the specific chemical analytes, $K_3[Fe(CN)_6]$ for Fe^{2+} and KI for Cu^{2+}, were deposited in the spot's center by drop-casting. The wax channels prevented the typical diffusion phenomenon of the liquid substances deposited on the paper. Five microliters of $FeCl_2.4H_2O$ and $CuSO_4.5H_2O$ (25, 50, 100, 200 µg/mL) were used. Chelation (1) and redox (2) chemical reactions produced a blue and brown color, respectively. The two reactions were the following:

(1) $Fe^{2+} + Fe(CN)^{3-}{}_6 \rightarrow Fe_3[Fe(CN)_6]^2$;

(2) $Cu^{2+} + 2I^- \rightarrow CuI_2 \rightarrow 2CuI_2 = 2CuI + I_2$.

After color formation, we acquired the images by a digital scanner to perform ImageJ analysis using the adjustment of Threshold Color both for Fe^{2+} (Figure 7) and Cu^{2+} (Figure 8). The analysis was repeated in three identical test zones.

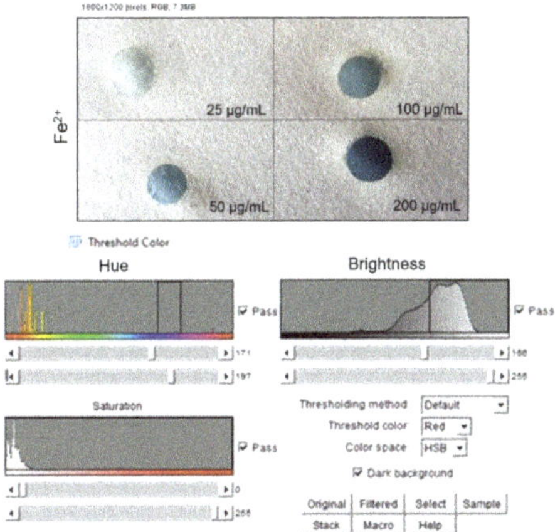

Figure 7. Top: Image acquisition of PAD after $FeCl_2.4H_2O$ deposition at different concentrations. Down: threshold analysis, saturation and brightness adjustment by ImageJ software.

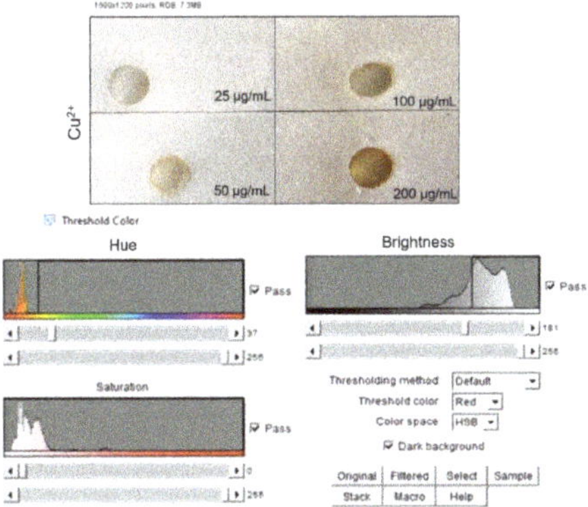

Figure 8. Top: Image acquisition of PAD after $CuSO_4.5H_2O$ deposition at different concentrations. Down: threshold analysis, saturation and brightness adjustment by ImageJ software.

As shown in Figure 9a, the color changed from light blue to dark blue, proportionally to the Fe^{2+} concentration increase. The corresponding calibration curve was obtained plotting the MGV values analyzed by ImageJ analysis after standards solution deposition, showing an R^2 value of 0.98 (Figure 9b).

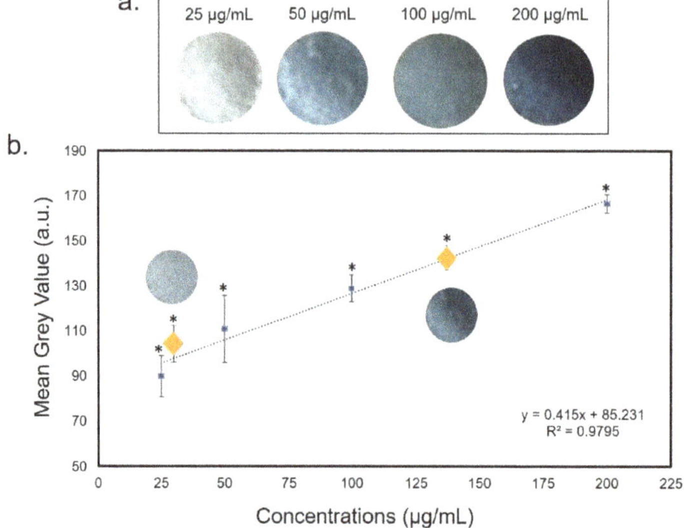

Figure 9. (**a**) Color change of filter paper after exposure to Fe(II) at different concentrations. (**b**) The yellow rhombuses represented the interpolation of Fe^{2+} intermediate concentrations (37 and 150 µg/mL). Data reported were the average of three independent experiments ± SD. The difference between as-prepared paper and colored papers was considered statistically significant performing a Student's *t*-test with $p < 0.05$ (<0.05 *).

A similar R^2 value was reported for the Cu^{2+} calibration curve; in the latter case, the color changed from light brown to dark brown (Figure 10a). As demonstrated for NH_3 and C_2H_4O we used two average concentrations between 25 and 50 µg/mL and between 100 and 200 µg/mL (37 and 150 µg/mL) to test the device reliability. The MGV values acquisitions revealed that the corresponding concentrations were on the calibration curve thus confirming the effectiveness and stability of the PAD (Figure 10b). The LOD for Fe^{2+} was 3.8 µg/mL and 3.2 µg/mL for Cu^{2+}. For both metals, the values were greatly below the maximum acceptable concentrations in drinking water stipulated by the World Health Organization (WHO) [54].

The LOD values of each device are summarized in Table 1.

Table 1. LOD values of NH_3, C_2H_4O, Fe^{2+} and Cu^{2+} PADs.

PADs	Concentrations Range	Limit of Detection (LOD)
NH_3	100–1000 ppm	7.64 ppm
C_2H_4O	100–1000 ppm	11.08 ppm
Fe^{2+}	25–200 µg/mL	3.8 µg/mL
Cu^{2+}	25–200 µg/mL	3.2 µg/mL

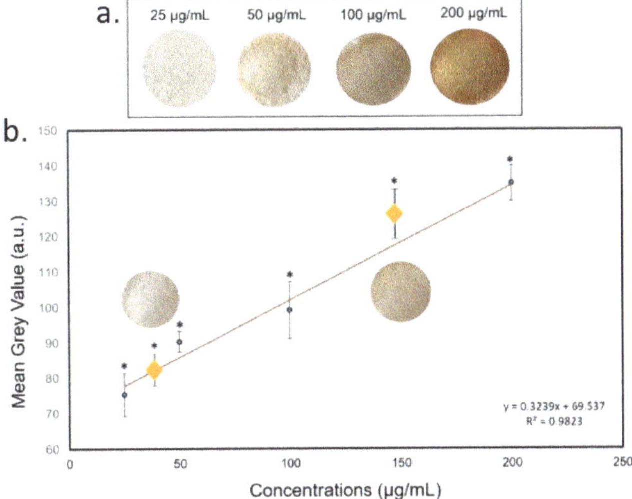

Figure 10. (**a**) Color change of filter paper after exposure to Cu(II) at different concentrations. (**b**) The yellow rhombuses represented the interpolation of Cu^{2+} intermediate concentrations (37 and 150 µg/mL). Data reported were the average of three independent experiments ± SD. The difference between as-prepared paper and colored papers was considered statistically significant performing a Student's *t*-test with $p < 0.05$ (<0.05 *).

In order to test the selectivity of the different PADs used in this study, several metal and gaseous solutions at 100 µg/mL and 100 ppm, respectively, were used. No significant visual color change had been observed in all the tested cases. For gaseous molecules, the PAD was exposed to methylamine, ethylamine, triethylamine, benzene, toluene, ethyl benzene, formaldehyde and ethanol at 100 ppm for ca. 15 min. Any noticeable effects on filter paper were recorded. This suggested the high selectivity of PAD to the NH_3 and C_2H_4O only (Figure 11a,b). Similar results were obtained analyzing the interferences of different metal ions after PAD exposure for 15 min. It was observed that 100 µg/mL of Na^+, K^+, Mg^{2+}, Ca^{2+}, Al^{3+}, Mn^{2+}, Fe^{3+}, Co^{2+}, Ni^{2+}, Zn^{2+}, Cd^{2+}, and Pb^{2+} highlighted negligible colorimetric effects on the PAD due to the small affinity with the analytes deposited on filter paper (Figure 11c,d).

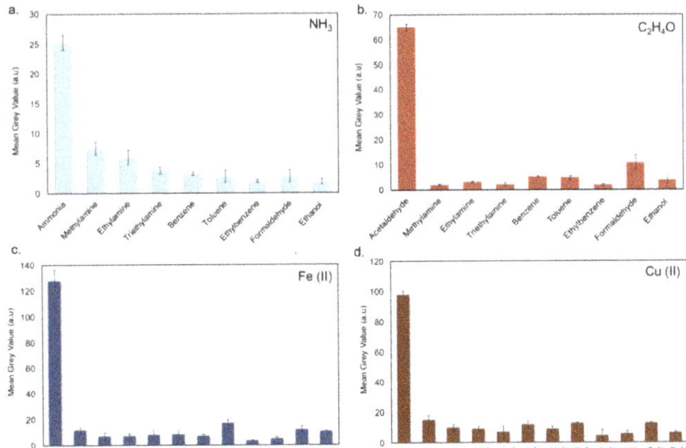

Figure 11. Interferences assay for NH_3 and C_2H_4O (**a**,**b**) and Fe(II) and Cu (II) (**c**,**d**). The values were expressed as MGV. Data reported were the average of three independent experiments ± SD.

4. Conclusions

The use of paper as a platform for sensing devices offers considerable advantages in terms of affordability and availability of functionalization processes; in fact, the hydrophilic nature of paper makes it a suitable tool due to the fast adsorption of different chemical solutions through its porous structure. Since only small volumes of reagents are needed to functionalize the paper device, it is very inexpensive. In addition, this technology does not require qualified personnel to collect and analyze the data. We developed an easy and versatile PAD that is able to measure different pollutant agents, namely NH_3, C_2H_4O, Fe^{2+} and Cu^{2+}, in two different mediums, air and water. The device architecture is a "proof of concept" of a new class of colorimetric sensors. In fact, it could be implemented in several environmental detection technologies, such as drones or aquatic sensors as well as individual protection devices or wearable technologies, by an electronic integration. In addition, the PAD can be used by different citizens of particular geographic areas to map the possible contaminations, with the aim to collect the global data and to build a database to monitor the pollution.

Author Contributions: Conceptualization and methodology V.D.M.; experimental V.D.M., G.F., M.C., L.M.; validation V.D.M.; formal analysis V.D.M., M.C., G.F.; data curation V.D.M., M.C.; writing—original draft V.D.M.; review and editing V.D.M., M.C., G.F., R.R.; supervision R.R.; project administration R.R. All authors have read and agreed to the published version of the manuscript.

Funding: This research was funded by project PAPER (Paper Analyzer for Particulate Exposure Risk), funded within POR Puglia FESR-FSE 2014-2020—Asse prioritario 1—Azione 1.6—Bando Innonetwork—grant number PH3B166 and the project M.In.E.R.V.A funded within POR Puglia FESR-FSE 2014-2020—Asse prioritario 1—Azione 1.4—Bando Innolabs—grant number NG2WPK0-1.

Acknowledgments: V.D.M. kindly acknowledge Franco Mazzotta for the beneficial discussion.

Conflicts of Interest: The authors declare no conflict of interest.

References

1. Ali, H.; Khan, E.; Ilahi, I. Environmental chemistry and ecotoxicology of hazardous heavy metals: Environmental persistence, toxicity, and bioaccumulation. *J. Chem.* **2019**, 1–14. [CrossRef]
2. Jaishankar, M.; Tseten, T.; Anbalagan, N.; Mathew, B.B.; Beeregowda, K.N. Toxicity, mechanism and health effects of some heavy metals. *Interdiscip. Toxicol.* **2014**, *7*, 60–72. [CrossRef] [PubMed]
3. Tchounwou, P.B.; Yedjou, C.G.; Patlolla, A.K.; Sutton, D.J. Heavy Metal Toxicity and the Environment. In *Experientia Supplementum*; Springer: Basel, Switzerland, 2012; Volume 101. [CrossRef]
4. Kumar, A.; Kumar, A.; MMS, C.-P.; Chaturvedi, A.K.; Shabnam, A.A.; Subrahmanyam, G.; Mondal, R.; Gupta, D.K.; Malyan, S.K.; Kumar, S.S.; et al. Lead toxicity: Health hazards, influence on food chain, and sustainable remediation approaches. *Int. J. Environ. Res. Public Health* **2020**, *17*. [CrossRef] [PubMed]
5. Kim, B.-E.; Nevitt, T.; Thiele, D.J. Mechanisms for copper acquisition, distribution and regulation. *Nat. Methods* **2008**, *4*, 176–185. [CrossRef]
6. Jan, A.T.; Azam, M.; Siddiqui, K.; Ali, A.; Choi, I.; Haq, Q.M.R. Heavy metals and human health: Mechanistic insight into toxicity and counter defense system of antioxidants. *Int. J. Mol. Sci.* **2015**, *16*, 29592–29630. [CrossRef]
7. Waldvogel-Abramowski, S.; Waeber, G.; Gassner, C.; Buser, A.; Frey, B.M.; Favrat, B.; Tissot, J.-D. Physiology of iron metabolism. *Transfus. Med. Hemother.* **2014**, *41*, 213–221. [CrossRef]
8. Keller, A.A.; Adeleye, A.S.; Conway, J.R.; Garner, K.L.; Zhao, L.; Cherr, G.N.; Hong, J.; Gardea-Torresdey, J.L.; Godwin, H.A.; Hanna, S.; et al. Comparative environmental fate and toxicity of copper nanomaterials. *NanoImpact* **2017**, *7*, 8–40. [CrossRef]
9. International Maritime Organization. Marine Environmental Protection Committee. In *Harmful Effects of the Use of Antifouling Paints for Ships*; 40th Session, Agenda Item 11, Annex 2nd; United Nations Economic and Social Council: London, UK, 1997.
10. Hall, L.W.; Anderson, R.D. A deterministic ecological risk assessment for copper in european saltwater environments. *Mar. Pollut. Bull.* **1999**, *38*, 207–218. [CrossRef]
11. Hall, L.W., Jr.; Scott, M.C.; Killen, W.D. *A Screening Level Probabilistic Ecological Risk Assessment of Copper and Cadmium in the Chesapeake Bay Watershed*; US EPA, Chesapeake Bay Program Office: Annapolis, MD, USA, 1997.

12. Taylor, A.A.; Tsuji, J.S.; Garry, M.R.; McArdle, M.E.; Goodfellow, W.L., Jr.; Adams, W.J.; Menzie, C.A. Critical review of exposure and effects: Implications for setting regulatory health criteria for ingested copper. *Environ. Manag.* **2019**, *65*, 131–159. [CrossRef]
13. Hébert, C.D. *NTP Technical Report on Toxicity Studies of Cupric Sulphate (CAS N°7758-99-8) Administered in Drinking Water and Feed to F344/N Rats and B6C3F1 Mice*; Technical Report No. 7758-99-8; United States Department of Health and Hum: Washington, DC, USA, 1993.
14. Eid, R.; Arab, N.T.; Greenwood, M.T. Iron mediated toxicity and programmed cell death: A review and a re-examination of existing paradigms. *Biochim. Biophys. Acta Mol. Cell. Res.* **2017**, *1864*, 399–430. [CrossRef]
15. Phippen, B.; Horvath, C.; Nordin, R.N.N. *Ambient Water Quality Guidelines for Iron: Overview*; Water Stewardship Division, Ministry of Environment Province of British Columbia: Cranbrook, BC, Canada, 2008.
16. Grazuleviciene, R.; Nadisauskiene, R.; Buinauskiene, J.G.T. Effects of elevated levels of manganese and iron in drinking water on birth outcomes. *Pol. J. Environ. Stud.* **2009**, *18*, 819–825.
17. Sutton, M.A. Introduction. In *Atmospheric Ammonia*; Sutton, M.A., Reis, S., Howard, C., Eds.; Springer: Dordrecht, The Netherlands, 2009.
18. Van Damme, M.; Clarisse, L.; Whitburn, S.; Hadji-Lazaro, J.; Hurtmans, D.; Clerbaux, C.; Coheur, P.-F. Industrial and agricultural ammonia point sources exposed. *Nature* **2018**, *564*, 99–103. [CrossRef] [PubMed]
19. Agency for Toxic Substances and Disease Registry (ATSDR). *Toxicological Profile for Ammonia*; U.S. Department of Health and Human Services, Public Health Service: Atlanta, GA, USA, 2004.
20. National Research Council. Acute Exposure Guideline Levels for Selected Airborne Chemicals. In *Acute Exposure Guideline Levels for Selected Airborne Chemicals*; The National Academies Press: Washington, DC, USA, 2008.
21. Michaels, R.A. Emergency planning and the acute toxic potency of inhaled ammonia. *Environ. Health Perspect.* **1999**. [CrossRef] [PubMed]
22. Goldmacher, V.S.; Thilly, W.G. Formaldehyde is mutagenic for cultured human cells. *Mutat. Res. Toxicol.* **1983**, *116*, 417–422. [CrossRef]
23. Fassett, D.W. Aldehydes and acetals. In *Indus Trial Hygiene and Toxicology*, 2nd ed.; Patty, F.A., Ed.; Interscience: New York, NY, USA, 1963; Volume 2, pp. 1959–1989.
24. Woutersen, R.; Appelman, L.; Feron, V.J.; Van Der Heijden, C. Inhalation toxicity of acetaldehyde in rats II. Carcinogenicity study: Interim results after 15 months. *Toxicology* **1984**, *31*, 123–133. [CrossRef]
25. Feron, V.J.; Kruysse, A.; Woutersen, R.A. Respiratory tract tumors in hamsters exposed to acetaldehyde vapour alone or simultaneously to benzo[a]pyrene or dimethylnitrosamine. *Eur. J. Cancer Clin. Oncol.* **1982**, *18*, 13–31. [CrossRef]
26. Pandey, S.K.; Mohanta, G.C.; Kumar, P. Development of Disposable Sensor Strips for Point-of-Care Testing of Environmental Pollutants. *Adv. Nanosens. Biol. Environ. Anal.* **2019**, *6*, 95–118. [CrossRef]
27. Cunningham, J.C.; DeGregory, P.R.; Crooks, R.M. New Functionalities for Paper-Based Sensors Lead to Simplified User Operation, Lower Limits of Detection, and New Applications. *Annu. Rev. Anal. Chem.* **2016**, *9*, 183–202. [CrossRef]
28. Nery, E.W.; Kubota, L.T. Sensing approaches on paper-based devices: A review. *Anal. Bioanal. Chem.* **2013**, *405*, 7573–7595. [CrossRef]
29. Huang, Y.-Q.; You, J.-Q.; Cheng, Y.; Sun, W.; Ding, L.; Feng, Y.-Q. Frontal elution paper chromatography for ambient ionization mass spectrometry: Analyzing powder samples. *Anal. Methods* **2013**, *5*, 4105. [CrossRef]
30. Xiao-Wei, H.; Zou, X.; Ji-Yong, S.; Zhi-Hua, L.; Jie-Wen, Z. Colorimetric sensor arrays based on chemo-responsive dyes for food odor visualization. *Trends Food Sci. Technol.* **2018**, *81*, 90–107. [CrossRef]
31. Liu, B.; Zhuang, J.; Wei, G. Recent advances in the design of colorimetric sensors for environmental monitoring. *Environ. Sci. Nano* **2020**. [CrossRef]
32. Ratnarathorn, N.; Chailapakul, O.; Henry, C.S.; Dungchai, W. Simple silver nanoparticle colorimetric sensing for copper by paper-based devices. *Talanta* **2012**, *99*, 552–557. [CrossRef] [PubMed]
33. Cate, D.M.; Noblitt, S.D.; Volckens, J.; Henry, C.S. Multiplexed paper analytical device for quantification of metals using distance-based detection. *Lab Chip* **2015**, *15*, 2808–2818. [CrossRef]
34. Irvine, G.W.; Tan, S.N.; Stillman, M.J. A Simple Metallothionein-Based Biosensor for Enhanced Detection of Arsenic and Mercury. *Biosensors* **2017**, *7*, 14. [CrossRef] [PubMed]
35. Li, Z.; Suslick, K.S. Ultrasonic Preparation of Porous Silica-Dye Microspheres: Sensors for Quantification of Urinary Trimethylamine N-Oxide. *ACS Appl. Mater. Interfaces* **2018**, *10*, 15820–15828. [CrossRef] [PubMed]

36. Xin, X.; Dai, F.; Li, F.; Jin, X.; Wang, R.; Sun, D. A visual test paper based on Pb(ii) metal–organic nanotubes utilized as a H2S sensor with high selectivity and sensitivity. *Anal. Methods* **2017**, *9*, 3094–3098. [CrossRef]
37. Francis, A.P.; Thiyagarajan, D. Toxicity of carbon nanotubes: A review. *Toxicol. Ind. Health* **2018**, *34*, 200–210. [CrossRef]
38. Maity, A.; Ghosh, B. Fast response paper based visual color change gas sensor for efficient ammonia detection at room temperature. *Sci. Rep.* **2018**, *8*, 16851. [CrossRef]
39. Babayigit, A.; Thanh, D.D.; Ethirajan, A.; Manca, J.; Muller, M.; Boyen, H.-G.; Conings, B. Assessing the toxicity of Pb- and Sn-based perovskite solar cells in model organism *Danio rerio*. *Sci. Rep.* **2016**, *6*, srep18721. [CrossRef]
40. Khattab, T.A.; Dacrory, S.; Abou-Yousef, H.; Kamel, S. Development of microporous cellulose-based smart xerogel reversible sensor via freeze drying for naked-eye detection of ammonia gas. *Carbohydr. Polym.* **2019**, *210*, 196–203. [CrossRef] [PubMed]
41. Lu, Y.; Shi, W.; Qin, J.; Lin, B. Fabrication and characterization of paper-based microfluidics prepared in nitrocellulose membrane by wax printing. *Anal. Chem.* **2010**, *82*, 329–335. [CrossRef] [PubMed]
42. Shen, L.; Hagen, J.A.; Papautsky, I. Point-of-care colorimetric detection with a smartphone. *Lab Chip* **2012**, *12*, 4240–4243. [CrossRef] [PubMed]
43. Wang, S.; Zhao, X.; Khimji, I.; Akbas, R.; Qiu, W.; Edwards, D.; Cramer, D.W.; Ye, B.; Demirci, U. Integration of cell phone imaging with microchip ELISA to detect ovarian cancer HE4 biomarker in urine at the point-of-care. *Lab Chip* **2011**, *11*, 3411–3418. [CrossRef] [PubMed]
44. Garcia, A.; Erenas, M.M.; Marinetto, E.; Abad, C.A.; De Orbe-Payá, I.; Palma, A.J.; Capitán-Vallvey, L. Mobile phone platform as portable chemical analyzer. *Sens. Actuators B* **2011**, *156*, 350–359. [CrossRef]
45. Sengupta, P.P.; Barik, S.; Adhikari, B. Polyaniline as a gas-sensor material. *Mater. Manuf. Process.* **2006**, *21*, 263–270. [CrossRef]
46. ImageJ. Available online: http://imagej.nih.gov/ij/ (accessed on 24 September 2020).
47. Yu, P.; Deng, M.; Yang, Y. New single-layered paper-based microfluidic devices for the analysis of nitrite and glucose built via deposition of adhesive tape. *Sensors* **2019**, *19*, 4082. [CrossRef]
48. Joshi, B.P.; Park, J.; Lee, W.I.; Lee, K.H. Ratiometric and turn-on monitoring for heavy and transition metal ions in aqueous solution with a fluorescent peptide sensortle. *Talanta* **2009**, *78*, 903–909. [CrossRef]
49. Long, G.L.; Winefordner, J.D. Limit of detection a closer look at the IUPAC definition. *Anal. Chem.* **1983**, *55*, 712A–724A. [CrossRef]
50. Manisalidis, I.; Stavropoulou, E.; Stavropoulos, A.; Bezirtzoglou, E. Environmental and health impacts of air pollution: A review. *Front. Public Health* **2020**, *8*. [CrossRef]
51. Vhahangwele, M.; Muedi, K.L. Environmental Contamination by Heavy Metals. *Heavy Met.* **2018**. [CrossRef]
52. Tuomisto, H.; Hodge, I.; Riordan, P.; Macdonald, D. Does organic farming reduce environmental impacts?—A meta-analysis of European research. *J. Environ. Manag.* **2012**, *112*, 309–320. [CrossRef] [PubMed]
53. Sharma, S.; Bhattacharya, A. Drinking water contamination and treatment techniques. *Appl. Water Sci.* **2016**, *7*, 1043–1067. [CrossRef]
54. WHO chronicle. *Guidelines for Drinking-Water Quality*, 4th ed.; World Health Organization: Geneva, Switzerland, 2011; ISBN 978924154815 1.

Article

Development and Evaluation of a New Spectral Disease Index to Detect Wheat Fusarium Head Blight Using Hyperspectral Imaging

Dongyan Zhang [1], Qian Wang [1], Fenfang Lin [1,2], Xun Yin [1], Chunyan Gu [3] and Hongbo Qiao [1,4,*]

1 National Engineering Research Center for Agro-Ecological Big Data Analysis & Application, Anhui University, Hefei 230601, China; zhangdy@ahu.edu.cn (D.Z.); p18201091@stu.ahu.edu.cn (Q.W.); linfenfang@126.com (F.L.); p17301129@stu.ahu.edu.cn (X.Y.)

2 School of Geography and Remote Sensing, Nanjing University of Information Science & Technology, Nanjing 210044, China

3 Institute of Plant Protection and Agro-Products Safety, Anhui Academy of Agricultural Sciences, Hefei 230031, China; guchunyan@aaas.org.cn

4 School of Information and Management Science, Henan Agricultural University, Zhengzhou 450002, China

* Correspondence: qiaohb@henau.edu.cn

Received: 30 March 2020; Accepted: 13 April 2020; Published: 16 April 2020

Abstract: Fusarium head blight (FHB) is a major disease threatening worldwide wheat production. FHB is a short cycle disease and is highly destructive under conducive environments. To provide technical support for the rapid detection of the FHB disease, we proposed to develop a new Fusarium disease index (FDI) based on the spectral data of 374–1050 nm. This study was conducted through the analysis of reflectance spectral data of healthy and diseased wheat ears at the flowering and filling stages by hyperspectral imaging technology and the random forest method. The characteristic wavelengths selected were 570 nm and 678 nm for the late flowering stage, 565 nm and 661 nm for the early filling stage, 560 nm and 663 nm for the combined stage (combining both flowering and filling stages) by random forest. FDI at each stage was derived from the wavebands of each corresponding stage. Compared with other 16 existing spectral indices, FDI demonstrated a stronger ability to determine the severity of the FHB disease. Its determination coefficients (R^2) values exceeded 0.90 and the *RMSEs* were less than 0.08 in the models for each stage. Furthermore, the model for the combined stage performed better when used at single growth stage, but its effect was weaker than that of the models for the two individual growth stages. Therefore, using FDI can provide a new tool to detect the FHB disease at different growth stages in wheat.

Keywords: hyperspectral imaging; spectral indices; random forest; growth stage; Fusarium head blight

1. Introduction

The production of wheat plays important social and economic roles, and the quality and safety issues related to these functions have been the focus of research at the national level and abroad. Fusarium head blight (FHB) is a wheat disease caused by the fungus *Gibberella zeae* (*Fusarium graminearum*) and often severely affects wheat yield and quality. Wheat infected with FHB accumulates a large amount of toxins in its grains, thereby seriously threatening public health. These bacterial toxins can contaminate flour and persist in the food chain for long periods, producing carcinogens. Therefore, FHB has become one of the crop diseases of great concern worldwide [1,2].

Conventional crop disease detection methods range from the naked eye to random monitoring, which have the disadvantages of strong subjectivity, high labor intensity, and time consumption. With the rapid development of spectral technology, hyperspectral imaging has been gradually applied to non-destructive detection of plant diseases and insect pests [3]. Hyperspectral images can provide hundreds of thousands of continuous narrow band data points and are very sensitive to changes in the physical and chemical parameters of plants caused by disease infection. These changes have gradually developed into effective features for expressing plant growth information and have proven to be effective in identifying plant diseases and insect pests [4]. Zheng et al. [5] used wavelengths of 570 nm, 525 nm, 705 nm, 860 nm, 790 nm, and 750 nm to identify yellow rust successfully in the early and middle stages of wheat growth. Huang [6] and others, based on the Relief-F algorithm, proposed that 515 nm, 698 nm, and 738 nm are key wavelengths for distinguishing wheat powdery mildew from other diseases. Bauriegel [7] believe that 550–560 nm and 665–675 nm are the best bands for field identification of FHB. However, with the increase in spectral and image spatial resolution, the simultaneous increase of data dimensions, noise, and redundant spectra pose considerable challenges to data storage, processing, and analysis [8].

The vegetation index is an effective method often used in the field of optical remote sensing to reflect changes in plant physiological and biochemical parameters. This index is a simple and efficient spectral data processing method that combines a few characteristic bands in a certain mathematical form. This method greatly eliminates the redundancy of hyperspectral data, has a small amount of calculation, and is widely used to estimate crop yields [9], pigment content [10], canopy structure [11], and changes in water status [12]. In recent years, exclusive spectral indexes have been proposed and demonstrated unique advantages in plant disease detection. Zhang et al. [8] developed a hyperspectral index based on hyperspectral microscopic images to identify FHB ears with classification accuracy of 0.898. Devadas et al. [13] observed that healthy and susceptible (yellow rust, leaf rust, stem rust) wheat can be distinguished based on the anthocyanin reflectance index. Rumpf et al. [14] combined the spectral index and support vector machine to identify beet leaf spot, leaf rust, and powdery mildew at an early stage, and the classification accuracy was above 0.65. The results of these studies indicate that a spectral index calculated by spectral reflectance at a special wavelength position has high potential for applications in the fields of crop diseases and insect pests. However, these proposed spectral indexes do not clearly indicate the applicable growth stage or only consider a certain growth period of the crop.

The pathological characteristics of wheat after being infected with FHB differ at separate stages, which may cause inconsistent relationships between the spectral index and the status of FHB during different growth periods. In many studies on spectral indexes, researchers have usually pooled observation data at different stages of the entire growth stage to explore characteristic bands and construct spectral indices, thereby weakening the inconsistencies of FHB status in different growth stages [15]. FHB usually occurs during the flowering and filling stages of wheat. At maturity, the damage caused by FHB to wheat yield and quality has been determined. Currently, conducting research on FHB identification is of minimal importance. Therefore, this study focuses on the accuracy and stability of the disease severity monitoring model for FHB at late flowering and early filling stages, with the goal of providing assistance for the scientific control of the disease. The main research objectives are as follows: (1) Based on the difference in spectral responses between healthy and infected ears, the most suitable characteristic wavelengths for identifying FHB were selected and determined by the random forest (RF) technique in the late flowering stage and early filling stage. (2) A Fusarium disease index (FDI) was constructed in the form of the normalized wavelength difference and compared with classical disease index to evaluate the accuracy and stability of FDI.

2. Materials and Methods

2.1. Wheat Material

This study was carried out at the experimental base of the Anhui Academy of Agricultural Sciences (31°89′ N, 117°1′ E) in China from 2017 to 2018 (Figure 1). The tested wheat variety was Xinong 979, which is moderately susceptible to FHB. A 10 × 10 m experimental plot was divided into an inoculation area (50 m^2) and a control area (50 m^2). In the early flowering stage, a small sprayer was used to spray a freshly prepared spore suspension (*F. graminearum*) on the ears of wheat in the inoculation area. The control area was sprayed with pesticide (Carbendazim, 750 g/hm^2) once between the full heading stage and the early flowering stage (18 April 2018) to prevent FHB and ensure a sufficient number of healthy samples for comparative research. Other field management techniques such as fertilization and irrigation were carried out in the two experimental plots according to local agronomic measures. In this study, 149 and 229 wheat ears were collected at the late flowering (3 May 2018) and early filling (9 May 2018) stages for a total of 378 samples.

Figure 1. Experimental field plots.

2.2. Inoculum Production

Under the bench with a sterile environment, the infected wheat grains were treated twice with mercury dichloride–alcohol–sterilized water. The treated grains were added to potato dextrose agar medium, and the culture was grown at 25 °C for three days. Five mycelium plugs were picked at the edge of the colony and placed in 100 mL carboxymethyl cellulose medium for four days. The conidia were filtered with two pieces of filter paper and centrifuged at 5000 rpm for 5 min. The concentration of the spore suspension was adjusted to 1×10^5/mL with sterile water.

2.3. Data Acquisition and Processing

2.3.1. Spectral Measurements

The spectral reflectance of the ears was measured using an SOC710E spectrometer (Surface Optics Corporation, San Diego, CA, USA). The spectral range of this instrument is 374–1050 nm, and the spectral resolution is 2.3 nm. After picking wheat ears in the field, the samples were quickly sent to the laboratory in a fresh-keeping box, and the spectral data were collected in a dark room. Wheat ears were placed on a black platform, and the exposure time and the distance between the platform and the lens were adjusted so that the wheat ears could be clearly imaged. Two 75-Watt halogen lamps were placed on both sides of the dark room to illuminate the sample. The hyperspectral imaging system is shown in Figure 2. Measurements on the whiteboard (with a reflectance of approximately one) and dark current (with a reflectance of approximately zero) were performed for spectral correction. The reflectance value of the dark current was recorded by covering the lens with a black cloth. The correction formula is as follows:

$$R = \frac{R_{original} - R_{dark}}{R_{white} - R_{dark}} \quad (1)$$

where R_{original} represents the original spectral reflectance, R_{dark} is the reflectance value of the dark current, R_{white} is the reflectance value of the reference whiteboard, and R is the corrected spectral reflectance of the image.

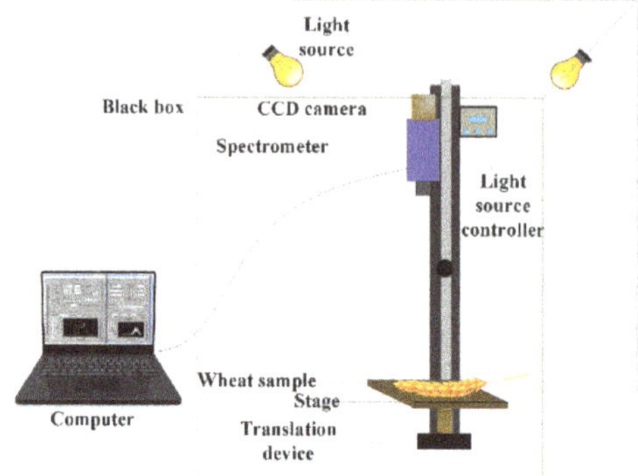

Figure 2. Hyperspectral imaging system.

2.3.2. Calculation of Disease Severity

With respect to ear scale, the disease severity is mainly quantified by the ratio of the diseased area on the ear to the whole ear area. Therefore, Fusarium head blight lesions were segmented from the whole ear to measure relative lesion area on ears. First, the third channel image of the original red green blue (RGB) image (Figure 3a) was processed with binarization and morphological corrosion and expansion to remove the tip of wheat and stalks in the image (Figure 3b).

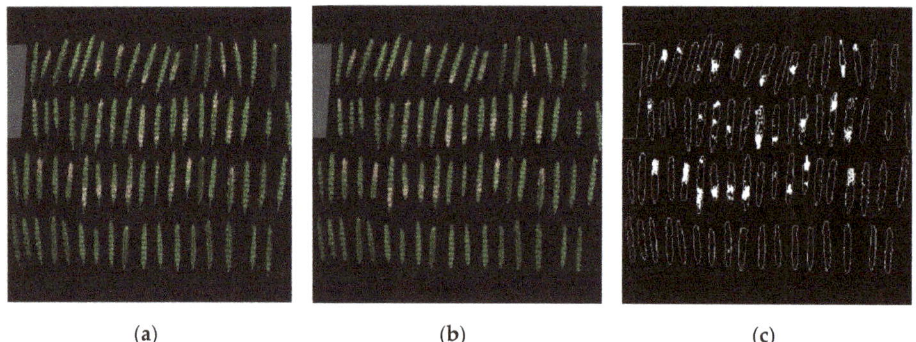

Figure 3. Extraction of diseased spots from wheat ears, (**a**) original image; (**b**) image of wheat tip and stalk removal; (**c**) image of diseased spots extraction.

Because the three components in the RGB image were represented by a three-dimensional Cartesian coordinate system, they were highly correlated and relatively heterogeneous, resulting in a small difference between the healthy area and the diseased area that was difficult to segment. The color space of YDbDr was used to separate the brightness and color difference, which was more suitable for distinguishing between green and red yellow susceptible areas. Therefore, the RGB image after the wheat tip and stalk removal was transferred to the YDbDr color space, and a threshold segmentation method was adopted to extract the ear disease spots (Figure 3c).

The severity of the FHB is expressed by the ratio of the number of pixels in the disease spot region to the pixel number in the whole wheat ear region, as shown in Equation (2):

$$SI = \frac{S_{lesion}}{S_{all}} \tag{2}$$

SI represents the severity of FHB, S_{lesion} is the number of pixels in the disease spot area, and S_{all} is the number of pixels in the whole wheat ear region.

2.3.3. Characteristic Band Selection

The RF algorithm was used to select the characteristic wavelengths that are sensitive to FHB. This algorithm is an ensemble learning algorithm based on multiple classification and regression trees (CARTs) proposed by Breiman [16] and is often used for characteristic wavelength selection in hyperspectral data analysis [17,18]. In this algorithm, the bootstrap resampling method is used to generate the training set; attributes are measured according to the minimum Gini index principle, and CART is gradually established. Subsequently, the classification of samples is determined by combining the voting of each decision tree. At the same time, the samples that do not appear in the training set are designated as "bag data" and are used to predict the accuracy of the algorithm.

The Gini index is an attribute splitting method based on impurity. The smaller the impurity, the worse the dispersion degree of the variables and the more information that is obtained [19]. The formula for calculating the impurity Gini index **G** is shown in Equation (3):

$$\mathrm{G(a)} = 1 - \sum_{\mathrm{i}=1}^{\mathrm{c}} \mathrm{P_i^2} \tag{3}$$

where c is the number of sample categories, and $\mathrm{P_i}$ is the probability that the sample corresponding to an attribute a belongs to category $\mathrm{c_i}$ ($\mathrm{c_i}$ represents the i-th category).

Because the Gini impurity index is negatively related to the available information, this study used the Gini purity index to convert the purity and available useful information into a positive correlation to more intuitively reflect the impact of features on the classification effect. The calculation formula is as follows:

$$G_{purity}(\mathrm{a}) = \sum_{\mathrm{i}=1}^{\mathrm{c}} \mathrm{P_i^2} \tag{4}$$

Through the converted formula, the Gini purity index of characteristic f can be obtained as follows:

$$\mathrm{G(f)} = \sum_{\mathrm{i}=1}^{\mathrm{k}} \frac{\mathrm{n_i}}{\mathrm{N}} G_{purity}(\mathrm{a_i}) \tag{5}$$

where N is the number of samples, k is the number of categories of a certain attribute a, $\mathrm{a_i}$ is a certain category of attributes, and $\mathrm{n_i}$ is the number of samples corresponding to a certain category. The greater the purity of a feature, the stronger the ability of the feature to recognize the sample. The calculation formula for the importance measurement of the feature is as follows:

$$\mathrm{S(v)} = \frac{1}{\mathrm{t}} \sum_{\mathrm{u}=1}^{\mathrm{t}} \mathrm{G}(f_{\mathrm{uv}}) \tag{6}$$

where t is the number of training datasets in the RF, $\mathrm{G}(f_{\mathrm{uv}})$ is the purity of the v-th dimension eigenvector in the u-th training dataset (v = 1, 2, 3, ..., k), and k is the overall dimension of the sample. Finally, the required characteristic wavelengths were obtained according to the positive maximum value and the negative minimum value of the importance score.

2.4. Construction of Proposed New Spectral Disease Index for Indentifing Wheat FHB

Previous studies [6,20] have shown that the disease spectral index in the form of the normalized wavelength difference is very sensitive to spectral changes caused by powdery mildew, stripe rust, and aphids. Therefore, this study used the normalized wavelength difference in combination with characteristic wavelengths to construct the exclusive FDI for each period. The calculation is carried out via Equation (7):

$$\text{FDI} = \frac{R_{\lambda 1} - R_{\lambda 2}}{R_{\lambda 1} + R_{\lambda 2}} \tag{7}$$

where $R_{\lambda 1}$ represents the reflectance at the $\lambda 1$ wavelength and $R_{\lambda 2}$ represents the reflectance at the $\lambda 2$ wavelength.

2.5. Traditional Spectral Indices for Wheat FHB Detection

Pigment content can provide information about the physiological state of leaves; consequently, a spectral index that can characterize the plant pigment content is highly related to plants' physiological and biochemical changes and is often used for non-destructive detection of plant diseases and insect pests. Sixteen commonly used spectral indexes (Table 1) were selected and compared with the FDI proposed in this study to evaluate FDI's ability to identify and distinguish infected ears.

Table 1. Traditional spectral indices tested in the study.

Full Name of Spectral Index	Spectral Index Abbreviation	Calculation Formula
nitrogen reflectance index [21]	NRI	$(R_{570} - R_{670})/(R_{570} + R_{670})$
photochemical reflectance index [22]	PRI	$(R_{531} - R_{570})/(R_{531} + R_{570})$
transformed vegetation index [23]	TVI	$0.5 \times [120 \times (R_{750} - R_{550}) - 200 \times (R_{750} + R_{550})]$
transformed chlorophyll absorption in the reflectance index [24]	TCARI	$3 \times [(R_{700} - R_{670}) - 0.2 \times (R_{700} - R_{550}) \times (R_{700}/R_{670})]$
modified chlorophyll absorption in the reflectance index [25]	MCARI	$[(R_{700} - R_{670}) - 0.2 \times (R_{700} - R_{550})] \times R_{700}/R_{670})$
red-edge vegetation stress index [26]	RVSI	$[(R_{712} + R_{752})/2] - R_{732}$
plant senescence reflectance index [27]	PSRI	$(R_{678} - R_{500})/R_{750}$
green index [28]	GI	R_{554}/R_{677}
structural independent pigment index [29]	SIPI	$(R_{800} - R_{445})/(R_{800} + R_{680})$
normalized pigment chlorophyll ratio index [30]	NPCI	$(R_{680} - R_{430})/(R_{680} + R_{430})$
normalized difference vegetation index [31]	NDVI	$(R_{840} - R_{675})/(R_{840} + R_{675})$
optimized soil-adjusted vegetation index [32]	OSAVI	$1.16 \times [(R_{800} - R_{670})/(R_{800} + R_{670} + 0.16)]$
Lichtenthaler's indices [33]	Lic1	$(R_{800} - R_{680})/(R_{800} + R_{680})$
Lichtenthaler's indices [34]	Lic2	R_{400}/R_{690}
anthocyanin reflectance index [35]	ARI	$(R_{550})^{-1} - (R_{700})^{-1}$
physiological reflectance index [22]	PHRI	$(R_{550} - R_{531})/(R_{550} + R_{531})$

2.6. Linear Regression Model and Validation

A linear regression model was used to model the relationship between spectral indices (FDI and existing spectral indices) and the severity index (SI) at different growth stages. The evaluation indexes of the model included the root mean square error (RMSE) and the coefficient of determination (R^2). RMSE represents the standard deviation of the difference between the predicted value and the measured value. R^2 used to measure the proportion of variation in the dependent variable that can be explained by the independent variable. The closer R^2 is to 1, the closer the regression line is to each observation point, and the better the regression fit.

To make the distribution of samples more uniform, the SI values of the samples were arranged in descending order and then divided into a training dataset and test dataset in a 3:1 proportion.

Specifically, one sample was taken from a group of four samples as the test dataset, and the remaining three were used as the training dataset. In the model, using FDI as the independent variable and SI as the dependent variable, the relationship between FDI and SI in different periods was determined by regression analysis. In a single growing period, the FDI of the sample in the training set was calculated, and the linear regression equation between it and the corresponding SI was established to obtain the R^2 and RMSE in the training dataset. The SI of each sample in the test dataset was predicted by a linear regression equation and FDI, and the R^2 and RMSE of the prediction set were obtained by comparing the actual SI with the predicted SI. In addition, the samples in the combined stage were modeled and predicted, and the linear regression equation of the combined stage was used to predict the test dataset samples of the late flowering and early filling stages.

3. Results and Analysis

3.1. Spectral Response Differences of Wheat FHB with Different Disease Severities

In the process of using hyperspectral data to accurately identify diseases, the study of spectral signatures under different disease severities is the basis for screening and identifying sensitive bands of diseases. Figure 4 shows the spectral signature of wheat ears with different infection levels. Generally, in the range of the 550–720 nm band, the spectral reflectance of healthy ears is lower than that of infected samples, with an obvious green peak and red valley; accordingly, these two spectral features disappear in the severely infected ears. Conversely, in the range of the 721–1000 nm band, the more severely infected the sample, the lower its reflectivity. The difference in the responses of wheat ears with different severities in the 550–720 nm and 721–1000 nm bands may be related to the difference in the pigment content and moisture content in mesophyll tissue [9]. Furthermore, with the increase in the severity of wheat diseases, a clear red edge moved in the short-wave direction. The above obvious spectral signature differences provide an important optical basis for analyzing and constructing the relationship between the spectral index and FHB severity in this study.

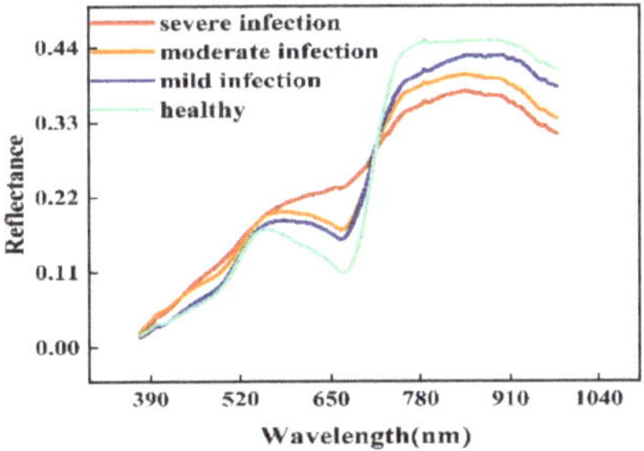

Figure 4. Spectral reflectance curves of wheat ears with different disease severities.

3.2. Construction of Proposed New Spectral Disease Index for Identifying Wheat FHB

3.2.1. Characteristic Bands for Identifying FHB at Different Growth Stages

RF was used to select characteristic wavelengths in samples during the late flowering stage, early filling stage, and combination of both. The weight coefficients of all wavelengths were calculated in the spectral range of 374–1050 nm. Except for the extreme points, the weight coefficients of adjacent wavelengths were similar, thus indicating that the information of adjacent wavelengths was highly correlated. To reduce the redundant information and maximize the effective spectral information, this study selected the wavelength corresponding to the positive highest weight coefficient and the negative lowest weight coefficient as the characteristic wavelengths. As shown in Figure 5, the characteristic wavelengths were 570 nm and 678 nm at the late flowering stage, 565 nm and 661 nm at the early filling stage, and 560 nm and 663 nm at the combined stage.

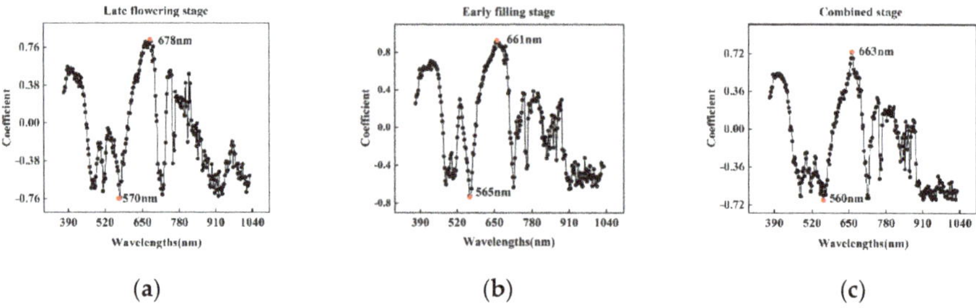

Figure 5. Weight coefficients calculated by RF at the late flowering stage (**a**), early filling stage (**b**) and combined stage (**c**).

Characteristic wavelengths selected from the different stages were all in the range of 565–680 nm. Combined with Figure 5, this band range shows a significant difference between healthy and infected wheat ears. Furthermore, the characteristic wavelengths selected in the two growth stages were different, from 678 nm and 570 nm in the late flowering stage to 661 nm and 565 nm in the early filling stage. With the development of FHB, the position of the characteristic wavelength moved in the direction of the short wave, which was consistent with the spectral change of the previously reported disease stress plants; specifically, the red edge shifted in the direction of the blue wave. Notably, the characteristic wavelengths of the combined stage are closer to those of the early filling period, which may be due to the sample size in the early filling stage (229) being larger than that in the late flowering stage (149); moreover, the incidence characteristics of FHB were obvious in this growth period.

3.2.2. Construction of New Fusarium Disease Index for Identifying Wheat FHB

In this study, with FDI as the independent variable and SI as the dependent variable, the relationship between FDI and SI in different stages was evaluated by linear regression analysis (Figure 6). FDI made an accurate prediction of the SI of wheat ears at the late flowering stage, early filling stage, and combined stage (R^2 was greater than 0.90, RMSE was less than 0.08). At each stage, the R^2 and RMSE of the training and test datasets were close, indicating that the model had a strong generalization ability. From the results of the training and test datasets, the FDI prediction was the most accurate in the early filling stage, followed by the late flowering period, and the lowest in the combined stage. (0.96, 0.94, and 0.90, respectively).

In this study, the regression model obtained from the combined stage was applied to the test set of the late flowering and early filling stages (Figure 7). The results obtained by applying the regression model established through the combined stage to the test datasets of the late flowering and early filling stages (R^2 = 0.91 and 0.94, respectively) were slightly lower than those of the regression models of the late flowering and early filling stages (R^2 = 0.94 and 0.96, respectively), especially in the late flowering stage.

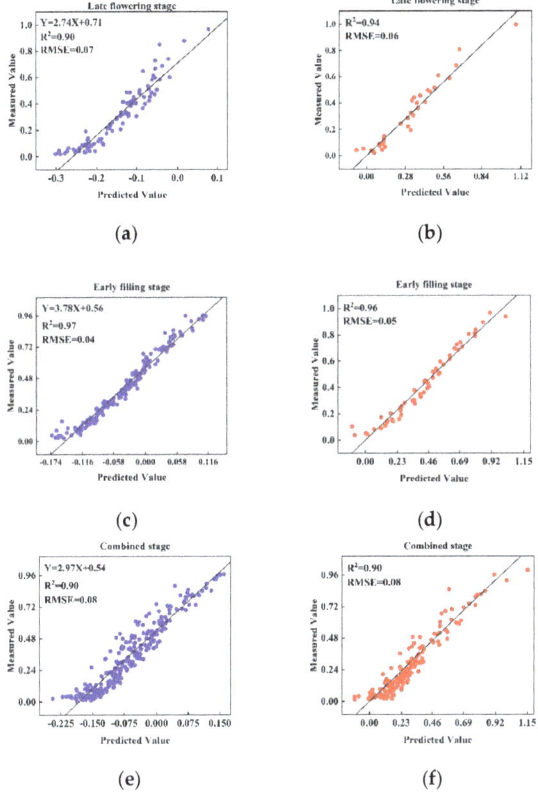

Figure 6. Evaluation of regression models in the training and test datasets at the late flowering stage (**a**,**b**), the early filling stage (**c**,**d**) and the combined stage (**e**,**f**).

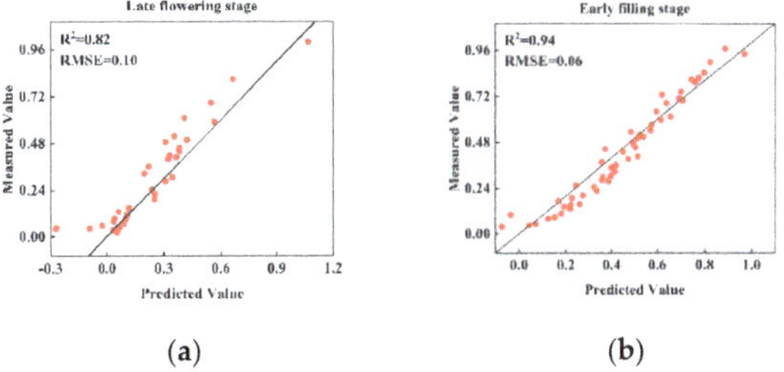

Figure 7. Evaluation of regression models at the combined stage used at the late flowering stage (**a**) and early filling stage (**b**).

3.3. Comparison of FDI and Traditional Spectral Indices

To verify the application potential of the FDI for detection of FHB, the results were compared with 16 other published spectral indexes at different stages (Tables 2–4). In the late flowering stage and combined stage, only the FDI proposed in this study had an R^2 above 0.9 in the training and the test datasets. In the early filling stage, only FDI and NRI had an R^2 above 0.9 in both the training and test datasets. The characteristic wavelengths of FDI (661 nm, 565 nm) and NRI (670 nm, 570 nm) were also close. The prediction results of FDI were higher than those of other spectral indexes in different stages, especially in the late flowering stage, which indicated that the FDI had excellent monitoring accuracy in the early stage of FHB infection. Furthermore, the detection capabilities of the selected spectral indices were different at separate stages, but the R^2 of FDI at every stage was greater than 0.9. Among these indices, nitrogen reflectance index (NRI), transformed vegetation index (TVI), and green index (GI) performed relatively well at different growth stages. What they all have in common is that they have better predictions during a single growth period than during the combined growth period. The diversity of samples in the combined stage may increase the difficulty of model prediction in this period because the characteristic wavelengths are dynamically changing in different growth stages [5]. At the same time, the performance of some indices in different growth stages will be very different. For example, modified chlorophyll absorption in the reflectance index (MCARI) has much better predictive power in the early filling period (R^2 = 0.67) than in the late flowering period (R^2 = 0.41); normalized pigment chlorophyll ratio index (NPCI) has much better predictive power in the combined period (R^2 = 0.77) than in the early filling period (R^2 = 0.17). The same index responds differently to diseases in different growth stages, which may be affected by the pathogenic mechanism of vegetation [13].

Table 2. Comparison of proposed FDI and traditional spectral indices at the late flowering stage.

Spectral Indices	Late Flowering Stage				
	Regression Equation	Training Set		Test Set	
		R^2	RMSE	R^2	RMSE
NRI	y = −2.77x + 0.68	0.87	0.08	0.86	0.09
PRI	y = −5.68x + 0.02	0.08	0.22	−7.9	0.21
TVI	y = −0.06x + 1.36	0.86	0.09	0.91	0.07
TCARI	y = −4.01x + 1.13	0.75	0.11	0.78	0.11
MCARI	y = −21.73x + 1.27	0.38	0.18	0.41	0.19
RVSI	y = 6.97x + 0.48	0.13	0.21	−4.46	0.22
PSRI	y = 4.96x + 0.01	0.73	0.12	0.79	0.12
GI	y = −0.85x + 1.48	0.86	0.09	0.88	0.07
SIPI	y = −3.39x + 2.40	0.74	0.12	0.68	0.12
NPCI	y = 2.58x − 0.61	0.41	0.18	−0.1	0.18
NDVI	y = −2.34x + 1.52	0.82	0.1	0.86	0.08
OSAVI	y = −2.79x + 1.55	0.81	0.1	0.81	0.1
Lic1	y = −2.40x + 1.52	0.82	0.1	0.86	0.08
Lic2	y = −0.81x + 0.63	0.02	0.23	−34.83	0.24
ARI	y = 0.36x − 0.12	0.2	0.21	−2.24	0.21
PHRI	y = −17.22x + 1.04	0.4	0.18	−1.82	0.23
FDI	y = 2.74x + 0.17	0.90	0.07	0.94	0.06

Table 3. Comparison of proposed FDI and traditional spectral indices at the early filling stage.

Spectral Indices	Early Filling Stage				
	Regression Equation	Training Set		Test Set	
		R^2	RMSE	R^2	RMSE
NRI	y = −3.2x + 0.55	0.93	0.07	0.92	0.07
PRI	y = −9.52x − 0.30	0.19	0.23	−2.14	0.23
TVI	y = −0.05x + 1.02	0.89	0.08	0.89	0.08
TCARI	y = −3.35x + 1.02	0.87	0.09	0.85	0.10
MCARI	y = −16.72x + 1.14	0.76	0.13	0.67	0.14
RVSI	y = 5.09x + 0.47	0.07	0.25	−14.73	0.26
PSRI	y = 3.60x − 0.15	0.88	0.09	0.85	0.10
GI	y = −1.25x + 1.79	0.89	0.09	0.85	0.09
SIPI	y = −3.74x + 2.44	0.78	0.12	0.71	0.12
NPCI	y = 3.48x − 1.16	0.52	0.18	0.17	0.19

Table 3. *Cont.*

Spectral Indices	Regression Equation	Early Filling Stage			
		Training Set		Test Set	
		R^2	RMSE	R^2	RMSE
NDVI	y = −2.45x + 1.32	0.87	0.09	0.85	0.10
OSAVI	y = −2.74x + 1.32	0.89	0.09	0.87	0.09
Lic1	y = −2.45x + 1.32	0.87	0.09	0.85	0.10
Lic2	y = −2.44x + 1.28	0.09	0.25	−6.40	0.24
ARI	y = 0.39x − 0.18	0.17	0.23	−4.87	0.25
PHRI	y = −13.41x + 1.13	0.17	0.23	−4.19	0.25
FDI	y = 3.78x + 0.56	0.97	0.04	0.96	0.05

Table 4. Comparison of FDI and traditional spectral indices at the combined stage.

Spectral Indices	Regression Equation	Combined Stage			
		Training Set		Test Set	
		R^2	RMSE	R^2	RMSE
NRI	y = −2.69x + 0.56	0.82	0.11	0.78	0.11
PRI	y = −7.56x − 0.16	0.46	0.19	0.03	0.18
TVI	y = −0.05x + 1.05	0.73	0.13	0.64	0.13
TCARI	y = −4.02x + 1.13	0.81	0.11	0.75	0.13
MCARI	y = −20.80x + 1.29	0.63	0.16	0.48	0.18
RVSI	y = 5.75x + 0.45	0.11	0.24	−8.03	0.26
PSRI	y = 2.57x − 0.05	0.72	0.14	0.63	0.14
GI	y = −0.97x + 1.53	0.79	0.12	0.73	0.12
SIPI	y = −3.71x + 2.51	0.34	0.21	−1.03	0.22
NPCI	y = 1.67x − 0.36	0.79	0.12	0.77	0.12
NDVI	y = −2.24x + 1.31	0.66	0.15	0.51	0.15
OSAVI	y = −2.54x + 1.32	0.66	0.15	0.56	0.14
Lic1	y = −2.24x + 1.32	0.66	0.15	0.51	0.15
Lic2	y = −2.45x + 1.27	0.73	0.13	0.70	0.13
ARI	y = 0.56x − 0.43	0.69	0.14	0.49	0.16
PHRI	y = −6.66x + 0.01	0.08	0.25	−9.57	0.24
FDI	y = 2.97x + 0.54	0.90	0.08	0.90	0.08

4. Discussion

4.1. Analysis of Spectral Characteristics for Identifying Wheat FHB

Previous studies have demonstrated that changes in crop physiological and biochemical parameters lead to changes in spectral reflectance, which is the basis for optical technology used to diagnose the severity of FHB [5]. As the symptoms of FHB are different in separate growth stages of wheat, characteristic wavelengths used to identify the severity of FHB are also different, so it is necessary to extract these wavelengths for different stages. In addition, there are some differences in the spectra of wheat ears with different degrees of infection in specific bands. Figure 4 shows the destruction of chloroplasts in the ear tissue of FHB which causes the chlorophyll in the cells to continuously degrade, and the reflectivity of the spectrum in the chlorophyll band (560–675 nm and 682–733 nm) decreases rapidly. At the same time, the decrease of the chlorophyll content in these cells reduces the possibility of photon reemission and reabsorption in this wavelength range, resulting in an increase in spectral reflectivity and the blue shift of the "red edge". These changes become more obvious with the increase in the degree of infection. The characteristic wavelengths selected in this study are in the range of 560–680 nm, including the green reflection peak and red absorption valley, and can characterize the characteristics of wheat ears. According to the characteristic wavelengths of the late flowering stage (570 nm and 678 nm) and early filling stage (565 nm and 661 nm), the 570 nm and 565 nm wavelengths are near the green peak, while 678 nm and 661 nm are near the red valley. Therefore, the characteristic wavelengths selected in this study are key wavelengths for identifying wheat ears with FHB. Bauriegel [7] also confirmed the importance of this band in the early detection of wheat ear scabs.

4.2. Comparison of Application Effects between Proposed New FDI and Traditional Spectral Indices

The most common and serious symptom of wheat FHB, ear rot, often begins at the early flowering stage. Therefore, the monitoring of wheat FHB at the flowering stage is highly valuable. However, the FHB fungi in the flowering stage are in the stage of mass reproduction, and the physiological and biochemical characteristics of the infected ears are not obvious, which makes it more difficult for a conventional spectral index to detect the severity of FHB at the flowering stage. The 16 published spectral indices selected in this study were constructed by collecting sample data from different stages of multiple growth stages, so their accuracies in detecting the severity of FHB at the late flowering stage were relatively poor. In this study, the sample models of the late flowering stage and combined stage were applied to the test set of the late flowering stage. The results show that the model obtained from the late flowering stage sample was more suitable for the detection of FHB at the late flowering stage than the model obtained from the combined stage sample. The characteristic wavelength selected from the samples at the late flowering stage (570 nm and 678 nm) was therefore more suitable for FHB detection than the characteristic wavelength selected during the combined stage (560 nm and 663 nm).

In the sample verification of the combined stage, the accuracies of the 16 published spectral indexes were not satisfactory because none of them were able to achieve an R^2 above 0.85 in both the training and test datasets. These spectral indices are mainly based on the leaves or canopies of crops rich in chlorophyll, and few diagnostic studies have used these indices to evaluate the severity of FHB in wheat ears, which is a special part with a low chlorophyll content. In the early filling period, in addition to the FDI, the NRI, plant senescence reflectance index (PSRI), GI, normalized difference vegetation (NDVI), and optimized soil-adjusted vegetation index (OSAVI) also had accurate detection results. This may be because the morphology and cell structure of wheat ears caused by FHB are more obvious than those at the late flowering, so other spectral indices may be more accurate in this context. Notably, among these spectral indexes, NRI and GI both performed strongly at the late flowering and early filling stages but performed relatively poorly in the combined stage. This shows that the diversity of samples has an important effect on predictions of FHB severity, which demonstrates the importance of subdividing the growth period when exploring the forecasts of FHB severity.

4.3. Analysis of Other Influential Factors

Hyperspectral data contain hundreds of narrow-band data points, but the adjacent wavelength information is often highly correlated, so the use of full-band information will only increase the complexity of data acquisition and calculation [5]. Usually, the most effective information is only contained in some specific bands, and the rest is redundant information [36]. In addition, the high price of hyperspectral imaging systems will also limit the application potential of the technology. This study explored the characteristic wavelengths at different stages to design a multispectral camera with a low price, a fast processing speed, and wide applications for specific identifications of FHB in different growth stages. In this study, the R^2 of predicted SI and FDI exceeded 0.90 in the late flowering, early filling, and combined stages. Considering the influence of man-made or natural environmental factors, this prediction accuracy is acceptable.

In addition, the spatial distribution and severity of FHB diseases are greatly affected by the genetic resistance of different varieties as well as environmental and agricultural management factors. Therefore, the factors that affect the biophysical and biochemical parameters of plants will affect the identification of FHB. This study was conducted under laboratory conditions, and its direct application in the field requires further verification. In the future, the growth stage will be further subdivided, especially in the early stages of FHB occurrence, such as mid-flowering, late flowering, early filling, and mid-filling, to achieve early detection, protection, and evaluation.

5. Conclusions

Monitoring wheat infection by FHB at different growth stages is important in making a decision on the use of pesticides to protect wheat from FHB and to evaluate yield losses. In this study, RF was used to select characteristic wavelengths for the late flowering stage, early filling stage, and combined stage of both. These wavebands were 570 nm and 678 nm for the late flowering stage, 565 nm and 661nm for the early filling stage, 560 nm and 663 nm for the combined stage. In the light of above wavelengths, FDI at each stage were constructed for establishing linear regression models with SI. Every model showed a high predictive accuracy with the test datasets, with their R^2 values exceeding 0.90. In addition, the R^2 of the model established at the late flowering stage and early filling stage was better than that of the combined stage, and the R^2 of applying the model of the combined stage to the test dataset at the late flowering stage and filling stage also decreased. Therefore, it is indicated that FHB shows different spectral characteristics at each growth stage, which provides a favorable basis for detecting the severity of the FHB disease at different growth stages in the future. However, additional studies are needed to verify the universality of FDI on different wheat varieties and in different field experiment settings.

Author Contributions: All authors contributed to the study conception and design. Methodology: F.L.; Writing—review and editing: H.Q.; Formal analysis and investigation, Writing—original draft preparation: D.Z.; Formal analysis and investigation, Writing—original draft preparation: Q.W.; Software: X.Y.; Conceptualization: C.G. All authors have read and agree to the published version of the manuscript.

Funding: This research was funded by the National Natural Science Foundation of China (Grant No. 41771463 and 41771469), Anhui Provincial Major Science and Technology Projects (Grant No. 18030701209), and National Key Research and Development Program of China (Grant No. 2016YFD0300700).

Conflicts of Interest: The authors declare no conflict of interest.

References

1. Gilbert, J.; Tekauz, A. Recent developments in research on Fusarium head blight of wheat in Canada. *Can. J. Plant Pathol.* **2000**, *22*, 1–8. [CrossRef]
2. Mcmullen, M.; Jones, R.; Gallenberg, D. Scab of wheat and barley: A Reemerging disease of devastating impact. *Plant Dis.* **1997**, *81*, 1340–1348. [CrossRef]
3. Lu, J.; Ehsani, R.; Shi, Y.; de Castro, A.I.; Wang, S. Detection of multi-tomato leaf diseases (late blight, target and bacterial spots) in different stages by using a spectral-based sensor. *Sci. Rep.* **2018**, *8*, 2793. [CrossRef]
4. Yuan, L.; Zhang, H.; Zhang, Y.; Xing, C.; Bao, Z. Feasibility assessment of multi-spectral satellite sensors in monitoring and discriminating wheat diseases and insects. *Int. J. Light Electron. Opt.* **2017**, *131*, 598–608. [CrossRef]
5. Zheng, Q.; Huang, W.J.; Cui, X.M.; Dong, Y.Y.; Shi, Y.; Ma, H.Q.; Liu, L.Y. Identification of Wheat Yellow Rust Using Optimal Three-Band Spectral Indices in Different Growth Stages. *Sensors* **2019**, *19*, 35. [CrossRef] [PubMed]
6. Huang, W.J.; Guan, Q.S.; Luo, J.H.; Zhang, J.C.; Zhao, J.L.; Liang, D.; Huang, L.S.; Zhang, D.Y. New optimized spectral indices for identifying and monitoring winter wheat diseases. *IEEE J. Sel. Top. Appl. Earth Obs. Remote Sens.* **2014**, *7*, 2516–2524. [CrossRef]
7. Bauriegel, E.; Giebel, A.; Geyer, M.; Schmidt, U.; Herppich, W.B. Early detection of fusarium infection in wheat using hyper-spectral imaging. *Comput. Electron. Agric.* **2011**, *75*, 304–312. [CrossRef]
8. Zhang, N.; Pan, Y.C.; Feng, H.K.; Zhao, X.Q.; Yang, X.D.; Ding, C.L.; Yang, G.J. Development of Fusarium head blight classification index using hyperspectral microscopy images of winter wheat spikelets. *Biosyst. Eng.* **2019**, *183*, 83–99. [CrossRef]
9. Weber, V.S.; Araus, J.L.; Cairns, J.E.; Sanchez, C.; Melchinger, A.E.; Orsini, E. Prediction of grain yield using reflectance spectra of canopy and leaves in maize plants grown under different water regimes. *Field Crop. Res.* **2012**, *128*, 82–90. [CrossRef]

10. Bannari, A.; Khurshid, K.S.; Staenz, K.; Schwarz, J.W. A Comparison of Hyperspectral Chlorophyll Indices for Wheat Crop Chlorophyll Content Estimation Using Laboratory Reflectance Measurements. *IEEE Trans. Geosci. Remote Sens.* **2007**, *45*, 3063–3074. [CrossRef]
11. Zhang, J.; Pu, R.; Huang, W.; Yuan, L.; Luo, J.; Wang, J. Using in-situ hyperspectral data for detecting and discriminating yellow rust disease from nutrient stresses. *Field Crop. Res.* **2012**, *134*, 165–174. [CrossRef]
12. Fensholt, R.; Huber, S.; Proud, S.R.; Mbow, C. Detecting Canopy Water Status Using Shortwave Infrared Reflectance Data from Polar Orbiting and Geostationary Platforms. *IEEE J. Sel. Top. Appl. Earth Obs. Remote Sens.* **2010**, *3*, 271–285. [CrossRef]
13. Devadas, R.; Lamb, D.W.; Simpfendorfer, S.; Backhouse, D. Evaluating ten spectral vegetation indices for identifying rust infection in individual wheat leaves. *Precis. Agric.* **2009**, *10*, 459–470. [CrossRef]
14. Rumpf, T.; Mahlein, A.K.; Steiner, U.; Oerke, E.C.; Dehne, H.W.; Plümer, L. Early detection and classification of plant diseases with Support Vector Machines based on hyperspectral reflectance. *Comput. Electron. Agric.* **2010**, *74*, 91–99. [CrossRef]
15. Feng, W.; Shen, W.; He, L.; Duan, J.; Guo, B.; Li, Y.; Wang, C.; Guo, T. Improved remote sensing detection of wheat powdery mildew using dual-green vegetation indices. *Precis. Agric.* **2016**, *17*, 608–627. [CrossRef]
16. Breiman, L. Random forests. *Mach. Learn.* **2001**, *45*, 5–32. [CrossRef]
17. Zhang, C.; Jiang, H.; Liu, F.; He, Y. Application of near-infrared hyperspectral imaging with variable selection methods to determine and visualize caffeine content of coffee beans. *Food Bioproc. Technol.* **2016**, *1*, 213–221. [CrossRef]
18. Zhao, Y.R.; Yu, K.Q.; He, Y. Hyperspectral imaging coupled with random frog and calibration models for assessment of total soluble solids in mulberries. *J. Anal. Methods Chem.* **2015**, *2015*, 343782. [CrossRef]
19. Fan, W.L.; Hu, P.; Liu, Z.G. Multi-attribute node importance evaluation method based on Gini-coefficient in complex power grids. *IET Gener. Transm. Distrib.* **2016**, *10*, 2027–2034.
20. Mahlein, A.K.; Rumpf, T.; Welke, P.; Dehne, H.W.; Plümer, L.; Steiner, U.; Oerke, E.C. Development of spectral indices for detecting and identifying plant diseases. *Remote Sens. Environ.* **2013**, *128*, 21–30. [CrossRef]
21. Filella, I.; Serrano, L.; Serra, J.; Penuelas, J. Evaluating wheat nitrogen status with canopy reflectance indices and discriminant analysis. *Crop Sci.* **1995**, *35*, 1400–1405. [CrossRef]
22. Gamon, J.A.; Penuelas, J.; Field, C.B. A narrow-waveband spectral index that tracks diurnal changes in photosynthetic efficiency. *Remote Sens. Environ.* **1992**, *41*, 35–44. [CrossRef]
23. Broge, N.H.; Leblanc, E. Comparing prediction power and stability of broadband and hyperspectral vegetation indices for estimation of green leaf area index and canopy chlorophyll density. *Remote Sens. Environ.* **2001**, *76*, 156–172. [CrossRef]
24. Haboudane, D.; Miller, J.R.; Pattey, E.; Zarco-Tejada, P.J.; Strachan, I.B. Hyperspectral vegetation indices and novel algorithms for predicting green LAI of crop canopies: Modeling and validation in the context of precision agriculture. *Remote Sens. Environ.* **2004**, *90*, 337–352. [CrossRef]
25. Daughtry, C.S.; Walthall, C.L.; Kim, M.S.; De Colstoun, E.B.; McMurtrey Iii, J.E. Estimating corn leaf chlorophyll concentration from leaf and canopy reflectance. *Remote Sens. Environ.* **2000**, *74*, 229–239. [CrossRef]
26. Merton, R.; Huntington, J. Early simulation results of the ARIES-1 satellite sensor for multi-temporal vegetation research derived from AVIRIS. In Proceedings of the Eighth Annual JPL Airborne Earth Science Workshop, Pasadena, CA, USA, 9–11 February 1999; pp. 9–11.
27. Merzlyak, M.N.; Gitelson, A.A.; Chivkunova, O.B.; Rakitin, V.Y. Non-destructive optical detection of pigment changes during leaf senescence and fruit ripening. *Physiol. Plant.* **1999**, *106*, 135–141. [CrossRef]
28. Zarco-Tejada, P.J.; Berjón, A.; López-Lozano, R.; Miller, J.R.; Martín, P.; Cachorro, V.; González, M.R.; De Frutos, A. Assessing vineyard condition with hyperspectral indices: Leaf and canopy reflectance simulation in a row-structured discontinuous canopy. *Remote Sens. Environ.* **2005**, *99*, 271–287. [CrossRef]
29. Peñuelas, J.; Baret, F.; Filella, I. Semi-empirical indices to assess carotenoids/chlorophyll a ratio from leaf spectral reflectance. *Photosynthetica* **1995**, *31*, 221–230.
30. Peñuelas, J.; Gamon, J.A.; Fredeen, A.L.; Merino, J.; Field, C.B. Reflectance indices associated with physiological changes in nitrogen-and water-limited sunflower leaves. *Remote Sens. Environ.* **1994**, *48*, 135–146. [CrossRef]
31. Schell, J.A. Monitoring Vegetation Systems in the Great Plains with ERTS. *NASA Spec. Publ.* **1973**, *351*, 309.

32. Rondeaux, G.; Steven, M.; Baret, F. Optimization of soil-adjusted vegetation indices. *Remote Sens. Environ.* **1996**, *55*, 95–107. [CrossRef]
33. Lichtenthaler, H.K.; Gitelson, A.; Lang, M. Non-destructive determination of chlorophyll content of leaves of a green and an aurea mutant of tobacco by reflectance measurements. *J. Plant Physiol.* **1996**, *148*, 483–493. [CrossRef]
34. Lichtenthaler, H.K.; Lang, M.; Sowinska, M.; Heisel, F.; Miehe, J.A. Detection of vegetation stress via a new high resolution fluorescence imaging system. *J. Plant Physiol.* **1996**, *148*, 599–612. [CrossRef]
35. Gitelson, A.A.; Merzlyak, M.N.; Chivkunova, O.B. Optical properties and nondestructive estimation of anthocyanin content in plant leaves. *Photochem. Photobiol.* **2001**, *74*, 38–45. [CrossRef]
36. Zhang, D.; Xu, Y.; Huang, W.; Tian, X.; Xia, Y.; Xu, L.; Fan, S. Nondestructive measurement of soluble solids content in apple using near infrared hyperspectral imaging coupled with wavelength selection algorithm. *Infrared Phys. Technol.* **2019**, *98*, 297–304. [CrossRef]

Article

Damage Indexing Method for Shear Critical Tubular Reinforced Concrete Structures Based on Crack Image Analysis

Yuan-Sen Yang [1,*], Chia-Hao Chang [1] and Chiun-lin Wu [2]

1 Department of Civil Engineering, National Taipei University of Technology, Taipei 106, Taiwan; changchiahoa@gmail.com
2 National Center for Research on Earthquake Engineering, Taipei 106, Taiwan; clwu@ncree.narl.org.tw
* Correspondence: ysyang@ntut.edu.tw; Tel.: +886-2-2771-2171 (ext. 2641)

Received: 14 September 2019; Accepted: 3 October 2019; Published: 4 October 2019

Abstract: Image analysis techniques have been applied to measure the displacements, strain field, and crack distribution of structures in the laboratory environment, and present strong potential for use in structural health monitoring applications. Compared with accelerometers, image analysis is good at monitoring area-based responses, such as crack patterns at critical regions of reinforced concrete (RC) structures. While the quantitative relationship between cracks and structural damage depends on many factors, cracks need to be detected and quantified in an automatic manner for further investigation into structural health monitoring. This work proposes a damage-indexing method by integrating an image-based crack measurement method and a crack quantification method. The image-based crack measurement method identifies cracks locations, opening widths, and orientations. Fractal dimension analysis gives the flexural cracks and shear cracks an overall damage index ranging between 0 and 1. According to the orientations of the cracks analyzed by image analysis, the cracks can be classified as either shear or flexural, and the overall damage index can be separated into shear and flexural damage indices. These damage indices not only quantify the damage of an RC structure, but also the contents of shear and flexural failures. While the engineering significance of the damage indices is structure dependent, when the damage indexing method is used for structural health monitoring, the damage indices safety thresholds can further be defined based on the structure type under consideration. Finally, this paper demonstrates this method by using the results of two experiments on RC tubular containment vessel structures.

Keywords: image-based measurement; crack measurement; shear cracks; flexural cracks; damage index

---

1. Introduction

Sensing and quantifying damage plays a critical role in the process of structural health monitoring, which aims to detect structural damage and provide early warnings when a possible risk of failure is detected. Many structural health monitoring systems employ accelerometers, displacement sensors, or piezoelectric sensors located at selected locations to monitor changes in the structure's deformation, natural frequencies, and modal shapes [1,2]. These systems then evaluate possible failure modes, damage levels, and locations. While accelerometers are typically employed for beam-column-based structures such as buildings, these are not the optimal sensors for structures whose failure modes are insensitive to the structure's vibration characteristics. For some types of structures such as dams, tunnels, and reinforced concrete vessels, or shear-critical components such as reinforced concrete (RC) walls, the detection and evaluation of cracks is a relatively practical approach for safety assessment and monitoring.

Several structural damage indices have been proposed. Park et al. [3] proposed a damage index for a structural system according to its largest system displacement, ultimate displacement, accumulated strain energy, cyclic loading effect, and system yield force and displacement. Based on the calculated damage index, the structural system can be classified into one of the following damage levels: slight, minor, moderate, severely damaged, and collapsed. Roufaiel and Meyer [4] proposed a damage index that uses the initial stiffness, current stiffness, and failure stiffness. Powell and Allahabadi [5] proposed an index based on the current displacement, yield displacement, and ultimate displacement. These damage indices consider a structure as a single-degree-of-freedom system to simplify damage level estimations. However, in practical applications, these damage indices are difficult to use, as the stiffness and the displacement of a structure is sometimes difficult to measure for real, multiple degrees of freedom, and partially damaged structures. Detailed structural performance and safety may require advanced structural analyses based on finite element analysis tools [6,7] or structural experiments [8,9] which are specific to a certain type of structure. For the purpose of structural health monitoring, the displacements of certain locations can be monitored by pre-installed displacement devices; however, current stiffness and other structural properties are difficult to accurately measure or estimate.

Alternatively, for easy to implement and quick structural safety assessments of reinforced concrete (RC) structures, several evaluation methods have been proposed that instead consider the surface cracks of concrete structures. The Japan Building Disaster Prevention Association (JBDPA) provides a guide based on the visible cracks in the concrete surface of beams, columns, or walls, and categorizes damage into five classes according to the maximum opening width of the cracks [10]. According to the JBDPA criterion, structures with a maximum crack width larger than 0.2 mm, 1 mm, and 2 mm are categorized as showing light damage, moderate damage, and heavy damage classes, respectively. The International Atomic Energy Agency (IAEA) uses a more conservative standard that categorizes cracks with an opening width larger than 0.2 mm and 1 mm as moderate and severe damage, respectively [11]. The bridge inspector's reference manual, published by US Department of Transportation [12], categorizes cracks into structural cracks, flexural cracks on a tee beam, shear cracks on a slab, temperature cracks, shrinkage cracks, longitudinal cracks, etc.

For surface damage detection and evaluation, image-based measurement is an automatic and cost-efficient method in terms of hardware cost. As the aforementioned structural health monitoring or damage detection methods have different features, advantages, and limits, no single method can be used to replace another, nor can it be used as the sole means of structural health monitoring or damage detection. Image-based measurements, and their potential for damage detection, are not intended to replace any of the aforementioned methods. Instead, the image-based method aims to provide an area-based measurement method to measure or monitor cracks [13], strain fields [14,15], multi-axial displacement [16], or structural vibrations [17], where technology for conventional displacement measurements is inadequate [18]. The hardware cost may be relatively low [19], and may even employ existing surveillance cameras in the structure, thus eliminating the need to install additional cameras [20]. With recent dramatic improvements in digital image processing techniques, image analysis algorithms, accuracies, reliability, and computing speed have improved as well; thus, image measurement has a strong potential for practical structural health monitoring applications [21].

This work develops an image analysis-based damage indexing method following a previously developed image-based crack measurement method. This method is tested using two cyclic tests of RC containment vessels [22]. The vessels are shear critical with a large number of shear cracks induced by only a small displacement. A fractal dimension method [23] is modified and employed in this work to quantify the number of cracks. Based on the number of cracks, as well as their opening widths and orientations, a method for calculating damage indices is proposed. This method modifies the previous image analysis method [24], such that concrete surface crack orientations can be determined automatically. In addition, the fractal dimension crack analysis method [23] is modified so that the damage index can be separated into a shear damage index and a flexural damage index to distinguish between the different types of failure. The combination of these methods will make it possible to carry out structural health monitoring in an automatic manner in practical applications in the future. This paper further demonstrates the image measurement and damage indices calculation procedure based on the aforementioned RC containment vessel experiments.

2. Image Measurement of Cracks on Concrete Surfaces

Image-based monitoring and damage identification consists of two major procedures: image measurements and damage quantification. Image measurements analyze the image(s) of the measurement regions of interest and provide details, such as locations, lengths, opening widths, sliding displacements, and the orientation of the cracks. The damage evaluation procedure estimates the damage level or index of the measurement region according to the analyzed results from the image measurement.

Many image measurement algorithms and methods have been proposed to detect cracks on measurement regions, such as on concrete surfaces or pavements. These methods can be classified into two groups: (1) edge detection-based methods, and (2) displacement field-based methods. Edge detection-based methods are capable of finding cracks that appear as dark lines in an image. The cracks need to be of sufficient width to appear as dark lines, which is theoretically the width of a pixel. Edge detection methods [25–28] or machine learning methods [29–32] are typically employed to identify the locations or widths of cracks. A review of crack detection methods can be found in [33].

Alternatively, the displacement field-based method identifies cracks according to the displacement field of the measurement region, where the displacement field is analyzed by image analysis techniques [24,34]. Due to the high precision of image-based displacement field measurements, displacement field-based methods are capable of detecting thin cracks with widths of much less than one pixel. Yang et al. [34] detected cracks as thin as 0.2 pixels in photos in an outdoor experiment where images contained environmental light noise. The same image analysis technique detected thin cracks whose width was equivalent to 0.03 pixels in photos in a structural laboratory [24]. This type of method estimates the cracks' opening widths, sliding displacements, and orientations, according to the change in the displacement field between each set of photos taken before and after cracks occurred, respectively. Thus, the first set of photos is used as a reference for the displacement field. Compared with edge detection-based methods, displacement field-based crack detection methods are suitable for thin crack detection, monitoring the early stages of crack development, or monitoring large regions where pixels are relatively coarsened. However, it should be mentioned that most edge detection-based methods used are tailored for inspection, rather than health monitoring. They are more suitable for that purpose than displacement field methods. In addition, displacement field methods tend to be more computationally expensive.

This work employs a displacement field-based method for crack measurement. However, this does not mean that edge detection-based methods cannot be applied to the damage evaluation method proposed in this work. The displacement field-based method is employed here because it is capable of detecting thin cracks that occur in the early stages of structural damage. In addition, the image measurement software, ImPro Stereo, is publicly available on the internet [35], and is further integrated with the damage evaluation computer codes developed in this work.

The displacement field-based method for crack measurement includes five main steps: camera calibration, measurement region positioning, metric rectification, displacement field analysis, and crack analysis. A detailed procedure can be found in [34,36]. This work only focuses on the analysis results as related to the follow-up damage evaluation procedure which is proposed herein.

Camera calibration is the process of finding the intrinsic and extrinsic parameters of the camera. The intrinsic parameters are essentially its optical properties, such as the fields of view and optical distortion coefficients. The extrinsic parameters describe the camera position and its orientation. Typically, the camera calibration process is only carried out once, by taking more than 10 pairs of photos of calibration objects (such as a chessboard of known size) during camera installation (see Figure 1).

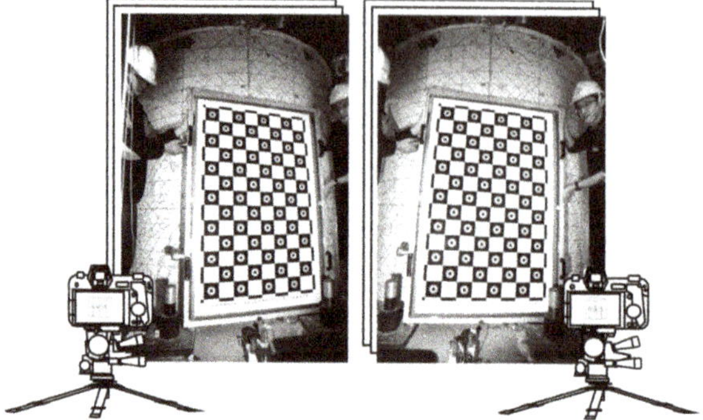

Figure 1. Stereo calibration of two cameras.

Measurement region positioning tracks the updated position of the measurement by precisely tracking the 3D positions of the control points that are used to define the measurement surface. Defining an ideal planer rectangle measurement requires at least three control points, while a cylindrical measurement region requires at least four, as shown in Figure 2. The positions of control points P1 to P4 describe the movement and deformation of the overall measurement region. Details of the process can be found in [24].

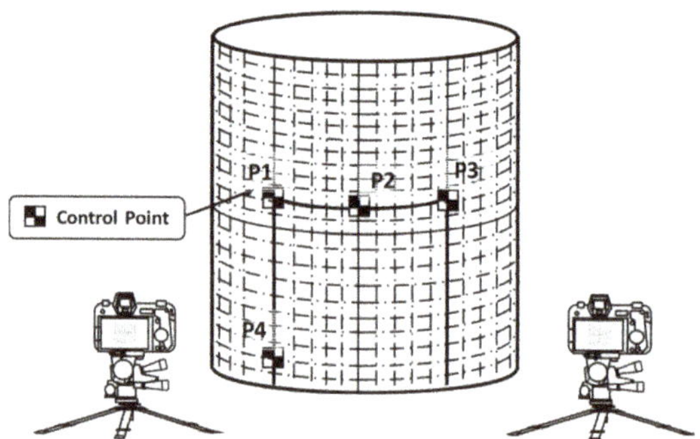

Figure 2. Measurement region positioning by tracking control points.

The image rectification process generates a rectangular image that represents the image pattern on the measurement region. The perspective and lens distortion effects are removed during this process. The metric rectified image can be seen as an expanded planer surface of the measurement region so that the ratio of a pixel to its physical length is constant over the entire measurement region; thus, it is essentially an image that represents the unfolded plane from the measurement region. The constant pixel-to-physical length ratio is an important property for the subsequent displacement field-based crack analysis. The rectified image is generated pixel-by-pixel, while the image intensity of each pixel is estimated by mathematically projecting a 3D point onto the surface to its image position in the photo according to the intrinsic and extrinsic parameters of the camera. Its image intensity is acquired through the numerical interpolation of neighboring pixels, as shown in Figure 3.

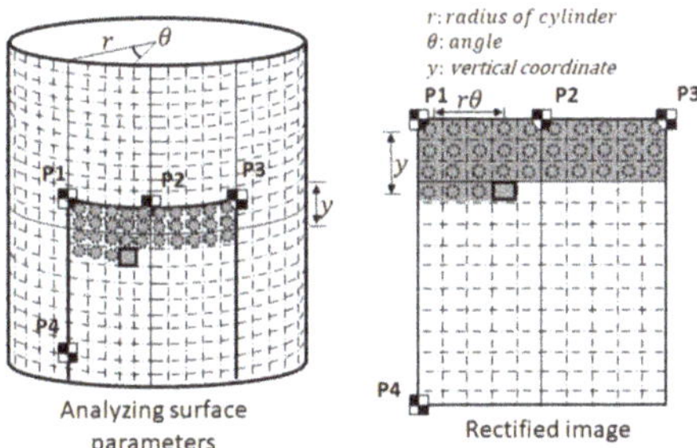

Figure 3. Metric rectification of the region of interest on a cylindrical structural component.

The displacement fields of the measurement region can be estimated by comparing the initial and current rectified images (see Figure 4a,b) using an object tracking method, such as template matching, digital image correlation, an enhanced correlation coefficient, or the optical flow method. Details of the process can be found in [24]. The example presented in Figure 4 was obtained from an experiment that had a measurement region of approximate dimensions of 1.4 m × 0.9 m. Each rectified image in Figure 4a is approximately 2400 × 1600 pixels. The displacement field in Figure 4b is a vector field with 90 × 60 cells, that is, each cell is represented by a sub-image with a size of 27 × 27 pixels (rounded from 2400 / 90 = 26.67). The crack opening in Figure 4c is a scalar field with the same refinement. The refinement is assigned by users, and should be tuned according to the image quality of photos when this method is being applied in practical applications. The displacement field of the rectified images is obtained by optical flow analysis [37]. The resolution of the rectified images and the refinement of the displacement and crack fields are adjusted by the user, and typically depend on the resolution and quality of the experimental photos.

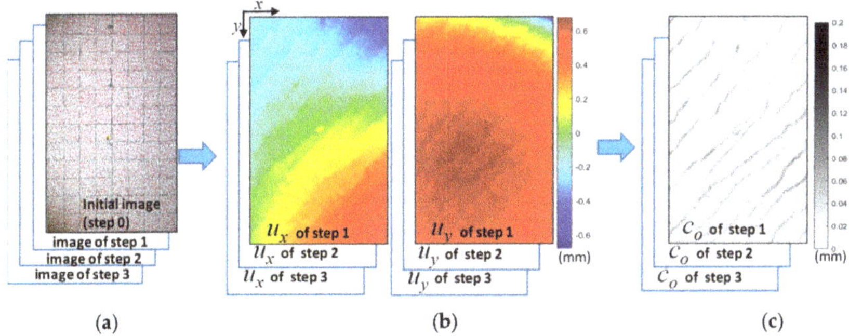

Figure 4. Estimating a displacement field by comparing initial and current rectified images. (**a**) Rectified images; (**b**) Displacement fields $\boldsymbol{u}$ (u_x and u_y); (**c**) Crack opening (c_o).

Crack analysis converts a displacement field to a crack distribution. Crack analysis is suitable for thin cracks that are too thin to display as a dark line in photos, thus requiring the use of the displacement field to estimate the crack's opening width. Each cell of the crack opening width c_o (see Figure 4c) and crack sliding displacement c_s of any arbitrary cell in the grid is estimated according to the displacement of its four neighboring cells. Crack sliding is the relative displacement of part A with respect to part B, i.e., parallel to the crack orientation. By using the formulation presented in [24], as shown in Equations (1)–(3), the crack distribution can be estimated by a displacement field. The crack analysis method is only suitable for brittle materials such as concrete, as it assumes that the deformation in the displacement field is mainly caused by cracks, rather than strains [34]. In addition, since the image is the appearance of the material surface, it does not represent the crack opening or sliding under beneath the surface; these are the limitations of this method. The crack distribution is a field of crack opening widths, sliding displacements, and crack orientations. It is discretized to a grid with the same grid density as the displacement. Each cell of the crack opening width c_o and crack sliding displacement c_s of any arbitrary cell in the grid can be calculated by Equations (1)–(3).

$$\begin{pmatrix} c_o \\ c_s \end{pmatrix} = \begin{pmatrix} \cos\theta & \sin\theta \\ -\sin\theta & \cos\theta \end{pmatrix} (u_A - u_B) \tag{1}$$

where

$$u_A = \begin{cases} \frac{u_U \cdot |\cos\theta| + u_L \cdot |\sin\theta|}{|\cos\theta| + |\sin\theta|}, & \text{if } 0 \le \theta < 0.5\pi \\ \frac{u_D \cdot |\cos\theta| + u_L \cdot |\sin\theta|}{|\cos\theta| + |\sin\theta|}, & \text{if } 0.5\pi \le \theta < \pi \end{cases} \tag{2}$$

$$u_B = \begin{cases} \frac{u_D \cdot |\cos\theta| + u_R \cdot |\sin\theta|}{|\cos\theta| + |\sin\theta|}, & \text{if } 0 \le \theta < 0.5\pi \\ \frac{u_U \cdot |\cos\theta| + u_R \cdot |\sin\theta|}{|\cos\theta| + |\sin\theta|}, & \text{if } 0.5\pi \le \theta < \pi \end{cases} \tag{3}$$

u_U, u_D, u_L, and u_R are the displacement vectors of the upper, lower, left, and right neighboring cells of any arbitrary cell in the displacement field, respectively (see Figure 5). The orientation of the crack of the analyzed cell is determined by iteratively testing θ within 0 and 180 degrees with a step of 15 degrees (i.e., 0, 15, 30, 45, ... , 165 degrees). To be conservative, the θ which leads to the largest crack opening is selected in this method. If there is no crack on the cell, c_s and c_o would be very small compared with those with cracks. Small values of c_s and c_o are caused by either noise, image analysis errors, or relatively small strains, and are ignored in the crack analysis. Figure 4c demonstrates the discretized grid of a crack pattern estimated from its displacement in Figure 4b. It should be noted that the size scale in Figure 5 is only for demonstration. A crack is typically much thinner than the size of a cell. The cracks shown in Figure 4c are actually as thin as 0.02–0.2 mm, i.e., much thinner than the size of a cell in Figure 4b,c. In Figure 4c, the size of a cell is equivalent to a 27 × 27-pixel sub-image. While a 0.02-mm crack can be recognized by the naked eye at a close distance when inspecting damage in structural experiments, it cannot be recognized by most of the edge detection-based methods, as the crack is typically too thin to appear as a dark line in photos. In addition, human inspection is not practical for automatic structural health monitoring.

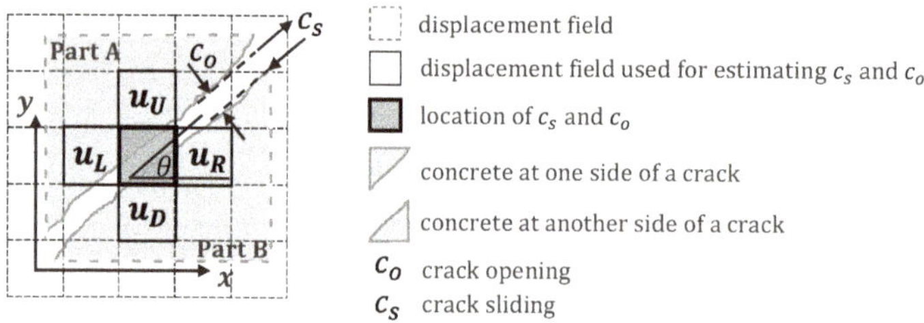

Figure 5. Crack opening calculation according to the analyzed displacement field.

3. Damage Indices based on Image Analysis of Cracks

The quantification of cracks in this work is based on a Fractal Analysis of Cracks (FAC) [23]. The quantification of the total length of cracks within a measurement region can be scale dependent; the smaller the scale and the more refined the crack pattern, the more likely it is that a longer total length of cracks would be measured. A typical scale-dependent example is the measurement of a coastline, which depends on the measurement scale. This method aims to quantify the number of cracks in a more objective and scale-invariant manner, rather than directly measuring the total lengths of cracks. The FAC method adopts a fractal analysis as a benchmark method to quantify a crack by estimating its fractal dimension. While mathematically, a line is one-dimensional and a filled rectangle is two-dimensional, the dimensions of a crack distribution over a measurement region are typically a real number between 1 and 2, and do not need to be an integer. The FAC method quantifies a crack by its fractal dimension. The details of FAC can be found in [23].

The crack analysis method proposed in this paper modifies the FAC method. The main modifications made in this work include the following:

(1) The crack data for FAC is based on a hand-sketched crack pattern. The crack data for the modified FAC is based on an image analyzed crack pattern.
(2) The modified FAC is capable of differentiating between the damage induced by shear cracks and that of flexural cracks according to crack orientation. In this work, the crack orientation is automatically determined by finding the orientation that results in the largest crack opening.

In this work, a framework for determining the damage indices by image analysis is proposed. In this framework, the damage indices include a flexural damage index d_F and a shear damage index d_S. The modified FAC method to determine these damage indices is composed of seven steps. All steps have been implemented in a public software implementation developed by the authors [35].

a. Analyze the crack opening pattern (see Figure 6a) using the image analysis approach described above, as shown in Figure 4. In this step, the crack opening field c_o is generated.
b. Define a threshold of crack opening width, such as 0.05 mm, and convert the crack opening pattern to a binary crack pattern (see Figure 6b). The crack opening width threshold is subjective and must be determined on the basis of the actual situation. While the image analysis method in this work is capable of observing cracks as thin as 0.02 mm (see cracks shown in Figure 4c, while some of the shown cracks are as thin as thin as 0.02 mm), a threshold of 0.05 mm was chosen in this work as it is the minimum crack width in a typical crack width ruler.
c. Analyze the fractal dimension by the FAC method. The FAC method is a multi-level discretization of the binary crack pattern. In each level, the crack pattern is discretized into a mesh composed of many square cells, with the number of cells that contain cracks then being counted (N). The width of each cell is ε. At each level, $log(1/\varepsilon)$ and $log(N)$ can be calculated, as shown in Figure 6c. Further details of calculating the fractal dimension can be found in [26]. Note that, typically, the actual meshes in FAC analyses are more refined, and the number of discretization levels is greater (e.g., 4 levels or higher) than as shown in Figure 6.
d. By applying multi-level mesh refinements (i.e., different sizes of ε), $log(N)$ versus $log(1/\varepsilon)$ can be plotted on a 2D plot. The fractal dimension f of the crack pattern is the slope of the line found by linear regression. Since the dimension of surface crack f is between 1 (that is, an ideal line) and 2 (a filled area), the damage index is estimated by $f-1$ in the FAC method. A damage index d, defined by Equation (4), is calculated, with a value between zero and one (see Figure 6d).

$$d = f - 1 \tag{4}$$

e. According to the crack orientation of each crack field cell, separate the crack opening field into a shear crack opening field and a flexural crack opening field, as shown in Figure 6e,f. The crack orientation is the angle of the crack. A crack orientation of zero degree means a horizontal crack; An orientation of 45 or 135 degrees means a diagonal crack. The range of the angle is from 0 to 180 degrees. The crack orientation of each crack field cell was calculated during the crack image analysis, as shown in Figure 4. In this work, the horizontal cracks, whose orientation is between 0 to 22.5 degrees or 157.5 to 180 degrees, are classified into flexural cracks and are assigned to the flexural crack opening field, while the remaining cracks are assigned to the shear crack opening field.
f. Separately calculate the total crack areas in the flexural crack opening field A_F and the shear crack opening field A_S. Since the crack opening field represents the crack opening widths, A_F is

the summation of all values in the flexural crack opening field multiplied by the width of each cell. A_S is calculated in the same manner.

g. Calculate the flexural damage index d_F and a shear damage index d_S using Equations (5) and (6).

$$d_F = d \cdot \frac{A_F}{A_S + A_F} \tag{5}$$

$$d_S = d \cdot \frac{A_S}{A_S + A_F} \tag{6}$$

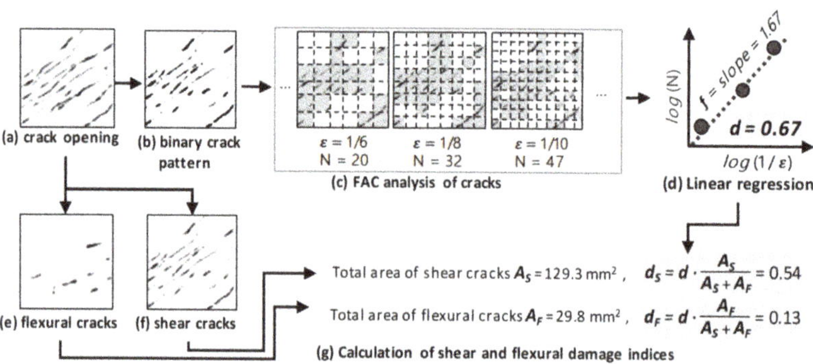

Figure 6. Demonstration of the proposed modified fractal analysis of cracks method. (**a**) crack opening; (**b**) binary crack pattern; (**c**) FAC analysis of crack; (**d**) linear regression; (**e**) flexural cracks; (**f**) shear cracks; (**g**) calculation of shear and flexural damage indices.

In most RC structures or components, crack orientation is a typical factor used to classify a crack as either flexural or shear. For RC columns or components that are subjected to bending and horizontal shear forces, horizontal cracks are typically classified as flexural, while the remaining cracks are classified as shear. This classification method is followed here. Furthermore, since the displacement field-based image analysis method provides not only the positions, opening widths, and sliding displacements of cracks, but also their orientations, it is practical to classify cracks according to their orientations. It should be noted that the classification of flexural and shear cracks by orientation is one of several classification methods, and is not necessarily applicable to all structure types. More details can be found in [10,12].

The proposed method not only integrates the previous crack image analysis [24] and FAC methods [23], but also makes some modifications. While the previous crack image analysis method requires analyzers to assign a crack orientation, the proposed method determines the crack orientation of each analyzed cell by finding the orientation that leads to the largest opening crack. While this is a conservative way to estimate crack orientation and opening width, it makes this method automatic, and does not require the orientation to be input manually. In addition, while the FAC method was originally designed for manually plotted cracks, this method uses automatically analyzed crack data for the FAC method. In the proposed method, the analyzed damage index is further separated into shear and flexural parts, providing more information on the failure mode for further safety evaluation. The integration of these methods and modifications makes it possible to carry out structural health monitoring based on crack information in practical applications.

4. Experiments

The proposed image-based shear and flexural damage indices were tested using two RC structural experiments [22]. The specimens were reduced-scale RC containment vessels (RCCVs), i.e., relatively short and wide tubular structures. They are denoted as RCCV #1 (Figure 7a) and RCCV #2 (Figure 7b), respectively. The specimens were identical in terms of geometry. The specimens were subjected to a constant vertical force of 160 kN, and a cyclic horizontal displacement history imposed through hydraulic controlled actuators, as shown in Figure 7c. The outer and inner diameters were 2500 mm and 2200 mm, respectively. The height of the structures was 2250 mm. The concrete strengths of the two specimens were 37.0 and 43.4 MPa, respectively. The yields and ultimate strength of steel rebars were 379 MPa and 572 MPa, respectively. Four cameras were set up to take photos of the measurement regions, as shown in Figure 7d. The photos from the two northern cameras were used in this work.

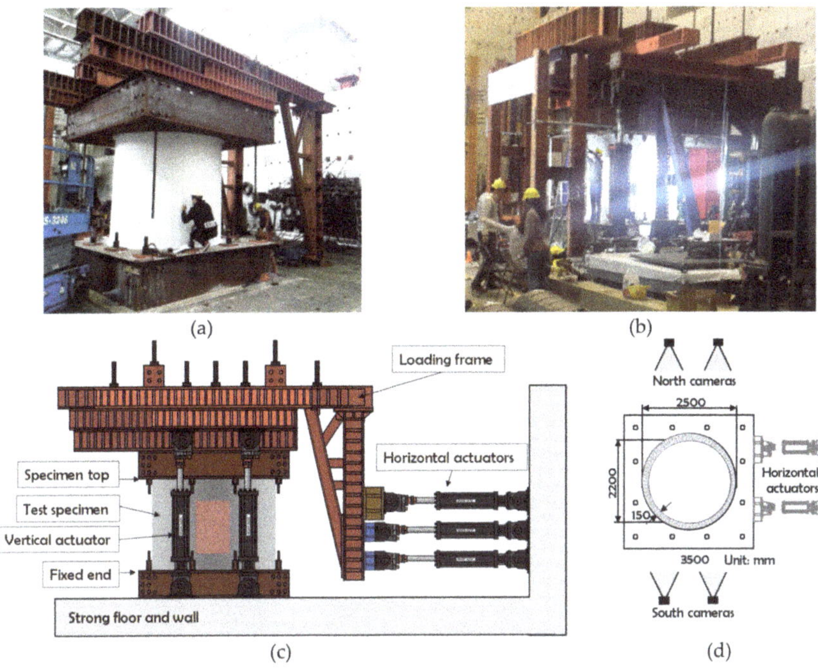

Figure 7. Experimental configuration and photos of both RCCV #1 and RCCV #2. (**a**) Photo of RCCV #1; (**b**) Photo of RCCV #2; (**c**) Elevation; (**d**) Plan.

The two RCCVs had slightly different rebar designs. Four cylindrical layers of rebars were constructed in the concrete tubular structures. Each layer contained up to 90 rebars. The steel ratio of RCCV #1 was 0.02 with reinforcement extending into the top and bottom for strong interfaces between the roof, the specimen, and the foundations. RCCV #2 had gradually increasing vertical steel ratios ρ_v near the top and bottom, as shown in Figure 8. The increased vertical steel reinforcement in RCCV #2 was designed to prevent sliding shear failure at the boundaries between the tubular structures and the top/bottom of the RC blocks, which occurred in the RCCV #1 test.

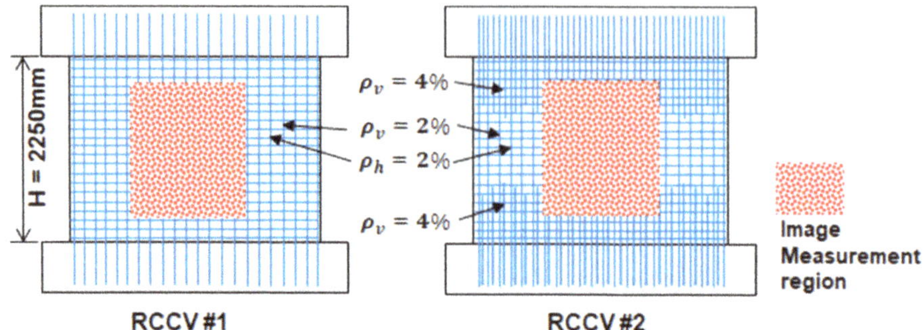

Figure 8. Steel rebar ratios in RCCV # 1 and RCCV # 2.

Four cameras were set up in both experiments; two were positioned to the north side and two to the south, as shown in Figure 7d. Two cameras were set up for each image measurement region, because stereo image analysis was employed, as described in the previous section. The measurement regions were painted with randomly striped patterns that provided image features for the displacement fields. The lightening conditions at the top and button regions of the specimens were not as good as those in the middle regions. In addition, the middle regions had better focal conditions in the experiments. Figure 9 shows the initial photos taken by the north cameras in both experiments.

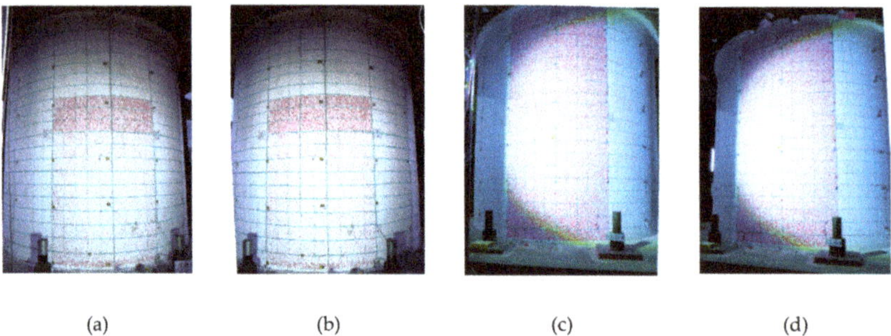

(a) (b) (c) (d)

Figure 9. Initial photos of the two RC containment vessels (RCCV) taken from the north cameras. (**a**) RCCV #1 left photo; (**b**) RCCV #1 right photo; (**c**) RCCV #2 left photo; (**d**) RCCV #2 right photo.

The experimental results show that the shear strength of the RCCV #2 was slightly higher than that of the RCCV #1 (see Figure 10a,b). The shear strengths of RCCV #1 and RCCV #2 were 5805 kN and 5580 kN, respectively. In addition, RCCV #1 and RCCV #2 had different ductilities. While both vessels reached their shear strengths for a displacement cycle of 16.9 mm (i.e., a drift ratio of 0.75% with respect to the specimen height of 2250 mm), RCCV #1 rapidly lost its shear strength after the 16.9 mm displacement cycle. In contract, RCCV #2 retained its shear capacity to 22.5 mm (i.e., a drift ratio of 1%), which was significantly higher because of the increased reinforcement at the top and bottom, as shown in Figure 10. The hysteresis loops of these specimens (see Figure 10c,d) show that the tangential stiffness did not significantly change until the cyclic displacements reached +/−3 mm. Details of the experimental results and explanations can be found in [22].

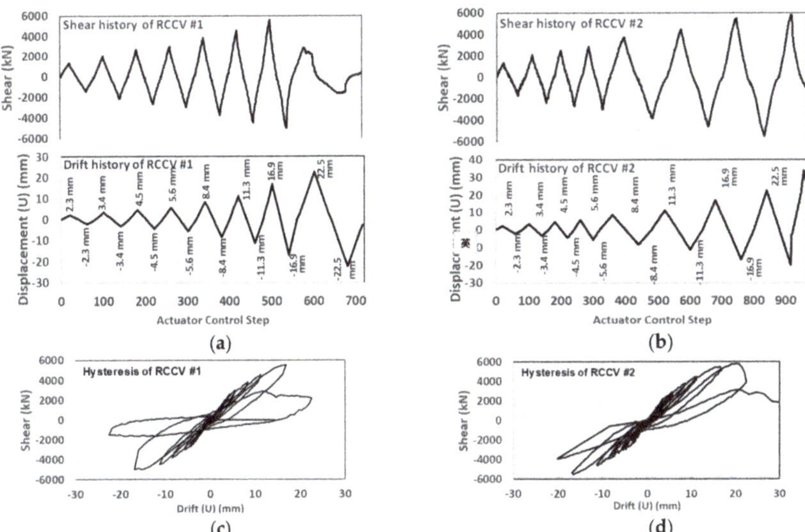

Figure 10. Shear/drift histories and hysteresis of RC containment vessels (RCCV) RCCV#1 and RCCV #2. (**a**) Shear and displacement history of RCCV #1; (**b**) Shear and displacement history of RCCV #2; (**c**) Hysteresis of RCCV #1; (**d**) Hysteresis of RCCV #2.

There were 163 and 1399 pairs of photos taken by the north cameras in the RCCV #1 and RCCV #2 experiments, respectively. Each pair of photos included a photo taken by the left camera and a photo taken by the right camera. The cameras were Canon EOS 5D Mark III with photo resolution of 3840 × 5760 pixels. Measurement regions were illuminated using a 100 W light-emitting-diode (LED). Figure 11 shows several north left camera photos of RCCV #1 and RCCV #2. The u in Figure 11 is the horizontal displacement at the top of the specimen. The displacements are so minor that the deformations are difficult to visually recognize in the figure. Since the RCCVs are shear-critical structures, a small displacement can cause significant shear failure. In addition to the experimental facilities and measurement devices, such as the load cells, the major way that we could observe the damage and the failure of the structure was to inspect the cracks on the surface. Diagonal (45-degree) shear cracks appeared on the north and south sides of the specimens, while the horizontal flexural cracks appeared at the top and bottom on the east and west sides. These cracks could be observed by human eyes only when we paused the testing, allowing people to get closer to the specimen to inspect the cracks. Details of the comparison of the manually plotted cracks and image analyzed cracks can be found in [24].

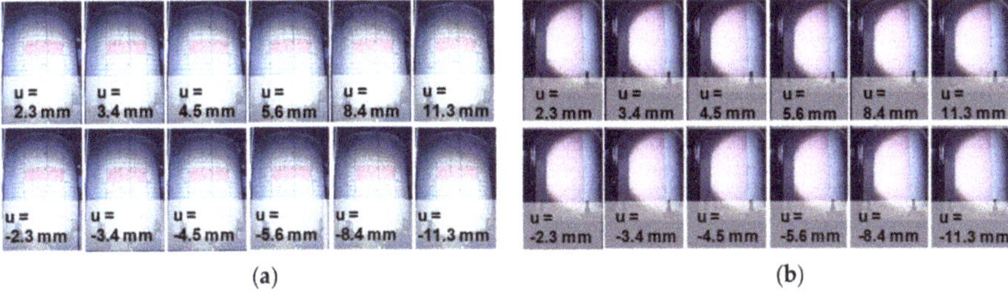

Figure 11. Selected experimental photos of RC containment vessels (RCCV) RCCV #1 and RCCV #2. (**a**) RCCV #1; (**b**) RCCV #2.

While both specimens underwent shear failures, different shear failure modes were observed for each vessel. RCCV #1 had a sliding shear mode at the top of the specimen, as shown in Figure 12a. A horizontal crack occurred at the top, where the shear stiffness dramatically changes, typically inducing a stress concentration. The red lines in Figure 12 represent the locations of the cracks. Sliding shear did not occur in RCCV #2 due to the gradual change in rebar density (the steel ratio was from 2% to 4%). RCCV #2 had a web shear failure in which the major shear crack passed through the specimen at a diagonal (45-degree) angle, as shown in Figure 12b.

(a)

(b)

Figure 12. Failure modes of RC containment vessel (RCCV) RCCV#1 and RCCV #2. (**a**) Sliding shear failure of RCCV #1; (**b**) Web shear failure of RCCV #2.

The crack patterns of the experimental photos, as shown in Figure 11, can be obtained by displacement field-based crack analysis. By using the displacement-based analysis, cracks as thin as 0.03 mm (approximately 0.06 pixels wide in the photos) that appeared at the very beginning of the failure could be detected. The crack patterns of the selected displacement peaks are shown in Figure 13. The crack patterns were analyzed and presented in a field discretized with a grid containing 90 × 60 cells. The size of each cell is equivalent to a sub-image with 27 × 27 pixels. In both cases, from the beginning of the tests, the cracks were distributed over almost the entire measurement region. The widths of the cracks then gradually increased from 0.03 mm (for the 2.3-mm displacement cycle) to up to 0.4 mm (for the 11.3-mm displacement cycle).

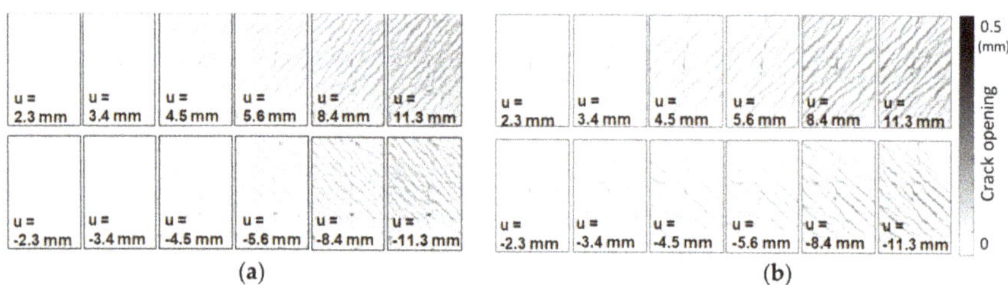

Figure 13. Displacement field-based crack analysis of RC containment vessel (RCCV) RCCV#1 and RCCV #2. (**a**) RCCV #1; (**b**) RCCV #2.

The proposed crack-based damage indices are calculated on the basis of the crack pattern obtained by the displacement field-based analysis (see Figure 14). In both experiments, the shear damage increased from 0 to approximately 0.75 for the displacement cycle of 8.4 mm (i.e., drift ratio of 0.375%), and did not significantly increase after that. The shear damage indices present a warning index that is capable of capturing the early stages of shear failure. Since both RCCVs #1 and #2 are shear critical, the cracks were mostly either at 45 degrees or 135 degrees, or typical shear cracks, with relatively fewer horizontal cracks observed in the measurement regions.

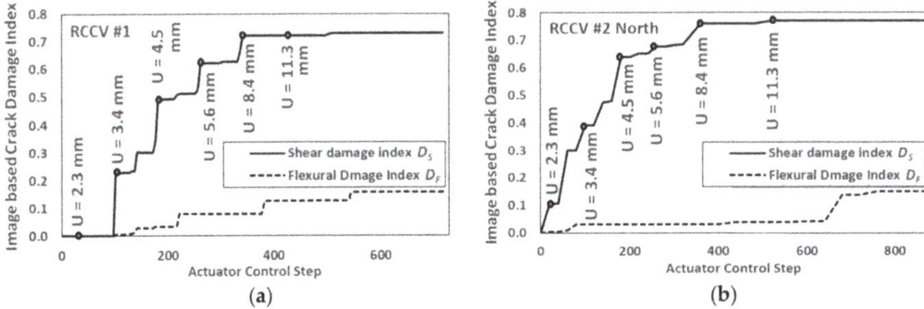

Figure 14. Image based damage index analysis of RC containment vessel (RCCV) RCCV#1 and RCCV #2. (**a**) RCCV #1; (**b**) RCCV #2.

This work examined the linear regression plots of several selected actuator control steps when analyzing the fractal dimension (as shown in Figure 6d). The plots showed that these points were very close to the line, and that the residual values were small. A selected plot of the linear regression of each specimen is shown in Figure 15. The crack pattern is a grid containing 90 × 60 cells, and is converted to different refinement of meshes with ε of 1, 2, 4, 8, 16, 32, 64, and 128 (while the most refined one is slightly more refined than the crack pattern), seven points were calculated in each of the fractal analyses.

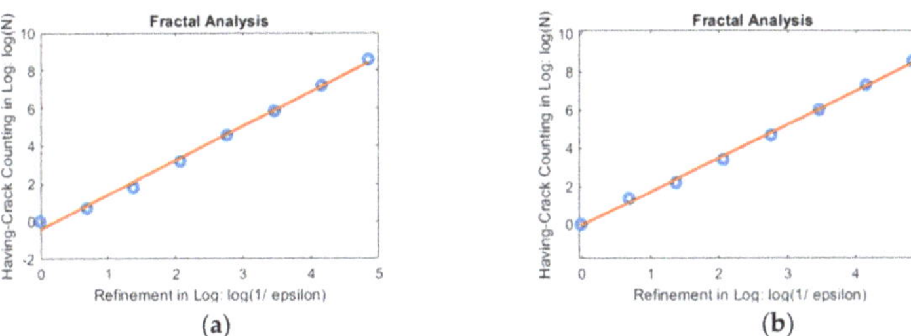

Figure 15. Selected fractal analysis plots of RC containment vessel (RCCV) RCCV#1 and RCCV #2. (**a**) RCCV #1; (**b**) RCCV #2.

The computing speed of the proposed method is great enough for static structural health monitoring, but still not sufficient for non-stop real-time dynamic analysis. For each step of the analysis, including image rectification, displacement field analysis, crack opening and orientation analysis, fractal analysis of cracks, and damage indices calculation, it takes about 40 seconds of computing time using a laptop equipped with an Intel i5-7300HQ 2.5 GHz processor and 32GB main memory. Sufficient computing speed may allow us to carry out automatic, non-stop crack detection and health monitoring with a sampling rate of 0.025 Hz, that is, once or twice per minute. It is still insufficient for detecting dynamic responses during a vibration event such as an earthquake, which typically requires a sampling rate of 200 Hz to 1000 Hz. To achieve non-stop dynamic analysis for structural health monitoring, this method requires not only a significant improvement in camera and computing hardware, but also further optimization of the algorithms and programming code.

5. Conclusions

This work proposed a damage indexing method based on crack image analysis, with the aim of indicating the early stage failure of shear critical RC structures. This method is based on a displacement field-based crack image analysis method, which is capable of detecting early stage, thin cracks on concrete surfaces. It is especially practical when displacement sensors and load cells are not applicable in real structures. Early stage, thin cracks can be detected when they are as thin as 0.03 mm, which is considerably thinner than the width of a pixel in a digital photo, and cannot be visually seen as a dark line. Based on the crack image analysis, a previously proposed fractal analysis of cracks was employed to estimate the overall damage index. According to the crack orientations, this method separates the fractal analysis damage index into a shear damage index and a flexural damage index to distinguish between the different types of failure. The software implementation method is publicly available.

The results of two RCCV experiments were used to verify the proposed damage indexing method. Since both RCCV specimens were shear critical structures, the analyzed damage indices showed that the shear cracks dominated the major failure. The flexural crack indices were relatively low throughout the experiments. In both experiments, the shear damage indices reached a relatively high value (i.e., 0.7) at a displacement of only 8.4 mm on the top of the specimen (i.e., a drift ratio of 0.375%). Earlier damage could be detected when the displacement was only 3.4 mm (i.e., a drift ratio of 0.15%) or even earlier, while the stiffness was still unchanged. This indicates that the crack image analysis-based damage indexing method is capable of indicating early stage failure in shear critical structures.

While this method estimates the damage indices of a structure, damage indices obtained from different types of structures are not comparable. The safety of a structure depends on many factors, including complicated design details such as the design of ties and stirrups, which are not visually observable. A non-ductile structure having a lower damage index does not mean it is safer than a ductile structure with a higher damage index. The practical health monitoring application of this method to other structures still requires sufficient experiments and investigations based upon the specific structure type.

Author Contributions: Data curation, C.-H.C.; Funding acquisition, C.-l.W.; Investigation, Y.-S.Y. & C.-l.W.; Methodology, Y.-S.Y. & C.-H.C.; Writing—original draft, Y.-S.Y.; Writing—review & editing, Y.-S.Y.

Funding: This work is partially funded by the Ministry of Science and Technology [MOST 107-2625-M-027-003 and MOST 108-2625-027-001] and the National Center for Research on Earthquakes in Taiwan.

Acknowledgments: The authors would like to acknowledge the National Center for Research on Earthquake Engineering in Taiwan for providing partial measurement resources and data of the shake table tests presented in this paper. We would like to thank Uni-edit (www.uni-edit.net) for editing and proofreading this manuscript.

Conflicts of Interest: The authors declare no conflict of interest.

References

1. Antunes, P.; Lima, H.; Varum, H.; André, P. Optical fiber sensors for static and dynamic health monitoring of civil engineering infrastructures: Abode wall case study. *Measurement* **2012**, *45*, 1695–1705. [CrossRef]
2. Karayannis, C.G.; Chalioris, C.E.; Angeli, G.M.; Papadopoulos, N.A.; Favvata, M.J.; Providakis, C.P. Experimental damage evaluation of reinforced concrete steel bars using piezoelectric sensors. *Constr. Build. Mater.* **2016**, *105*, 227–244. [CrossRef]
3. Park, Y.; Ang, A.H.; Wen, Y.K. Seismic Damage Analysis of Reinforced Concrete Buildings. *J. Struct. Eng.* **1985**, *111*, 740–757. [CrossRef]
4. Roufaiel, M.S.L.; Meyer, C. Analytical Modeling of Hysteretic Behavior of R/C Frames. *J. Struct. Eng.* **1987**, *113*, 429–444. [CrossRef]
5. Powell, G.H.; Allahabadi, R. Seismic Damage Prediction Deterministic Methods—Concepts and Procedures. *Earthq. Eng. Struct. Dyn.* **1998**, *16*, 719–734. [CrossRef]
6. Zhu, M.; McKenna, F.; Scott, M.H. OpenSeesPy: Python library for the OpenSees finite element framework. *SoftwareX* **2018**, *7*, 6–11. [CrossRef]
7. Yang, Y.; Wang, W.; Lin, J. Direct-Iterative Hybrid Solution in Nonlinear Dynamic Structural Analysis. *Comput. Civ. Infrastruct. Eng.* **2017**, *32*, 397–411. [CrossRef]
8. Yang, Y.S.; Tsai, K.C.; Elnashai, A.S.; Hsieh, T.J. An Online Optimization Method for Bridge Dynamic Hybrid Simulations. *Simul. Model. Pract. Theory* **2018**, *28*, 42–54. [CrossRef]
9. Ma, H.W.; Wang, J.W.; Liu, E.M.; Wan, Z.Q.; Wang, K. Experimental Study of the Behavior of Beam-Column Connections with Expanded Beam Flanges. *Steel Compos. Struct.* **2019**, *31*, 319–327.

10. Japan Building Disaster Prevention Association. *Standard for Seismic Evaluation of Existing Reinforced Concrete Buildings*; Japanese edition of 2001, English Version 1st; Japan Building Disaster Prevention Association: Tokyo, Japan, 2001.
11. International Atomic Energy Agency. *Guidebook on Non-Destructive Testing of Concrete Structures*; International Atomic Energy Agency: Vienna, Austria, September 2002.
12. Federal Highway Administration. *Bridge Inspector's Reference Manual*; FHWA NHI 12-049; Federal Highway Administration, U.S. Department of Transportation: Washington, DC, USA, 2012.
13. Szeląg, M. The Influence of Metakaolinite on the Development of Thermal Cracks in a Cement Matrix. *Materials* **2018**, *11*, 520. [CrossRef]
14. Castillo, E.D.R.; Allen, T.; Henry, R.; Griffith, M.; Ingham, J. Digital Image Correlation (DIC) for Measurement of Strains and Displacements in Coarse, Low Volume-Fraction FRP Composites Used in Civil Infrastructure. *Compos. Struct.* **2019**, *212*, 43–57. [CrossRef]
15. Dzaye, E.D.; Tsangouri, E.; Spiessens, K.; De Schutter, G.; Aggelis, D.G. Digital image correlation (DIC) on fresh cement mortar to quantify settlement and shrinkage. *Arch. Civ. Mech. Eng.* **2019**, *19*, 205–214. [CrossRef]
16. Yang, Y.-S. Measurement of Dynamic Responses from Large Structural Tests by Analyzing Non-Synchronized Videos. *Sensors* **2019**, *19*, 3520. [CrossRef] [PubMed]
17. Schumacher, T.; Shariati, A. Monitoring of Structures and Mechanical Systems Using Virtual Visual Sensors for Video Analysis: Fundamental Concept and Proof of Feasibility. *Sensors* **2013**, *13*, 16551–16564. [CrossRef]
18. Salhaoui, M.; Guerrero-González, A.; Arioua, M.; Ortiz, F.J.; El Oualkadi, A.; Torregrosa, C.L. Smart Industrial IoT Monitoring and Control System Based on UAV and Cloud Computing Applied to a Concrete Plant. *Sensors* **2019**, *19*, 3316. [CrossRef] [PubMed]
19. Feng, D.; Feng, M.Q. Experimental validation of cost-effective vision-based structural health monitoring. *Mech. Syst. Signal Process.* **2017**, *88*, 199–211. [CrossRef]
20. Cheng, C.; Kawaguchi, K. A preliminary study on the response of steel structures using surveillance camera image with vision-based method during the Great East Japan Earthquake. *Meas.* **2015**, *62*, 142–148. [CrossRef]
21. Ye, X.W.; Dong, C.Z.; Liu, T. A Review of Machine Vision-Based Structural Health Monitoring: Methodologies and Applications. *J. Sensors* **2016**, *2016*, 1–10. [CrossRef]
22. Wu, C.L.; Hsu, T.T.C.; Chang, C.Y.; Yang, H.C.; Chang, C.C.; Wang, K.J.; Yang, Y.S.; Mo, Y.L.; Lu, H.J.; Chen, Y.C. *Reversed Cyclic Tests of 1/13 Scale Cylindrical Concrete Containment Structures*; Technical Report 18-001; National Center for Research on Earthquake Engineering: Taipei, Taiwan, 2018.
23. Farhidzadeh, A.; Dehghan-Niri, E.; Moustafa, A.; Salamone, S.; Whittaker, A. Damage Assessment of Reinforced Concrete Structures Using Fractal Analysis of Residual Crack Patterns. *Exp. Mech.* **2013**, *53*, 1607–1619. [CrossRef]
24. Yang, Y.-S.; Wu, C.-L.; Hsu, T.T.; Yang, H.-C.; Lu, H.-J.; Chang, C.-C. Image analysis method for crack distribution and width estimation for reinforced concrete structures. *Autom. Constr.* **2018**, *91*, 120–132. [CrossRef]
25. Cho, H.-W.; Yoon, H.-J.; Yoon, J.-C. Analysis of Crack Image Recognition Characteristics in Concrete Structures Depending on the Illumination and Image Acquisition Distance through Outdoor Experiments. *Sensors* **2016**, *16*, 1646. [CrossRef] [PubMed]
26. Yu, L.; Tian, Y.; Wu, W. A Dark Target Detection Method Based on the Adjacency Effect: A Case Study on Crack Detection. *Sensors* **2019**, *19*, 2829. [CrossRef] [PubMed]
27. Dorafshan, S.; Thomas, R.J.; Maguire, M. Benchmarking Image Processing Algorithms for Unmanned Aerial System-Assisted Crack Detection in Concrete Structures. *Infrastructures* **2019**, *4*, 19. [CrossRef]
28. Hoang, N.-D.; Nguyen, Q.-L. Metaheuristic Optimized Edge Detection for Recognition of Concrete Wall Cracks: A Comparative Study on the Performances of Roberts, Prewitt, Canny, and Sobel Algorithms. *Adv. Civ. Eng.* **2018**, *2018*, 1–16. [CrossRef]
29. Cha, Y.J.; Choi, W.; Suh, G.; Mahmoudkhani, S.; Büyüköztürk, O. Autonomous Structural Visual Inspection using Region-Based Deep Learning for Detecting Multiple Damage Types. *Comput. Aided Civ. Infrastruct. Eng.* **2018**, *33*, 731–747. [CrossRef]
30. Rafiei, M.H.; Adeli, H. A novel unsupervised deep learning model for global and local health condition assessment of structures. *Eng. Struct.* **2018**, *156*, 598–607. [CrossRef]

31. Dorafshan, S.; Thomas, R.J.; Maguire, M. Comparison of deep convolutional neural networks and edge detectors for image-based crack detection in concrete. *Constr. Build. Mater.* **2018**, *186*, 1031–1045. [CrossRef]
32. Luo, Q.; Ge, B.; Tian, Q. A fast adaptive crack detection algorithm based on a double-edge extraction operator of FSM. *Constr. Build. Mater.* **2019**, *204*, 244–254. [CrossRef]
33. Mohan, A.; Poobal, S. Crack detection using image processing: A critical review and analysis. *Alex. Eng. J.* **2018**, *57*, 787–798. [CrossRef]
34. Yang, Y.-S.; Yang, C.-M.; Huang, C.-W. Thin crack observation in a reinforced concrete bridge pier test using image processing and analysis. *Adv. Eng. Softw.* **2015**, *83*, 99–108. [CrossRef]
35. Yang, Y.S. ImPro Stereo. National Taipei University of Technology. 2019. Available online: https://sites.google.com/site/improstereoen/ (accessed on 1 August 2019).
36. Yang, Y.S.; Huang, C.W.; Wu, C.L. A Simple Image-Based Strain Measurement Method for Measuring the Strain Fields in an RC-Wall Experiment. *Earthq. Eng. Struct. Dyn.* **2012**, *41*, 1–17. [CrossRef]
37. Kaehler, A.; Bradski, G. *Learning OpenCV3: Computer Vision in C++ with the OpenCV Library*; O'Reilly Media: Sebastopol, CA, USA, 2016.

MDPI
St. Alban-Anlage 66
4052 Basel
Switzerland
Tel. +41 61 683 77 34
Fax +41 61 302 89 18
www.mdpi.com

Sensors Editorial Office
E-mail: sensors@mdpi.com
www.mdpi.com/journal/sensors

www.ingramcontent.com/pod-product-compliance
Lightning Source LLC
LaVergne TN
LVHW070112170726
843515LV00007B/1667